Heathkit/Zenith Educational Systems

DIGITAL TECHNIQUES

A Step-by-Step Introduction

A SPECTRUM BOOK

Training in Computers & Electronics Series

Prentice-Hall, Inc. Englewood Cliffs, New Jersey 07632

Library of Congress Cataloging in Publication Data

Main entry under title:

Digital techniques.

(Training in computers & electronics series)
"A Spectrum Book."
"Adapted from a larger work entitled Digital techniques"—T.p. verso.
Includes index.
1. Digital electronics. I. Heathkit/Zenith Educational Systems (Group) II. Series.
TK7868.D5D49 1983 621.3815 83-3038
ISBN 0-13-214049-7
ISBN 0-13-214031-4 (pbk.)

A SPECTRUM BOOK

12 9 8 7 6 5 4 3 2

Printed in the United States of America

ISBN 0-13-214031-4 {PBK.}

ISBN 0-13-214049-7

Manufacturing buyer Patrick Mahoney

Prentice-Hall International, Inc., *London*
Prentice-Hall of Australia Pty. Limited, *Sydney*
Prentice-Hall of Canada Inc., *Toronto*
Prentice-Hall of India Private Limited, *New Delhi*
Prentice-Hall of Japan, Inc., *Tokyo*
Prentice-Hall of Southeast Asia Pte. Ltd., *Singapore*
Whitehall Books Limited, *Wellington, New Zealand*
Editora Prentice-Hall do Brasil Ltda., *Rio de Janeiro*

CONTENTS

Unit 1

INTRODUCTION TO DIGITAL TECHNIQUES

INTRODUCTION

The purpose of this first unit on digital techniques is to give you an overview of the subject and to introduce you to the basic concepts. You will learn what digital techniques are, how they are used and why they are used. You will learn about binary numbers and codes which are the basic language of all digital systems. And finally, you will see how digital techniques are implemented with hardware.

This unit provides you with the background and base upon which you will build your knowledge of digital techniques. This foundation will put the concept of digital techniques into perspective so that you can fully relate them to the field of electronics. With the information presented here you will clearly understand the need for and uses of digital techniques.

Digital techniques are so widely used today that it is almost impossible to think of electronic equipment without them. Digital techniques are used in virtually every area of electronics. They have greatly improved electronic methods and have given us practical electronic equipment with amazing capability. And, there is potential for further improvements and advances. As an electronic engineer, technician, or hobbyist you can benefit by knowing digital techniques. This program will provide you with a solid understanding of digital methods and a guide to their application.

DIGITAL TECHNIQUES

There are two basic types of electronic signals and techniques, analog and digital. Analog signals are the most familiar type. An analog signal is an ac or dc voltage or current that varies smoothly or continuously. It is one that does not change abruptly or in steps. An analog signal can exist in a wide variety of forms. Several types of analog signals are shown in Figure 1-1.

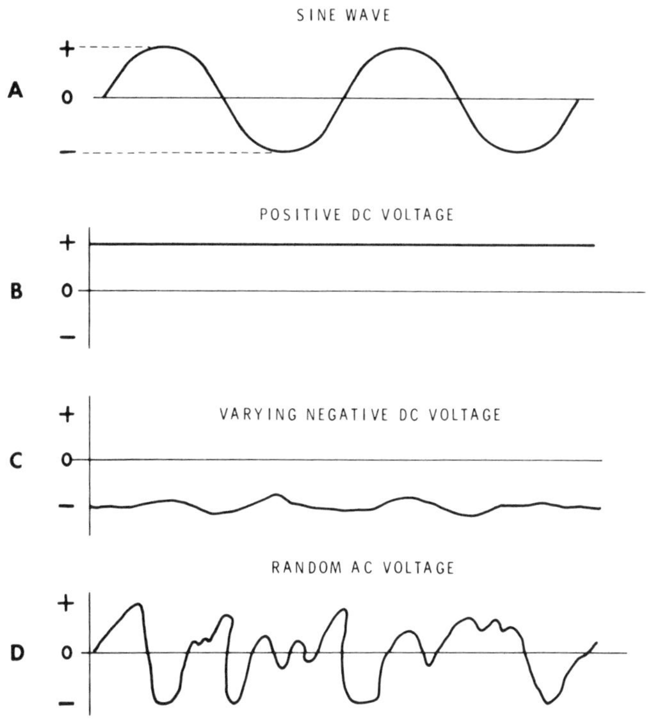

Figure 1-1
Types of Analog signals

Figure 1-1A shows the most common type of analog signal, a sine wave. A significant number of electronic signals are sinusoidal. Radio signals and audio tones are examples. A fixed dc voltage is also an analog signal. Figure 1-1B shows a constant positive dc voltage. Another type of analog signal is varying dc voltage or current. A changing negative dc voltage is illustrated in Figure 1-1C. Any random but continuously varying voltage waveform is considered to be analog. The signal shown in Figure 1-1D is only one of an infinite variety of such signals. Electronic circuits that process these analog signals are called linear circuits.

Digital signals are essentially a series of pulses or rapidly changing voltage levels that vary in discrete steps or increments. Digital signals are pulses of voltage that usually switch between two fixed levels. Figure 1-2 shows several types of digital signals. Notice how these signals switch between two distinct voltage levels. In Figure 1-2A, the two levels are 0 (ground) and +5 volts. In Figure 1-2B, the levels are 0 (ground) and −6 volts. In Figure 1-2C the signal alternates between the +3 and −3 volt levels. This two-level, off-on or up-down fast switching characteristic is fundamental of all digital signals. Electronic circuits that process these digital signals are called digital, logic, or pulse circuits.

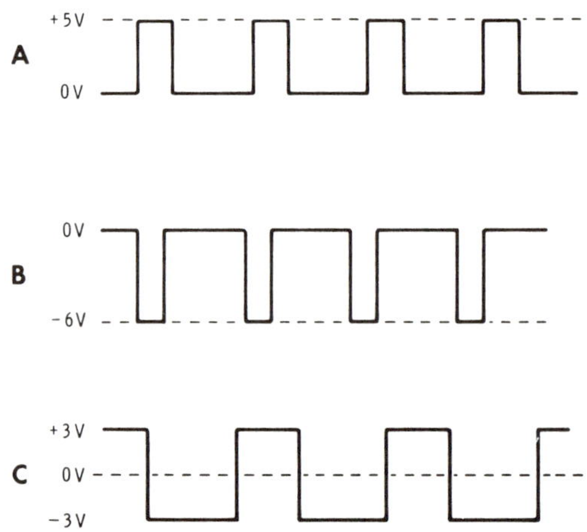

Figure 1-2
Types of Digital signals

Contrasting Analog and Digital Devices and Techniques

Now let's further define analog and digital methods in terms of devices and ideas that are already quite familiar to you. For example, a light bulb can be either an analog or digital device depending upon how it is used. The amount of current through a light bulb can be set to any level less than its maximum rated value. We can vary the current through it continuously and its brightness will vary. Used in this way the light bulb is an analog device. The brilliance of the lamp is proportional to the current through it. There are virtually an infinite number of brightness levels.

The lamp can also be used as a digital device where the current through it and its brightness varies in discrete steps. The most common way of using the light bulb as a digital device is to give it two brilliance levels, usually off and on. The important point is that the lamp has two states. Because of this off-on characteristic, we say that the lamp is binary in nature. The term binary designates any two-state device or signal.

Let's take some other examples to illustrate the concept of analog and digital techniques. The VHF channel selector switch on your television set is digital in nature because it can assume only discrete positions. It can be set to any one of thirteen unique states, channels 2 through 13 and UHF. Any type of switch is a digital device because it has two or more discrete positions.

The volume control on your television set is an analog device. You can vary the volume of the sound continuously over a wide range from completely off to extremely loud.

The speedometer on your car is an analog device. It tells you the speed of your car in miles per hour on a smooth and continuously varying basis. You read the speed from a dial that is usually calibrated in no smaller increments than 5 miles per hour. To determine the speeds between the markings you must interpolate, or guess at, the exact speed.

The odometer portion of your speedometer, the part that indicates the number of miles traveled, is a digital device. Since the odometer records mileage in increments of one mile (or in some cases one-tenth mile), it is digital in nature.

Another example of an analog device is a typical clock or watch. It indicates the time continuously by the positions of the hands on a calibrated dial. The second hand sweeps smoothly and continuously around in an analog fashion as do both hour and minute hands. To determine the exact time, you must estimate the positions of the hands. Your ability to read the time accurately is limited by the precision of the dial calibration increments. Digital clocks overcome this problem. On a digital clock, you read the time directly from decimal number display readouts in discrete increments of hours, minutes and seconds. The accuracy is greatly improved and you gain the added convenience of a direct number display.

A standard voltmeter is also an analog device. It reads or measures voltage and indicates its value by the position of a pointer on a meter scale. The pointer moves smoothly or continuously as the amplitude of the analog voltage being measured varies. Of course, digital voltmeters are also available. These instruments measure the voltage and display it as discrete digits on a decimal readout.

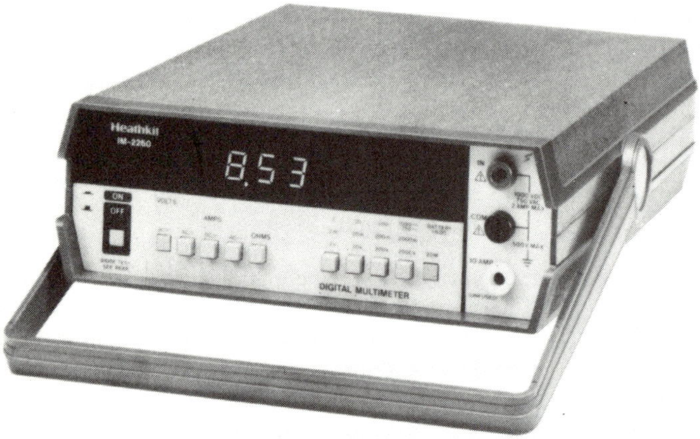

The low-priced Heathkit IM-2260 Portable Digital Multimeter features attractive styling, a wide measurement range, and highly accurate readings.

An analog multimeter for measuring voltage, current and resistance

Here are a few other analog quantities and devices.

temperature — thermometer
direction — compass
light intensity — light meter

Keep in mind that all of these variables could be monitored and displayed as a digital readout.

Further examples of digital variables are:

money
heart pulse rate

Where Are Digital Techniques Used?

Perhaps the greatest use of digital techniques today is in computers. Digital computers are used in virtually all areas of business and industry. They are extremely useful machines that can save man a tremendous amount of effort and greatly extend his capabilities.

Over the years digital computers have grown in capability, but have become smaller, cheaper and easier to use. As a result, their use has increased tremendously. The small, low-cost but very powerful minicomputer has put digital and computer techniques within the reach

A typical microcomputer.

of nearly everyone. A minicomputer can be quickly and easily designed into a system to replace more conventional equipment and circuitry for control, computation and automation.

Advanced semiconductor technology has recently given us a computer that is really a component. Known as microprocessors, these devices are complete digital computers in a single miniature integrated circuit package. Microprocessors can replace minicomputers in many applications and can often be used to replace conventional digital circuits. Like all digital computers, the microprocessor must be programmed to carry out its specified function.

But computers aren't the only application for digital techniques. Digital methods are being employed in almost every imaginable area of electronics. Here are just a few examples.

Communications. Instead of transmitting information over wire lines or by radio by analog methods, much data is now being transmitted in digital form. It has been found that pulse type signals are easier to work with and are less susceptible to noise and other problems common in communications systems. Digital computers can readily communicate with one another by transmitting information over the telephone lines by using digital techniques.

Telemetry systems. Those systems used for transmitting measurement data from a remote location, use digital techniques extensively. In an unmanned satellite, for example, sensors are used to monitor various environmental conditions such as temperature, light, and radiation.The analog voltages produced by these sensors, in response to the quantity being measured, can be transmitted back to the earth via radio by modulating a carrier using conventional analog methods. However, it has been found that by converting the analog variables into digital signals, an improvement in transmission reliability and accuracy can be achieved. Today, telemetry systems in satellites and missiles make extensive use of digital techniques.

Test Instruments. The trend in test and measurement equipment is clearly toward the use of digital techniques. Besides the convenience of a direct decimal display and the increased precision of measurement brought about by the use of digital techniques, it is possible to interface many digital instruments with a computer. This permits automatic monitoring, controlling, measuring and recording of data.

The most common electronic test instrument, the analog voltmeter, is gradually being replaced by the more sophisticated digital voltmeter (DVM). The DVM does the same job as its analog counterpart. The DVM measures voltage but instead of presenting the reading to the observer in the form of a pointer on a meter face, the voltage is a direct readout display of decimal numbers. Such an instrument, while generally higher in cost than an analog voltmeter, is extremely convenient to use and read. More important, it gives more accurate measurements of voltages. The digital multimeter (DMM), a DVM with refinements, is capable of measuring voltages as well as resistance and current like an analog VOM.

Another widely used digital test instrument is the electronic counter. This unit is widely used for measuring frequency and time intervals. Again the digital techniques provide a convenient decimal read-out of the exact quantity being measured, thereby eliminating man's need to interpolate continuous or analog meter scales to provide a reading. This result is greater accuracy and less error in measurement.

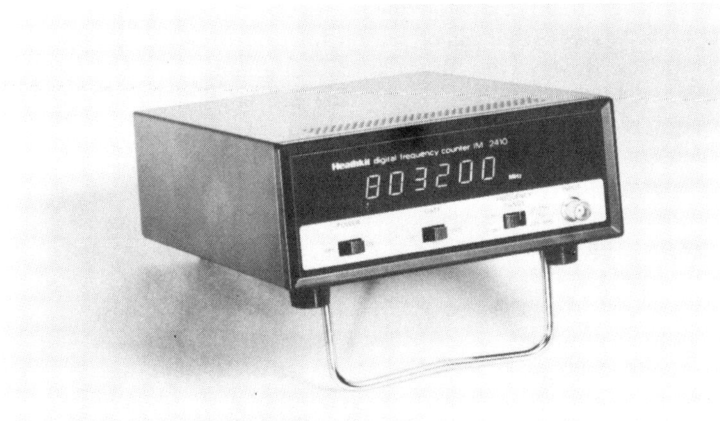

The new Heathkit IM-2410
Portable Frequency Counter
introduces new capabilities
to its price range.

Industrial Controls. Digital techniques are becoming more widely used in manufacturing plants and refineries where complex operations must be accurately controlled. These systems use sensors to monitor various phases of the operation, and the outputs of these sensors are then used to produce signals that will control the various operations that affect the process.

Industrial feedback control systems have traditionally used analog techniques. Today, many of these control systems are digital in nature. Most involve both analog and digital methods. Analog quantities like temperature, pressure, liquid level and flow rate are monitored by sensors to produce an analog signal. These are converted into digital values by analog-to-digital converters. Other system inputs are already digital in nature such as limit switches or sensors of an off-on or go no-go nature.

All of this digital information is fed into a digital computer which monitors the input variables and according to a predetermined program generates output signals to control the process.

Another popular industrial application for digital techniques is machine tool control. Here a digital computer controls the drilling, cutting, punching and stamping of materials to produce specific metal parts accurately and automatically.

Consumer Electronic Equipment. Perhaps the biggest holdout against the use of digital techniques has been common consumer items such as radio, TV, hi-fidelity and other products. The higher cost of some digital methods has been the major reason for the delay in adopting these techniques. However, as prices of digital circuits have dropped, even these products have given way to the use of digital techniques. An example is the circuitry for digital channel selection used in some of the newer color television sets. Some hi-fi receivers use digital frequency synthesizers for tuning.

An FM tuner using digital techniques for frequency selection and display (Heath Co.)

Other popular consumer devices using digital methods are the popular digital clocks and electronic calculators. Even a home digital thermometer is available.

Handheld electronic calculator using digital circuits (Heath Co.)

Digital clock and thermometer. (Heath Co.)

While consumer electronic devices will continue to use analog techniques, more digital controls and circuits will gradually be added to increase versatility and improve performance.

Where are digital techniques used? The answer is almost everywhere in electronics. And their use will increase. Why are digital techniques used? Digital methods are used basically to obtain greater resolution of measurement, control or calculation, and for convenience.

Why Use Digital Techniques?

The primary impetus for the widespread use of digital techniques has been the availability of low cost,digital integrated circuits (ICs). Advances in integrated circuit technology have produced many excellent low cost digital circuits. Such circuits are small, inexpensive and very reliable. The more recent medium scale and large scale (MSI and LSI) integrated devices can replace entire circuits and instruments. Electronic equipment designers recognize the availability of such devices and have begun to take advantage of them. While digital techniques have been known for years, it took integrated circuits to make them practical.

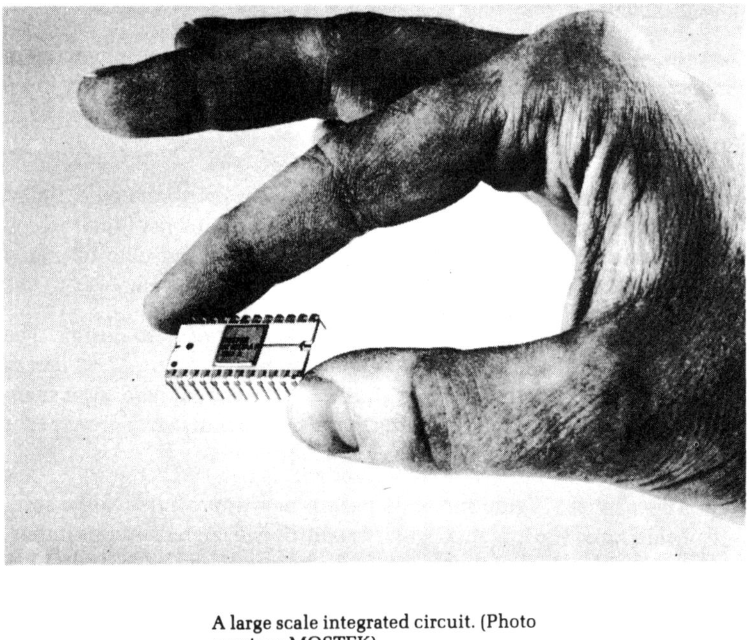

A large scale integrated circuit. (Photo courtesy MOSTEK).

By using digital IC's many equipment improvements have been made. Reductions in size, weight, cost and power consumption usually result when analog techniques are replaced by digital methods, thanks to digital IC's. Here are a few more reasons why digital techniques have become so popular.

Greater accuracy. Digital techniques permit greater precision and resolution in representing quantities or in making measurements than with analog methods.

Greater Dynamic Range. Dynamic range is the difference between the upper and lower data values that a system or instrument can handle. Analog systems are limited because of component capabilities and noise to a range of something less than 100,000 to 1. With digital techniques practically any desired dynamic range can be obtained.

Greater Stability. Analog or linear circuits are subject to the effects of drift and component tolerance problems. Temperature and other environmental factors affect resistor, capacitor and inductance values. Transistor biases vary causing non-linear operation and distortion to occur. Component imperfections and ageing cause drift and resultant problems. Digital methods greatly minimize or completely eliminate such problems.

Convenience. Digital techniques make instruments and equipment more convenient to use. The direct decimal display of data is not only more convenient, but the error of reading or interpolating analog meters or in setting analog dials is eliminated.

Automation. Many electronic processes can be fully automated if digital techniques are used. Special control circuits or a digital computer which are programmed can automatically, set up, control and monitor many operations. Data is readily recorded, stored and displayed.

Design Simplicity. Digital equipment is relatively easy to design. The available digital ICs make digital design a pleasure. Little or no breadboarding is required. In analog or linear circuits breadboarding is mandatory to ensure a workable circuit. Digital equipment can go from paper design to finished product in a very short time.

New Approaches. Digital methods permit new approaches to the solution of electronic equipment design. In addition, design solutions impossible with analog techniques are readily implemented with digital circuits. Digital circuits make it possible to do some things that have no analog equivalent.

THE BINARY NUMBER SYSTEM

All digital circuits, instruments and systems work with numbers that represent specific quantities. For example, the analog voltage measured by a digital voltmeter is converted into digital form and displayed as a specific decimal number. The number that you enter into an electronic calculator is stored and used in the calculation you specify. The digital computer that prints your payroll check works with numbers, specifically your salary, the number of hours you work and the various deductions. As you can see, numbers or quantities are the basic source for an end product of most digital equipment. Figure 1-3 shows how most digital equipment accepts input numbers, processes them and generates number outputs. The actual form of the input and output numbers depends on the application. They may be in binary or decimal form. In some applications the input and/or output may be in analog form despite the digital processing.

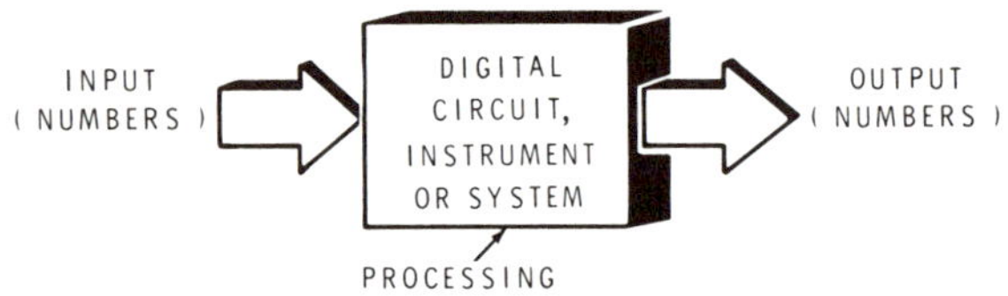

Figure 1-3 A general block diagram of any digital system

The type of numbers we are most familiar with are decimal numbers. In the decimal number system we combine the ten digits 0 through 9 in a certain way so that they indicate a specific quantity. In the binary number system, we use only two digits, 0 and 1. These binary digits, or bits, when appropriately arranged can also represent any decimal number. For example, the binary number 110101 represents the decimal quantity 53. All modern digital techniques are based on the binary number system.

The basic distinguishing feature of a number system is its base or radix. The base indicates the number of characters or digits used to represent quantities in that number system. The decimal number system has a base or radix of 10 because we use the ten digits 0 through 9 to represent quantities. The binary number system has a base of 2 since only the digits or bits 0 and 1 are used in forming numbers.

The decimal number system came about basically as a result of man having ten fingers. Man's earliest attempts to represent numbers, count and keep track of quantities involved the use of his fingers. Of course, the decimal number system is universally used and understood because it is our way of communicating information about quantities.

The binary number system, while simple, is inconvenient to use because we are not familiar with it. But once you learn it, you will find it easy to work with. It has special benefits when it comes to constructing the hardware used in digital equipment.

Digital systems can be implemented with either the decimal or binary number systems. However, the advantages of the binary number systems over the decimal number system in terms of hardware implementation are many. If we use the decimal number system, our hardware must have ten discrete steps or states each representing one of the digits 0 through 9. Various types of electrical, electromechanical, or electronic components and circuits are available to do this. A ten position rotary switch or a stepping relay are examples. However, when the binary number system is used, we only need to represent two states. An electronic component or circuit that has only two states is significantly simpler, less expensive, faster and more reliable than one with ten. Each bit can be implemented with components of a simple off/on nature such as switch or relay contacts or a transistor that conducts or does not conduct.

Positional Number Systems

The decimal and binary number systems are positional or weighted number systems. This means that each digit or bit position in a number carries a particular weight in determining the magnitude of that number. For example, you know that a decimal number has positional weights of units, tens, hundreds, thousands, etc. Each position has a weight that is some power of the number system base, in this case ten. The positional weights are $10^0 = 1$ (units)*, 10^1 (tens), 10^2 (hundreds), etc. We evaluate the total quantity represented by considering the specific digits and the weights of their positions. Consider the decimal number 7438 in which there are 8 ones, 3 tens, 4 hundreds, and 7 thousands. The number can be written as indicated below.

$$(7 \times 10^3) + (4 \times 10^2) + (3 \times 10^1) + (8 \times 10^0) =$$
$$7000 + 400 + 30 + 8 = 7438$$

To determine the value of the number, you multiply each digit by the weight of its position and add your results.

Binary numbers work the same way. Each bit position carries a specific weight. As in the decimal number system, the position weights are some power of the base of the number system. These weights from right to left are $2^0 = 1$*, $2^1 = 2$, $2^2 = 4$, $2^3 = 8$, etc. The weight of each position is twice that of the weight of the number to the right. Consider the binary number 110101. This can be written as indicated below.

$$(1 \times 2^5) + (1 \times 2^4) + (0 \times 2^3) + (1 \times 2^2) + (0 \times 2^1) + (1 \times 2^0) =$$
$$32 + 16 + 0 + 4 + 0 + 1 = 53$$

The quantity represented by the number is determined by multiplying each bit by its position weight and obtaining the sum.

Fractional Numbers

So far we have only discussed whole numbers or integer quantities. But as you know it is often necessary to express quantities in terms of fractional parts of a whole.

Decimal fractions are numbers whose positions have weights that are negative powers of ten such as $10^{-1} = 1/10 = .1$, $10^{-2} = 1/100 = .01$, $10^{-3} = 1/1000 = .001$, etc. A decimal point separates the whole and fractional parts of a number. The integer or whole number portion is to the left of the decimal point and has weights of units, tens, hundreds, etc. The fractional part of the number is to the right of the decimal point and the positions have weights of tenths, hundredths, thousandths, etc. To illustrate this, the number 278.94 can be written as shown below.

$$(2 \times 10^2) + (7 \times 10^1) + (8 \times 10^0) + (9 \times 10^{-1}) + (4 \times 10^{-2}) =$$
$$200 + 70 + 8 + .9 + .04 = 278.94$$

In a fractional binary number, the weights of the fractional positions are negative powers of 2 or $2^{-1} = 1/2 = .5$, $2^{-2} = 1/4 = .25$, $2^{-3} = 1/8 = .125$, $2^{-4} = 1/16 = .0625$, etc. The position weight is one half of the weight of the position directly to the left. A binary point separates the whole and fractional parts of the number.

The binary number 1101.101 is evaluated as shown below.

$$(1 \times 2^3) + (1 \times 2^2) + (0 \times 2^1) + (1 \times 2^0) + (1 \times 2^{-1}) + (0 \times 2^{-2}) + (1 \times 2^{-3}) =$$
$$8 + 4 + 0 + 1 + .5 + 0 + .125 = 13.625$$

*Any number with an exponent of zero is equal to one.

Converting Between the Binary and Decimal Number Systems

In working with digital equipment, you will often need to determine the decimal value of binary numbers. In addition, you will also find it necessary to convert a specific decimal number into its binary equivalent. Let's see how such conversions are accomplished.

Binary to Decimal. To convert a binary number into its decimal equivalent you simply add together the weights of the positions in the number where binary 1's occur. The weights of the integer and fractional positions are indicated below.

INTEGER									FRACTIONAL		
2^7	2^6	2^5	2^4	2^3	2^2	2^1	2^0		2^{-1}	2^{-2}	2^{-3}
128	64	32	16	8	4	2	1	•	.5	.25	.125

Binary Point

As an example, let's convert the binary number 1010 into its decimal equivalent. Since no binary point is shown, the number is assumed to be a whole number where the binary point is to the right of the number. The right-most bit, called the least significant bit or LSB, has the lowest integer weight of $2^0 = 1$. The left-most bit is the most significant bit (MSB) because it carries the greatest weight in determining the value of the number. In this example, it has a weight of $2^3 = 8$. To evaluate the number we add together the weights of the positions where binary 1's appear. In this example, 1's occur in the 2^3 and 2^1 positions. The decimal equivalent is ten.

Binary Number	1	0	1	0	
Position Weights	(8)	(4)	(2)	(1)	
Decimal Equivalent	8 +	0 +	2 +	0	= 10

As a further illustration of this process, let's convert the binary number 101101.11 into its decimal equivalent.

Binary Number	1	0	1	1	0	1	.1	1	
Position Weights	(32)	(16)	(8)	(4)	(2)	(1)	(.5)	(.25)	
Decimal Equivalent	32+	0+	8+	4+	0+	1+	.5+	.25	= 45.75

Note that you can disregard the position weights where binary 0s occur since they add nothing to the number value.

After you solve a few practice problems, you will quickly catch on to this procedure.

Decimal to Binary. Converting a known decimal number into its binary equivalent can be accomplished by a simple trial and error method once you are familiar with the weighting sequence of binary numbers. Suppose that you wish to convert the decimal number 175 into its binary equivalent. To do this you first determine the highest positional weight that is equal to or less than the number being converted. This is 128. Subtract 128 from 175 and note the remainder.

$$\begin{array}{r} 175 \\ -\ 128 \\ \hline 47 \end{array}$$

Again determine the highest positional weight that does not exceed the remainder. This is 32. Next determine their difference and continue to repeat the process until no further subtractions are possible.

$$\begin{array}{r} 47 \\ -\ 32 \\ \hline 15 \end{array}$$

The highest positional weight less than 15 is 8.

$$\begin{array}{r} 15 \\ -\ 8 \\ \hline 7 \end{array}$$

The highest positional weight less than 7 is 4.

$$\begin{array}{r} 7 \\ -\ 4 \\ \hline 3 \end{array}$$

The highest positional weight less than 3 is 2.

$$\begin{array}{r} 3 \\ -\ 2 \\ \hline 1 \end{array}$$

And finally, the highest positional weight less than or equal to 1 of course is 1.

Now using this information you construct the equivalent binary number. You write a binary 1 for the weight positions you subtracted from the original number and the remainders. In this exmple, these were 128, 32, 8, 4, 2 and 1. Note that you did not use the 64 and 16 weights, so these positions will be binary 0. The number then is 10101111 = 175. You can check it by converting the binary version back into decimal form using the procedure discussed earlier.

Another method for converting decimal numbers into binary is to repeatedly divide the number by 2 and note the remainder. When dividing by 2, the remainder will always be either 1 or 0. The remainder forms the equivalent binary number.

As an example, convert the number 175 into its binary equivalent.

	REMAINDER
175 ÷ 2 = 87	1 ← LSB
87 ÷ 2 = 43	1
43 ÷ 2 = 21	1
21 ÷ 2 = 10	1
10 ÷ 2 = 5	0
5 ÷ 2 = 2	1
2 ÷ 2 = 1	0
1 ÷ 2 = 0	1 ← MSB

Binary Number Sizes

Binary numbers are also referred to as binary words. An 8 bit binary number is also an 8 bit word. You will also see the term byte used to refer to binary words. Most digital circuits and equipment use a fixed word size. The size of this word determines the maximum magnitude and resolution with which numbers can be represented. The number of bits in a word determine the number of discrete states that can exist and the maximum decimal number value that can be represented.

The formula below indicates the number of states that can be represented with a given number of bits.

$N = 2^n$ N = total number of states.
n = number of bits in the word.

For example, with a 4-bit word, we can represent a maximum of

$$N = 2^n = 2^4 = 2 \times 2 \times 2 \times 2 = 16$$

This means that by using 4-bit positions, we can create a total of 16 different binary bit patterns or number combinations. These are shown in Table I along with their decimal equivalents.

TABLE I

DECIMAL	BINARY
0	0000
1	0001
2	0010
3	0011
4	0100
5	0101
6	0110
7	0111
8	1000
9	1001
10	1010
11	1011
12	1100
13	1101
14	1110
15	1111

Binary and decimal number equivalents for a 4-bit word.

As Table I indicates we represent the numbers 0 through 15 using the binary number weighting system. The maximum decimal number that can be represented is one less than the total number of states. The largest decimal number value (N) that can be represented for a given number of bits (n) is expressed with the formula below.

$$N = 2^n - 1$$

For example, with a 6-bit number we can represent a maximum value of

$$N = 2^6 - 1 = 64 - 1 = 63$$

If you know the maximum decimal quantity (N) that you wish to represent with a binary number, you can determine the required number of bits (B) with the expression given below.

$$B = 3.32 \log_{10} N$$

The common logarithm can be obtained from a set of tables, a slide rule or an electronic calculator with log capability. For example, if the maximum decimal number that you need to represent is 500, the number of bits required is:

$$B = 3.32 \log_{10} 500$$

$$B = 3.32\ (2.69897)$$

$$B = 8.96$$

Of course you cannot implement fractional bits so the total number of bits required is the next highest whole number. Therefore, to represent the number 500, you would need a total of nine bits. Using the previously given expression, you can determine that, with a total of nine bits, the maximum number you can represent is

$$N = 2^n - 1 = 2^9 - 1 = 512 - 1 = 511$$

The Appendix at the end of this unit contains a table of numbers that are powers of 2. It will help you to quickly determine the relationship between decimal number size and binary word bit length.

Number Identification

When working with both binary and decimal numbers, it is often necessary to have some way of identifying whether a number is a decimal or a binary number. This is particularly true of numbers involving only ones and zeros. For example, the number 101 could represent a quantity of one hundred and one if it is a decimal number. However, if this number is in binary form it would represent a quantity of five.

To distinguish binary from decimal numbers a small subscript number is generally written after the number. The identifying subscript number is the base or radix of the number system being used. Several examples of this notation are indicated below.

$$101_2 = 5_{10}$$

$$101_{10} = 1100101_2$$

BINARY CODES

The general term given to the process of converting a decimal number into its binary equivalent is coding. We express a decimal number as a binary code or binary number. The binary number system, as discussed, is known as the pure binary code. We give it this name to distinguish it from other types of binary codes. In this section you will see some of the other types of binary codes used in digital systems.

Binary Coded Decimal

Because the decimal number system is so familiar, it is easy to use. The binary number system is less convenient to use because we are not as intimately familiar with it. It is difficult to quickly glance at a binary number and recognize its decimal equivalent. For example, the binary number 1010011 represents the decimal number 83. You certainly cannot tell immediately by looking at the number what its decimal value is. However, you know that within a few minutes, using the procedures described earlier, that you could readily calculate its decimal value. The amount of time that it takes you to convert or recognize a binary number quantity is a distinct disadvantage in working with this code despite the numerous hardware advantages. Digital engineers recognized this problem early and developed a special form of binary code that was more compatable with the decimal system. Because so many digital devices, instruments and equipment use decimal input and output, this special code has become very widely used and accepted. This special compromise code is known as binary coded decimal (BCD). The BCD code combines some of the characteristics of both the binary and decimal number systems.

The BCD code is a system that represents the decimal digits 0 through 9 with a four-bit binary code. This BCD code uses the standard 8421 position weighting system of the pure binary code. The standard 8421 BCD code and the decimal equivalents are shown in Table II. As with the pure binary code, you can convert the BCD numbers into their decimal equivalents by simply adding together the weights of the bit positions whereby the binary 1's occur. Note, however, that there are only ten possible valid four bit code arrangements. The 4-bit binary numbers representing the decimal numbers 10 through 15 are invalid in the BCD system.

TABLE II

8421 BCD CODE

DECIMAL	BCD
0	0000
1	0001
2	0010
3	0011
4	0100
5	0101
6	0110
7	0111
8	1000
9	1001

To represent a decimal number in BCD notation you simply substitute the appropriate four bit code for each decimal digit. For example, the number 834 in BCD would be 1000 0011 0100. Each decimal digit is represented by its equivalent 8421 four bit code. A space is left between each four bit group in order to avoid confusing the BCD format with the pure binary code.

The beauty of the BCD code is that the ten BCD code combinations are very easily remembered. Once you begin to work with binary numbers on a regular basis you will find that the BCD numbers will come to you as quickly and automatically as decimal numbers. For that reason by simply glancing at the BCD representation of a decimal number you can make the conversion almost as quickly as if it were already in decimal form.

While the BCD code does help to simplify the man-machine interface it is less efficient than the pure binary code. It takes more bits to represent a given decimal number in BCD than it does with pure binary notation. For example, the decimal number 83 in pure binary form is 1010011. In BCD code the decimal number 83 is written as 1000 0011. In the pure binary code it takes only a 7 bit word to represent the number 83. In BCD form it takes 8 bits. The inefficiency arises out of the fact that for each bit in a data word there is usually a certain amount of digital circuitry associated with it. The extra circuitry associated with the BCD code costs more, increases equipment complexity, and consumes more power. Arithmetic operations with BCD numbers are also more time consuming and complex than those with pure binary numbers. As you recall, with four bits of binary information we can represent a total of $2^4 = 16$ different states or the decimal number equivalents 0 through 15. In the BCD system we waste six of these states (10-15), thus the inefficiency. When we use the BCD number system we trade off some efficiency for the improved communications between the digital equipment and the human operator.

Special Binary Codes

Besides the standard pure binary coded form, the BCD numbering system is by far the most widely used digital code. You will find one or the other in most of the applications that you encounter. However, there are several other codes that are used for special applications. Let's consider some of these.

Excess 3 Code. The excess 3 code (sometimes abbreviated XS3) is not a weighted code. Here each four bit number is three greater than the standard 8421 code number. To obtain the excess 3 code of a specific decimal number, you simply add 3 to it then use the corresponding 8421 binary code (Table I). For example, the excess 3 code for the number 7 is the binary equivalent of $7 + 3 = 10$. The four bit excess 3 code for the number 7 then is 1010. To convert from XS3 to decimal, you simply write the decimal equivalent of each four bit group then subtract three from each digit. The excess 3 code was developed primarily because of its ability to greatly simplify the arithmetic computations with BCD numbers.

Gray Code. The Gray Code is a widely used non-weighted code system. Also known as the cyclic, unit distance or reflective code, the Gray code can exist in either the pure binary or BCD formats. The Gray code is shown in Table III. As with the pure binary code, the first ten codes are used in BCD operations. Notice that there is a change in only one bit from one code number to the next in sequence. You can get a better idea about the Gray code sequence by comparing it to the standard four bit 8421 pure binary code also shown in Table III. For example, consider the change from 7 (0111) to 8 (1000) in the pure binary code. When this change takes place all bits change. Bits that were 1's are changed to 0's and 0's are changed to 1's. Now, notice the code change from 7 to 8 in the Gray code. Here 7 (0100) changes to 8 (1100). Only the first bit changes.

TABLE III

THE GRAY CODE

DECIMAL	GRAY	PURE BINARY
0	0000	0000
1	0001	0001
2	0011	0010
3	0010	0011
4	0110	0100
5	0111	0101
6	0101	0110
7	0100	0111
8	1100	1000
9	1101	1001
10	1111	1010
11	1110	1011
12	1010	1100
13	1011	1101
14	1001	1110
15	1000	1111

The Gray code is generally known as an error-minimizing code because it greatly reduces the possibility of ambiguity in the electronic circuitry when changing from one state to the next. When binary codes are implemented with electronic circuitry, it takes a finite period of time for bits to change from 0 to 1 or 1 to 0. These state changes can create timing and speed problems. This is particularly true in the standard 8421 codes where many bits change from one combination to the next. When the Gray code is used, however, the timing and speed errors are greatly minimized because only one bit changes at a time. This permits code circuitry to operate at higher speeds with fewer errors.

The biggest disadvantage of the Gray code is that it is difficult to use in arithmetic computations. Where numbers must be added, subtracted or used in other computations, the Gray code is not applicable. In order to perform arithmetic operations the Gray code number must generally be converted into pure binary form.

ASCII Code. The ASCII code is a special form of BCD code that is widely used in digital computers and data communications equipment. This code is known as the American Standard Code for Information Interchange (ASCII). It is a 7 bit binary code that is used in transferring data between computers and their external peripheral devices and in communicating data by radio and telephone lines. With 7 bits we can represent a total of $2^7 = 128$ different states or characters. The ASCII code is used to represent the decimal numbers 0 through 9, the letters of the alphabet (both upper and lower case) plus other special characters used for controlling various computer peripheral devices and communications circuits. The standard ASCII code is shown in Table IV.

TABLE IV
AMERICAN STANDARD CODE FOR INFORMATION INTERCHANGE

ROW	COLUMN	0	1	2	3	4	5	6	7
	BITS 4321 ⇩ / 765 ⇨	000	001	010	011	100	101	110	111
0	0000	NUL	DLE	SP	0	@	P	`	p
1	0001	SOH	DC1	!	1	A	Q	a	q
2	0010	STX	DC2	"	2	B	R	b	r
3	0011	ETX	DC3	#	3	C	S	c	s
4	0100	EOT	DC4	$	4	D	T	d	t
5	0101	ENQ	NAK	%	5	E	U	e	u
6	0110	ACK	SYN	&	6	F	V	f	v
7	0111	BEL	ETB	'	7	G	W	g	w
8	1000	BS	CAN	(	8	H	X	h	x
9	1001	HT	EM	)	9	I	Y	i	y
10	1010	LF	SUB	*	:	J	Z	j	z
11	1011	VT	ESC	+	;	K	[	k	{
12	1100	FF	FS	,	<	L	\	l	¦
13	1101	CR	GS	–	=	M	]	m	}
14	1110	SO	RS	.	>	N	⌒	n	~
15	1111	SI	US	/	?	O	—	o	DEL

Explanation of special control functions in columns 0, 1, 2 and 7.

NUL *Null*
SOH *Start of Heading*
STX *Start of Text*
ETX *End of Text*
EOT *End of Transmission*
ENQ *Enquiry*
ACK *Acknowledge*
BEL *Bell (audible signal)*

BS *Backspace*
HT *Horizontal Tabulation (punched card skip)*
LF *Line Feed*
VT *Vertical Tabulation*
FF *Form Feed*
CR *Carriage Return*
SO *Shift Out*
SI *Shift In*
SP *Space (blank)*

DLE *Data Link Escape*
DCI *Device Control 1*
DC2 *Device Control 2*
DC3 *Device Control 3*
DC4 *Device Control 4*
NAK *Negative Acknowledge*
SYN *Synchronous Idle*
ETB *End of Transmission Block*
CAN *Cancel*
EM *End of Medium*
SUB *Substitute*
ESC *Escape*
FS *File Separator*
GS *Group Separator*
RS *Record Separator*
US *Unit Separator*
DEL *Delete*

The 7-bit ASCII code for each number, letter or control function is made up of a 4 bit group and a 3 bit group. Figure 1-7 shows the arrangement of these two groups and the numbering sequence. The 4-bit group is on the right and bit 1 is the LSB. Note how these groups are arranged in rows and columns in Table IV.

To determine the ASCII code for a given number letter or control operation, you locate that item in the table. Then you use the three and four bit codes associated with the row and column in which the item is located. For example, the ASCII code for the letter L is 1001100. It is located in column 4, row 12. The most significant 3 bit group is 100 while the least significant four bit group is 1100.

There are both 6 and 8 bit special versions of the ASCII code. In addition, the International Business Machines Corporation (IBM) uses another 8 bit coding system called Extended Binary Coded Decimal Interchange Code (EBCDIC) instead of ASCII, for its peripheral and data communications operations.

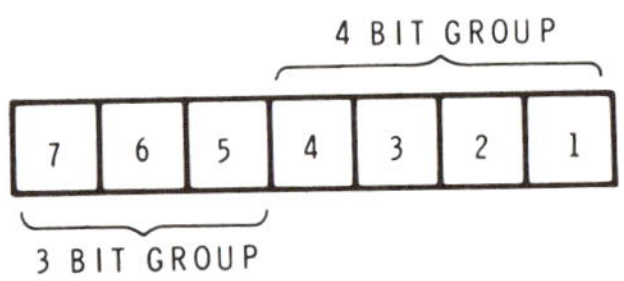

Figure 1-7
ASCII Code Word Format

DATA REPRESENTATION

Now that you understand the reason for using the binary number system and are familiar with some of the binary coding schemes used in digital equipment, you are ready to consider the actual hardware means of implementing these binary numbers. By hardware we mean the electronic components and circuits that are used to represent and manipulate the binary numbers used in the digital system. It is relatively easy to represent a binary number with electronic components. The component to represent a specific bit in a binary word must be capable of assuming two distinct states. One of the states will represent a binary 0 and the other a binary 1.

Electromechanical Devices

Switches and relays are ideal for representing binary data. A closed switch or relay contact can represent a binary 1 while the open switch or contact can represent a binary 0. Of course these logic representations can also be reversed. Switches and relays are still widely used to implement digital systems or parts of digital equipment. They are used in places where static binary conditions are required or very low speed operation can be tolerated.

Early digital equipment such as computers and test instruments used relays to represent binary numbers. But the relays were soon replaced by vacuum tubes in many applications. Each bit was represented by a vacuum tube that was either conducting or cut-off. When the tube was conducting it represented one binary state and when it was cut-off it represented the other binary state. Vacuum tubes worked well in digital applications. They achieved operating speeds significantly higher than that of relays. However, because of their large size, high power consumption and speed limitations they have been replaced by solid state devices.

Transistors

Today the most common way of representing binary data in digital equipment is with a transistor. A transistor can readily assume two distinct states, conducting and cut-off. When a transistor is cut-off it is essentially an open circuit. When a transistor is conducting heavily, it acts as a very low resistance and accurately simulates a closed switch. Most digital circuitry in use today uses saturated bipolar switching transistors for data representation. Non-saturated bipolar transistor switches are also used in many applications where high speed operation is desirable. Keep in mind that both discrete component and integrated circuit transistors are used in digital applications.

The enhancement mode MOS field effect transistor (MOS FET) is also widely used as a two state switch to represent binary data. This type of transisor is the key element in MOS and CMOS integrated circuits.

Magnetic Cores

Another common means of representing binary data is by the use of a magnetic core. This is a small donut shaped piece of magnetic material approximately one tenth inch in diameter or smaller. A wire passing through the center of the core carries a current that magnetizes the core in either of two directions depending upon the direction of current flow in the wire. One direction of magnetization represents a binary 0 and the other a binary 1. The permeability of the magnetic material is such that the core retains the magnetization once the current through the wire is removed. This means that the core can store one bit of binary data. It remembers in which direction it was magnetized so that the state of the core can be determined later. Such cores are the primary data storage medium for many digital computers.

Logic Levels

The basic element for representing a single bit of data is a switch: mechanical, electromechanical, electronic or magnetic. The on-off nature of a switch makes it perfect for binary data representation. The exact relationship between the state of the switch and the bit represented by this switch is arbitrary. In actual digital hardware we are not so much concerned with whether the transistor is off or on. Instead the bit assignments are generally represented by voltage levels. The switching element controls these voltage levels. For example, a binary 0 may be represented by 0 volts or ground. A binary 1 may be represented by +5 volts. Depending upon the equipment power supplies available, the exact circuitry used, and the application; almost any voltage level assignments can be used.

Figure 1-4 shows two ways in which a bipolar transistor can be used to produce two distinct voltage levels. In figure 1-4A, the transistor is connected as a shunt switch. This means that the transistor is in parallel with the output. When the transistor is not conducting, the output voltage is +5 volts as seen through collector resistor Rc. When the transistor is conducting it acts as a very low resistance or near short circuit. At this time the output is some low positive voltage level near ground or zero volts. The switching of the transistor of course is controlled by the application of the appropriate base signal. Switching times in the nanosecond (10^{-9} seconds) region are possible with modern transistors.

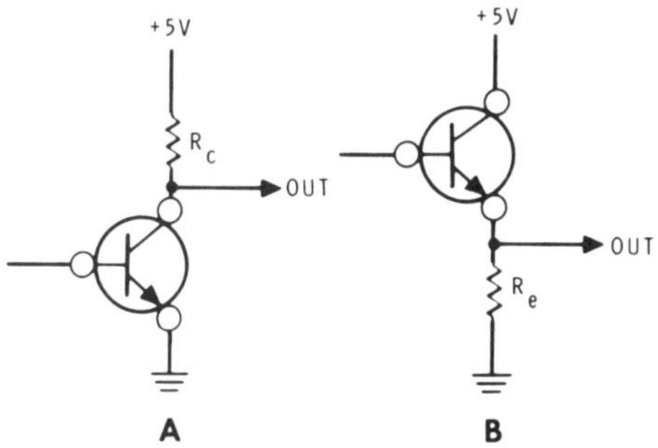

Figure 1-4
Bipolar transistor logic switches, (A) shunt switch (B) series switch.

In Figure 1-4B the transistor is connected as a series switch. When the transistor is cut off the output is zero volts or ground as seen through resistor R_E. When the transistor conducts, it acts as a very low impedance and connects the +5 volts supply line to the output. Again the operation of the transistor is controlled by applying the appropriate signal to the base. You will find both series and shunt transistor switches used in digital circuits.

Positive and Negative Logic

There are two basic types of logic level representation, positive logic and negative logic. When the most positive of two voltage levels is assigned the binary 1 state we say that positive logic is being used. When the negative or least positive of two voltage levels is assigned to the binary 1 state, we say that negative logic is being used. Indicated below are several examples of both positive and negative logic level assignments. Keep in mind that the assignments are strictly arbitrary and are selected by the designer when the circuit or equipment is designed.

Positive Logic	Negative Logic
binary 0 = +.2V. binary 1 = +3.4V.	binary 0 = +3.4V. binary 1 = +.2V.
binary 0 = −6V. binary 1 = 0V.	binary 0 = 0 V. binary 1 = −6V.
binary 0 = +1V. binary 1 = +15V.	binary 0 = +15V. binary 1 = +1V.

Parallel vs. Serial Data Representation

There are two basic ways in which digital numbers are transmitted, processed or otherwise manipulated. These methods are designated as serial and parallel. In the serial methods of data handling, each bit of binary word or number is processed serially one at a time. In a parallel system all bits of a word or a number are processed simultaneously.

Serial Data. Figure 1-5 shows a binary number represented in a serial data format. The binary number exists as a series of voltage levels representing the binary 1s and 0s. These voltage level changes occur at a single point in a circuit or on a single line. Each bit of the word exists for a specific interval of time. The time interval allotted to each bit is in this example, one millisecond. The most significant bit (MSB) is the one at the far left. It occurs first since time is considered to be increasing from left to right. Because this is an 8 bit binary word, it takes 8 milliseconds for the entire word to occur or be transmitted. Positive logic level assignments are used. By observing the voltage levels at the specific point or on the transmission line, the number can be determined. The number is 10110010. This is the binary equivalent of the decimal number 178.

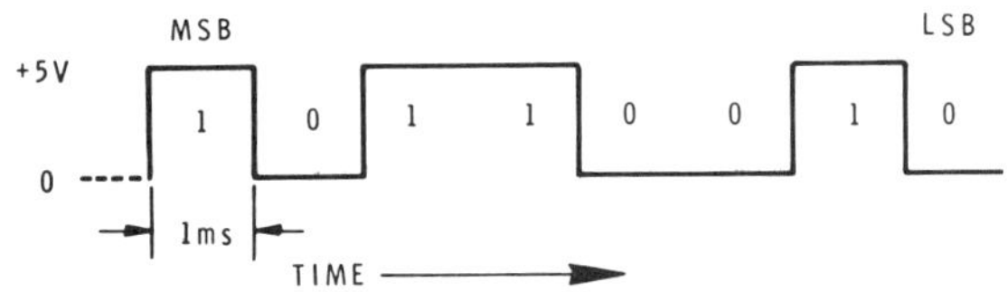

Figure 1-5
The serial binary word 10110010

The primary advantage of the serial binary data representation is that it requires only a single line or channel for transmitting it from one place to another. In addition, since each bit on the single line occurs separately from the others then only one set of digital circuitry is generally needed to process this data. For these reasons, serial data representation is the simplest and most economical of the two types. Its primary disadvantage is that the transmission and processing time required for a serial word is significant since the bits occur one after the other. Despite this time penalty, serial data representation is widely used because of its economy and simplicity.

Parallel Data. The other method of representing, transmitting and processing binary data is designated as parallel. The reason for this is that all bits of a binary word or number are transmitted or processed simultaneously. For this reason a separate line or channel is required for each bit of the word in transmitting that word from one point to another. Refer to Figure 1-6. Here the 8-bit digital word 10110010 is available as voltage levels on eight separate output lines. Since all of the bits of the word are available at the same time, digital circuitry must be provided to process or otherwise manipulate each of the bits in the word simultaneously. The transmission and processing of parallel data, therefore, is more complex and expensive than that required for serial data. However, the clear advantage of parallel data transmission is its speed. All bits are processed at the same time and, therefore, the time required for handling of the data is very short. For high speed applications requiring rapid processing, parallel digital techniques are preferred.

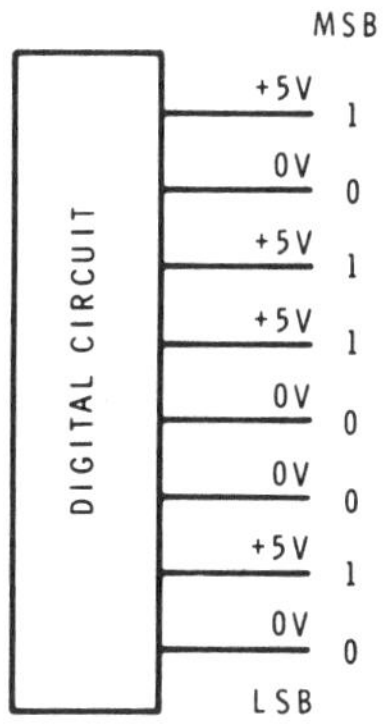

Figure 1-6
The parallel binary word 10110010

Logic Circuits

The transistor is by far the most common way of representing binary data in modern digital equipment. These transistors are combined with other electronic components to form digital circuits. Digital, or logic, circuits are used to process or manipulate the binary information in some way. The primary emphasis in this program is on digital logic circuits and how they are designed and used.

There are two basic types of logic circuits, decision-making and memory. Our ability to think logically requires both a capability for making decisions based on given information and the ability to remember facts. This holds true for digital circuits and equipment as well. Logic circuits accept input binary data and based upon that information generate output binary signals that represent the results of a built in decision making capability or they store the input data for later use.

Decision making logic circuits are called gates. Such circuits have two or more inputs and a single output. Both the inputs and outputs are binary signals. These gates are combined in a variety of ways to form combinational logic circuits that can perform a wide range of sophisticated decision making functions.

The memory circuits store binary data. Commonly called flip flops, these circuits remember a single bit of data. Flip flops are combined to form a class of logic circuits called sequential circuits. Such circuits store, count, and shift binary data and otherwise manipulate it.

All digital equipment is made up of both gates and flip flops that form functional combinational and sequential circuits. Such functional circuits perform the specific operations called for by the application. The remaining lessons in this program are concerned with the detailed design and application of such circuits.

The key components used in digital circuits are semiconductors such as diodes and transistors. These are combined with resistors, capacitors and other electronic components to form the circuits that produce the various logic functions. Early digital circuits were made up of discrete components, individual components interconnected to form the circuit. Today, most digital logic is in integrated circuit form. These are microminiature semiconductor devices that are complete logic circuits constructed in a single package.

Since semiconductors are the basis for all modern digital circuits, you will find a knowledge of their operation helpful in designing, using or troubleshooting digital equipment. The next Unit in this program covers this important subject.

APPENDIX

Table of Powers of 2

2^n	n	2^{-n}
1	0	1.0
2	1	0.5
4	2	0.25
8	3	0.125
16	4	0.062 5
32	5	0.031 25
64	6	0.015 625
128	7	0.007 812 5
256	8	0.003 906 25
512	9	0.001 953 125
1 024	10	0.000 976 562 5
2 048	11	0.000 488 281 25
4 096	12	0.000 244 140 625
8 192	13	0.000 122 070 312 5
16 384	14	0.000 061 035 156 25
32 768	15	0.000 030 517578 125
65 536	16	0.000 015 258 789 062 5

n = number of bits

Unit 2

SEMICONDUCTOR DEVICES FOR DIGITAL CIRCUITS

INTRODUCTION

At the heart of all modern digital circuits are semiconductor devices like diodes and transistors. Your ability to understand digital circuits and apply them to practical situations depends directly on a knowledge of semiconductors. The purpose of this Unit is to provide you with a solid background in semiconductor fundamentals as they apply to digital circuits, both discrete components as well as integrated circuits. For many of you, this Unit will be primarily a review. It will help you to remember the key points about semiconductor devices. For some of you the information may be mostly new. We have developed it to give you exactly what you need to understand digital circuits.

Since most engineers, technicians, and experimenters are users of semiconductor devices rather than designers, our discussion here will focus on practical operation and application rather than detailed coverage of internal device physics.

THE BIPOLAR TRANSISTOR SWITCH

The basic component used in implementing any digital logic circuit is a switch. Modern digital integrated circuits use a high speed transistor switch as the primary component. There are two basic types of transistor switches used in implementing digital integrated circuits, the bipolar transistor and the metal oxide semiconductor field effect transistor (MOS FET). An understanding of these two devices is pertinent to the operation, capabilities and limitations of the various types of digital integrated circuits.

The primary function of a transistor switch in a digital logic circuit is to alternately connect and disconnect a load to and from the circuit power supply. In doing this the transistor switch produces two distinct voltage levels across the load which represent the binary 0 and binary 1 states. The transistor switch should make and break these connections as quickly and efficiently as possible.

The most commonly used digital switch is the bipolar transistor. In digital applications, the bipolar transistor operates as an off/on or two state device. In one state the transistor is non-conducting or cut-off and acts as essentially an open circuit. In the other state, the transistor is conducting heavily and acts as a very low resistance approaching a short circuit. A two state logic input signal is applied to the transistor to produce this on/off operation.

Modes of Operation

A bipolar transistor has three basic regions or modes of operation: cut-off, linear or active and saturation. All three of these modes are used in digital circuits, the cut-off and linear mode in non-saturated bipolar circuits and the cut-off and saturation modes in saturated bipolar circuits.

Cut-Off. In the cut-off mode the transistor is nonconducting. Both the emitter-base (E-B) and collector base (C-B) junctions are reverse biased or not biased at all to produce the cut-off state. In theory, no emitter or collector current flows, and the transistor acts as an open circuit between emitter and collector. In most practical transistors, however, the cut-off is not perfect. Because of imperfections in the semiconductor material out of which the device is made, some leakage current flows. In most modern transistors this leakage current is extremely low and for most practical applications can be neglected. However, where very high temperature operation is expected the leakage current becomes a more important consideration. In silicon transistors, the leakage current nearly doubles for each 10°C rise in temperature.

Linear. The linear or active mode of operation is characterized by a forward biased emitter-base junction and a reverse biased collector-base junction. In this mode, the transistor does conduct. Emitter and collector current flows. The emitter and collector currents are directly proportional to the base current variations. The emitter and collector currents are simply amplified versions of the base current variation. In this mode of operation, the transistor functions as a variable resistance and is used to amplify or otherwise process analog signals. This region is of concern in non-saturating digital integrated circuits.

Saturation. The third mode of bipolar transistor operation is saturation. In this mode, both the emitter-base and collector-base junctions are forward biased. The transistor conducts heavily and acts as a very low resistance. The resistance between the emitter and collector is very low, approaching that of a short circuit.

In digital applications, the bipolar transistor is usually switched between the cut-off and saturation states. As it switches, the transistor passes quickly through the linear region. The primary responsibility of the designer of a digital circuit is to see that the bipolar transistor switches as quickly as possible between cut-off and saturation and that these two states are as stable as possible. In non-saturating digital circuits, the transistors switch between cut-off and the linear region. Again, high speed is the primary requirement.

Saturated Switching Circuits

Figure 2-1 shows the most common form of saturated bipolar transistor logic switch. Here, the transistor is connected as a shunt switch since it is in parallel with the output load R_L. This circuit is also known as a transistor logic inverter.

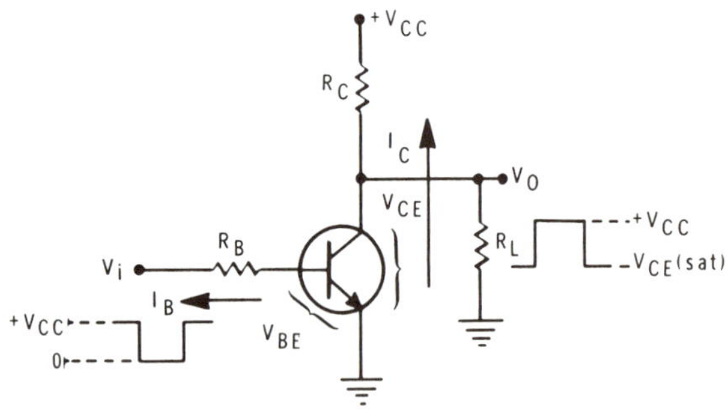

Figure 2-1 Basic transistor switching circuitry

With an input voltage V_i of zero volts or ground, the transistor is cut-off. The emitter-base junction is not forward biased, therefore the transistor does not conduct. The only collector current flowing at this time is a minute leakage current. The output voltage V_o, with no load, is equal to the supply voltage V_{CC} as seen through collector resistor R_C. If a finite resistive load R_L is connected between the output and ground V_o will be some value less than V_{CC} and dependent upon the division ratio between R_C and R_L.

$$V_O = V_{CC}\left(\frac{R_L}{R_C + R_L}\right)$$

When an input voltage V_i of sufficient amplitude is applied to base resistor R_B, the emitter-base junction will become forward biased and the transistor will conduct. The transistor will be in the linear or saturation regions depending upon the size of V_i, the value of R_B and the gain (Beta or h_{FE}) of the transistor.

Figure 2-2 shows the typical input and output waveforms of a transistor switching circuit. The input switches between zero volts (LOW) and V_i (HIGH). When the input is LOW, the transistor is cut-off so you see an output voltage V_O that is equal to V_{CC} or some value slightly less (HIGH). When the input is HIGH, the transistor saturates and acts as a low resistance. The output voltage V_O is the collector-emitter saturation voltage V_{CE} (sat) which is only a few tenths of a volt (LOW). As you can see, the output is HIGH when the input is LOW, and the output is LOW when the input is HIGH. The input and output signals are always opposite one another. This is the reason for the name inverter.

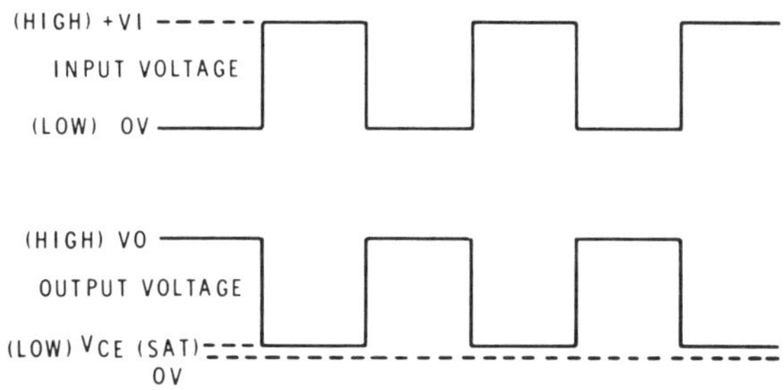

Figure 2-2 Input and output waveforms of a shunt transistor switching circuit

In a digital circuit the magnitude of V_i and R_B are such that sufficient base current flows in order to cause the transistor to saturate. This condition occurs when the actual base current I_B is greater than the ratio of the collector current I_C to the dc forward current transfer ratio h_{FE} (also known as the dc current gain or Beta).

$$I_B > \frac{I_C}{h_{FE}} \quad \text{(for saturation)}$$

The base current is a direct function of the applied input voltage V_i and the value of the base resistor R_B. This relationship is

$$I_B = \frac{V_i - V_{BE}}{R_B}$$

where V_{BE} is the voltage across the forward biased emitter-base junction.

As you can recall, the ratio of the collector current (I_C) to the base current (I_B) in a common emitter circuit is known as the dc forward current transfer ratio or h_{FE}. This ratio

$$h_{FE} = \frac{I_C}{I_B}$$

expresses the effective gain of the device or the ability of the base current to control the larger collector current. The greater this ratio the higher the gain.

By algebraically rearranging this expression, we can calculate the base current for a given collector current and gain.

$$I_B = \frac{I_C}{h_{FE}}$$

If we design the circuit so that the base current is greater than this ratio, the transistor will saturate. That is, both emitter-base and collector-base junctions will become forward biased. In this state the transistor is conducting heavily and the resistance between the emitter and collector is very low. For typical switching transistors this value of resistance is in the 5 to 30 ohms range. The voltage drop between the collector and emitter during saturation V_{CE} (sat) is only several tenths of a volt. This is very low compared to the supply voltage V_{CC} and therefore for most practical purposes is considered to be nearly zero volts.

During saturation the amount of emitter and collector current flowing becomes basically a function of the value of the supply voltage V_{CC} and the collector resistance R_C. Because the voltage drop across the transistor is essentially zero, the collector current can be found from the expression

$$I_C = \frac{V_{CC} - V_{CE}\,(\text{sat})}{R_C} \approx \frac{V_{CC}}{R_C}$$

This relationship holds true only if sufficient base current flows to saturate the transistor. If the value of the base current is less than the ratio of I_C/h_{FE}, the transistor will be operating in the linear region. The emitter-base junction will be forward biased but the collector base junction will be reverse biased. The collector-emitter voltage V_{CE} will be correspondingly higher.

There are two ways you can determine whether a transistor is saturated or operating in the linear region. The first method is to measure the junction potentials on the transistor. When measuring these voltages it is important to note the polarity of each transistor element with respect to the other. By knowing the magnitudes of the junction voltages and their relative polarities you can establish the state of the transistor.

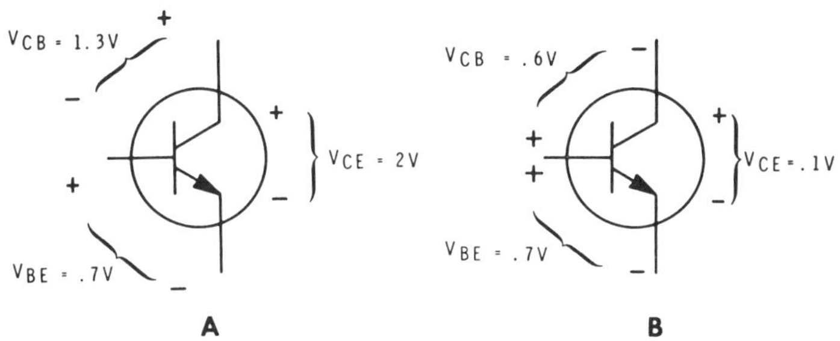

Figure 2-3 Polarity and voltage relationships in a conducting transistor (A) linear operation and (B) saturation

Figure 2-3 shows typical junction voltages for an NPN transistor both in the saturated and unsaturated states. The transistor in Figure 2-3A is operating in the linear or active region. The emitter-base junction is forward biased because junction voltage V_{BE} is of the proper polarity. A conducting junction in a silicon transistor typically has a voltage drop of approximately .7 volt. Observing the collector-base junction we see that the voltage across it (V_{CB}) is 1.3 volts with the collector being more positive than the base. This indicates a reverse biased condition on the collector-base junction. The voltage drop between the emitter and collector V_{CE} is two volts. Note the relationship between V_{CE}, V_{BE} and V_{CB}.

$$V_{CE} = V_{BE} + V_{CB}$$

In Figure 2-3B, the transistor is saturated. Again V_{BE} is approximately .7 volt with the polarity indicated. The big difference however is the polarity change in V_{CB}. Here the base is more positive than the collector indicating a forward biased condition on this junction. The junction voltage drop is approximately .6 volt. Again note that the collector-emitter voltage V_{CE} is the *algebraic* sum of (or difference between) V_{CB} and V_{BE} and in this case is only .7 − .6 = .1 volt. Because of this low voltage drop the effective resistance of the transistor is extremely low. Another beneficial characteristic of saturated operation is that the low collector-emitter voltage greatly minimizes the power dissipation in the conducting transistor. The power dissipation is

$$P = V_{CE} \cdot I_C$$

With V_{CE} so low during saturation, the power dissipation is also very low even though the collector current may be large.

The other method of determining whether or not a transistor is in saturation is to find both the base and collector current through actual circuit measurements and then determine if the base current is less than or greater than the ratio of the collector current and the dc current gain h_{FE}. As indicated before, if $I_B > I_C/h_{FE}$ then the transistor is saturated. If $I_B < I_C/h_{FE}$, the transistor is operating in its active region. The actual value of h_{FE} depends upon the type of transistor being used, the current level in the transistor, its temperature and other factors. The current gain value varies even among transistors of the same type. To ensure saturation in a switching transistor, designers generally make the base resistor small enough to produce a base current with a given minimum logic input voltage that is greater than the I_C/h_{FE} ratio. This safety factor is necessary in order to ensure saturation under all conditions. When the base current is greater than the I_C/h_{FE} ratio, we say that we are overdriving the base.

Switching Speed

One of the most important characteristics of a logic circuit is its ability to switch rapidly between the binary logic levels. This switching speed is affected by the transistor characteristics, the circuit component values, stray capacitance and inductance, the current and voltage levels in the circuit and the specific circuit configuration. When the input signal to a digital circuit changes from one logic level to the other, the output of the circuit does not change instantaneously in response. Instead there is a delay time existing between the change in the input signal and the corresponding change in the output. This time lag is generally referred to as propagation delay.

The turn on time of a transistor is primarily a function of the transistor characteristics and the amount of base drive applied to the circuit. A heavy base current helps to ensure a rapid turn on.

The turn off time delay is affected mainly by the transistor characteristics. The turn off of the transistor is delayed because of storage time. When a transistor is saturated, an excess of minority carriers (holes in an NPN and electrons in a PNP transistor) build up in the collector-base junction region. This charge storage keeps the transistor conducting even with the base drive removed. It takes a finite period of time for this charge to be removed so that the transistor begins to come out of saturation. This storage time is a function of the transistor characteristics and the amount of base drive.

By using the proper transistors, circuit configuration and component values, the switching delay times can be significantly reduced. Despite all of the factors that limit the switching speed, modern high speed switching transistors can change from one state to the next in only a matter of nanoseconds (10^{-9} seconds).

Non-Saturating Switching Circuits

The most serious limitation to the switching speed of a bipolar transistor is the storage time associated with the condition of saturation. The charge storage build up in the collector-base region during saturation takes a finite time to be cleared away in order to turn the transistor off. This storage time is the most significant part of the turn off time and therefore any means of minimizing it will greatly increase the switching speed of the transistor. Special bipolar switching transistors have been designed to help minimize this storage time effect and external circuitry can be adjusted to minimize it to some degree. The obvious way to increase switching speeds, therefore, is to avoid saturation. By keeping the transistor from going into saturation no charge storage occurs therefore very high switching speeds can be achieved. A number of logic circuits have been designed using non-saturating bipolar transistor switches. The transistors switch between the cut-off and linear regions. Such circuits are capable of switching at frequencies as high as 1 GHz.

DESIGNING A SATURATED SWITCH LOGIC INVERTER

While most modern digital equipment is implemented with integrated circuits it is sometimes necessary or desirable to supplement the IC with a descrete component circuit to perform a special function. This includes things such as logic level conversion, driving an indicator light such as an incandescent lamp or light emitting diode (LED) or operating a relay. All of these circuits use a saturated bipolar transistor switch. Several examples are given in Figure 2-4.

The circuit in Figure 2-4A is simply an inverter where the input and output logic levels may or may not be equal. Such a circuit is useful in interfacing different types of logic circuits. In Figure 2-4B, the LED will light when an input voltage V_i is applied. The transistor acts as an on/off switch controlled by V_i. This is also true in the relay driver circuit of Figure 2-20C. Contacts A and B of relay K1 are normally open (N.O.) when V_i is zero. When V_i becomes sufficiently positive, the transistor saturates and K_1 is energized. The magnetic field produced by the relay coil closes contacts A and B. Diode D_1 is used to protect the transistor when the input voltage is removed. When the transistor cuts off the magnetic field the coil collapses thereby inducing a high negative voltage spike that can damage the transistor. The voltage spike causes D_1 to conduct and clamp the collector voltage to a safe level.

Such circuits are so common that it is desirable to know how to design them. The procedure is simple as you will see here.

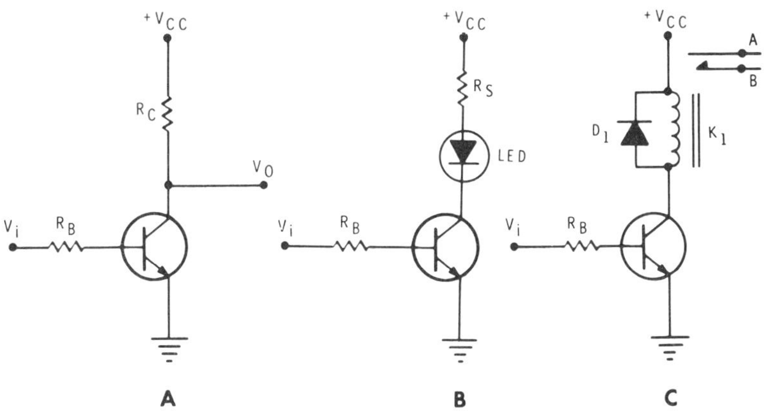

Figure 2-4
Bipolar transistor switch used as a logic inverter (A) an LED indicator driver (B), and a relay driver (C)

Procedure

1. Define the load.

   For a lamp or relay driver, the characteristics of the lamp or relay are given. These usually include a rated voltage (V_L), current and/or resistance. If the circuit is an inverter used to supply a signal to an external load, the load voltage and current or load resistance must be given.

2. Specify a supply voltage.

   The supply voltage (V_{CC}) will usually be equal to or higher than the desired load voltage. Normally, the circuit will operate from an existing supply of some standard value such as +5 volts, +12 volts, etc.

3. Select a suitable transistor.

   A wide variety of types are available. The exact application will guide you in its selection. A transistor designed for switching rather than linear applications is usually preferred. The voltage and current ratings will be determined by the load and supply characteristics. The collector current and voltage breakdown ratings should be at least twice the operating characteristics. Gain (h_{FE}) and operating speed requirements will be dictated by the application. The exact transistor characteristics can be determined from the manufacturer's data sheets.

4. Determine the value of any series dropping resistor.

   If the supply voltage is larger than the load (lamp, relay, etc.) voltage, then some series dropping resistor (R_S) will be needed. See Figure 2-20B. The voltage across this resistance will be the supply voltage (V_{CC}) less the load voltage (V_L) and the transistor saturation voltage V_{CE} (sat). The resistance can then be found with the expression

$$R_S = \frac{V_{CC} - V_L - V_{CE}\,(\text{sat})}{I_C}$$

   where I_C is the load current and the collector current V_{CE} (sat) is very low so can be considered zero. Use the closest standard resistor value. This step does not apply to an inverter with only a collector resistor.

5. Specify an output voltage.

If the circuit being designed as an inverter with a shunt load, the desired output voltage (V_o) should be specified. It will be some value less than V_{CC} depending upon the values of R_C and R_L. See Figure 2-5.

$$V_O = V_{CC} \frac{R_L}{R_C + R_L}$$

If there is no shunt load, $V_O = V_{CC}$.

6. Determine the value of the collector resistor.

If the circuit being designed is an inverter with a shunt load to ground, the collector resistor R_C value can be calculated if the supply voltage (V_{CC}), load resistance (R_L) and desired output voltage (V_O) are known. Since

$$V_O = V_{CC} \left(\frac{R_L}{R_C + R_L}\right)$$

The collector resistance can be found by rearranging this expression.

$$R_C = \frac{R_L (V_{CC} - V_O)}{V_O}$$

Use the closest standard value of resistance.

7. Determine the collector current.

This is the load current specified earlier for lamp or relay drivers. For the inverter arrangement described in Step 6, the current is

$$I_C = \frac{V_{CC} - V_{CE}\,(\text{sat})}{R_C} \approx \frac{V_{CC}}{R_C}$$

Since V_{CE} (sat) is usually only several tenths of a volt it can be considered negligible or zero in the above expression.

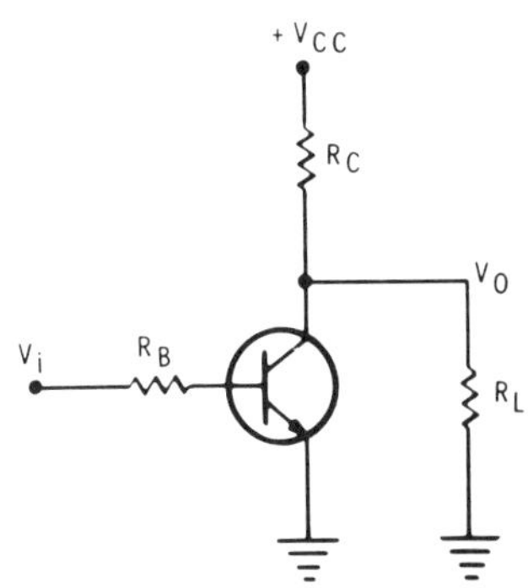

Figure 2-5
Logic inverter with shunt load

8. Calculate the base current.

Knowing the gain (h_{FE}) from the manufacturer's data sheet and the collector current (I_c), the base current (I_B) can be found.

$$h_{FE} = \frac{I_C}{I_B} \quad \text{therefore } I_B = I_C/h_{FE}$$

Use the minimum value of h_{FE} quoted in the data sheet. To provide some overdrive and ensure saturation it is desirable to provide some safety factor by derating the h_{FE} by a factor of 2 (or more if desired). Therefore

$$I_B = \frac{I_C}{h_{FE}/2} \quad \text{or} \quad I_B = \frac{2\ I_C}{h_{FE}\ (\text{min})}$$

9. Calculate the base resistor.

The base resistor (R_B) is found with the expression below.

$$R_B = \frac{V_i - V_{BE}}{I_B}$$

Here V_i is the lowest expected value of base drive voltage. It is the binary 1 voltage level for positive logic. V_{BE} of course is the emitter-base voltage drop which is typically .7 volts for a silicon transistor.

If the circuit being designed is to be driven by another inverter as in Figure 2-6 its supply voltage and collector resistance should be considered.

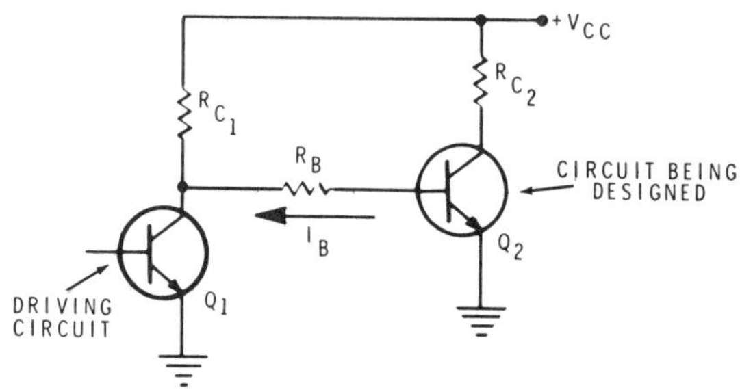

Figure 2-6

When Q_1 cuts off, the driver circuit will be activated by I_B flowing through R_B and the collector resistor R_{C1} of the driving circuit. In this case,

$$R_B = \frac{V_{CC} - I_B R_{C1} - V_{BE}}{I_B}$$

Use the closest lower standard resistance value.

The following examples will illustrate the procedure for two practical applications.

Example Application 1

Design a driver for an LED indicator. The steps below correspond to those in the procedure. See Figure 2-4B.

1. The load is a light emitting diode indicator whose normal brilliance is obtained with 20 ma of current. Its normal operating voltage, the load voltage drop (V_L), is 1.7 volts at this current value.
2. The supply voltage V_{CC} is +5 volts.
3. A type MPSA20 silicon transistor will be used. h_{FE} (min) = 100.
4. Since the LED voltage drop is less than the supply voltage a series dropping resistor is needed. The load (collector) current is 20 ma.

$$R_S = \frac{V_{CC} - V_L - V_{CE}\text{ (sat)}}{I_C}$$

If we consider V_{CE} (sat) negligible then

$$R_S = \frac{5 - 1.7}{.02} = \frac{3.3}{.02} = 165 \text{ ohms}$$

A standard 150 or 180 ohm 10 percent resistor or 160 ohm 5 percent resistor could be used.

5. Not applicable.
6. Not applicable.
7. I_C = 20 ma or .02 amperes.
8. $I_B = \frac{2\ I_C}{h_{FE}} = \frac{2\ (.02)}{100} = \frac{.04}{100} = .4 \text{ ma} = 400\ \mu A$

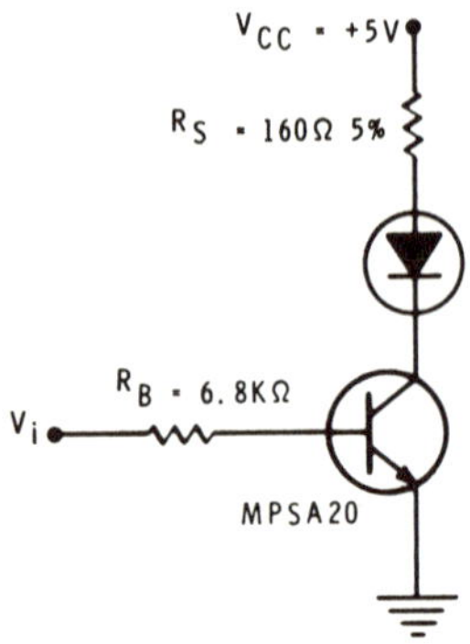

Figure 2-7 LED driver circuit: Example Application 1

9. Assume the input voltage V_i = +3.5 volts. V_{BE} = .7 volts for a silicon transistor.

$$R_B = \frac{V_I - V_{BE}}{I_B} = \frac{3.5 - .7}{.0004} = 7000 \text{ ohms}$$

A standard 6.8k 10 percent value can be used.

See Figure 2-7.

Example Application 2

Design an inverter circuit to apply +6 volts across a 600 ohm load. The supply voltage is +15 volts. The driving signal (V_i) is supplied by the +15 volt supply through a 1000 ohm collector resistor. See Figure 2-22.

1. R_L = 600 ohms V_L = 6 volts.
2. V_{CC} = +15 volts.
3. Use an MPSA20.
4. Not applicable.
5. V_O = 6 volts.
6. $$R_C = \frac{R_L(V_{CC} - V_O)}{V_O}$$

$$R_{C2} = \frac{600\ (15 - 6)}{6} = \frac{600\ (9)}{6} = 900 \text{ ohms}$$

A standard 910 ohm resistor could be used.

7. $$I_C = \frac{V_{CC}}{R_C} = \frac{15}{910} = .0165 \text{ or } 16.5 \text{ ma}$$

8. $$I_B = \frac{2\ I_C}{h_{FE}} = \frac{2\ (.0165)}{100} = .00033 \text{ amp} = .33 \text{ ma} = 330\ \mu A$$

9. $$R_B = \frac{V_{CC} - I_B\, R_{C1} - V_{BE}}{I_B}$$

$$R_B = \frac{15 - (.00033)\ (1000) - .7}{.00033} = \frac{13.97}{.00033} = 42{,}333 \text{ ohms}$$

A standard 38 k ohm resistor could be used. See Figure 2-8.

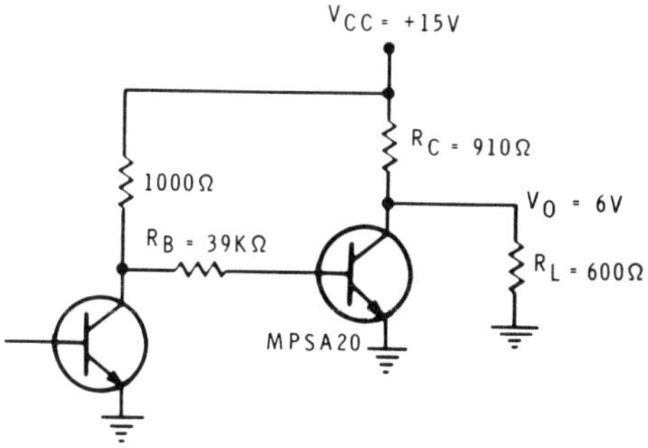

Figure 2-8 Inverter circuit: Example Application 2

MOS FIELD EFFECT TRANSISTORS

Another transistor widely used in digital integrated circuits is the enhancement mode metal oxide semiconductor field effect transistor (MOSFET). Also known as an insulated gate field effect transistor (IGFET), this device offers numerous advantages over the bipolar transistor particularly for digital work and integrated circuits. Because the MOSFET is basically a simple device, it is more easily constructed than a bipolar transistor. Since it can be made smaller than bipolar devices it permits more circuitry to be produced on a given area of semiconductor material. The cost of an integrated circuit is directly proportional to the area of the semiconductor material used to construct the circuit and the complexity of the devices used. Simple, low cost, high density circuits are readily constructed with MOSFET components.

The N-channel MOSFET

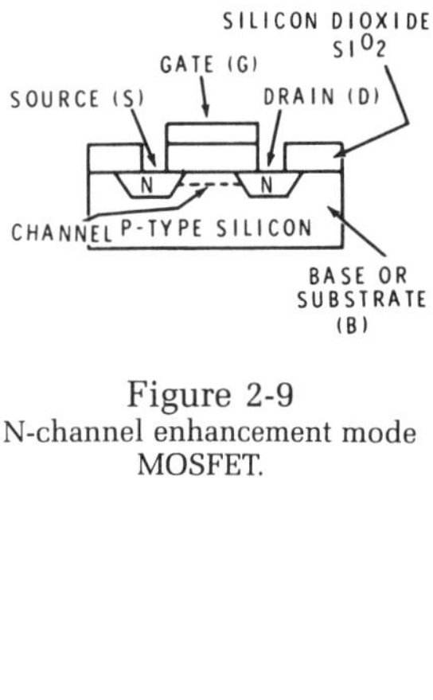

Figure 2-9
N-channel enhancement mode MOSFET.

Figure 2-9 shows the basic construction of an N-channel enhancement mode MOSFET. It is constructed of a P-type silicon base or substrate into which is diffused two N-type semiconductor material areas. These form the source and drain elements of the transistor. On top of this is diffused a thin layer of silicon dioxide, a glass insulator which isolates the source and drain regions from the remainder of the device. On top of the silicon dioxide insulator is formed a third element called the gate. The silicon dioxide insulates the gate terminal from the P-type silicon material. The gate is simply a metallized diffusion such as aluminum or a silicon conductive material that forms a capacitor with the P-type silicon base, the silicon dioxide layer acting as the dielectric. The area beneath the gate dielectric and between the source and drain is known as the channel. If the drain is made positive with respect to the source, current will flow between the source and the drain. It is the level and polarity of the voltage between the source and the gate that determines the conductivity of the channel.

With the gate source voltage equal to zero, no current flows between the source and the drain. The alternate N-type and P-type materials between source and drain effectively form two back-to-back diodes both of which are cut off. However, when the voltage applied between the gate and source exceeds a certain threshold level with the gate positive with respect to the source, an electric field will be established in the channel region. This causes the transistor to conduct and electrons will flow between the source and the drain.

The gate-source arrangement in the MOSFET acts as a capacitor. Applying a gate-source voltage charges this capacitor. The gate becomes positive and the area below the gate in the substrate becomes negative. The majority carriers in the P-type substrate (holes) will be depleted by the negative charge and the electron density will be enhanced. This negative charge in the P-type base establishes a channel for current flow between the two N-type regions. Removing the gate-source voltage or decreasing it below the threshold level will cause the conduction to cease.

The enhancement mode MOSFET is an excellent switch. When the gate voltage is below the threshold value, the resistance between the source and the drain is extremely high and very closely approximates an open circuit. When the gate voltage is above the threshold level, the transistor conducts and the resistance between the source and the drain is very low, approximating a short circuit. Such characteristics make the enhancement mode MOSFET ideal for digital circuits.

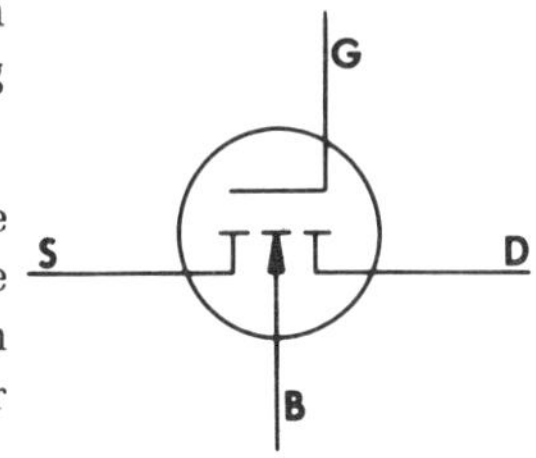

Figure 2-10
Schematic symbol for an N-channel enhancement mode MOSFET.

One of the primary benefits of the MOSFET over the standard bipolar transistor is that the input impedance between source and gate is extremely high. The input impedance between the source and gate is many thousands of megohms and, in effect, is more capacitive than resistive. This high input impedance minimizes the loading of one logic circuit on the next. Figure 2-10 shows the schematic symbol normally used to represent an N-channel enhancement mode MOSFET.

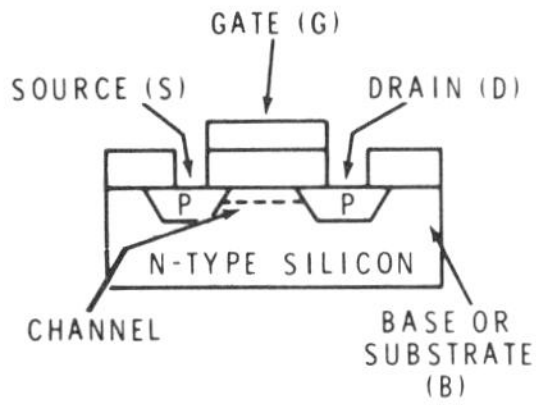

Figure 2-11
P-channel enhancement mode MOSFET.

The P-Channel MOSFET

Figure 2-11 shows the construction of a P-channel enhancement mode MOSFET. It is similar in construction to the N-channel device. However, the substrate is N-type material, while the source and drain diffusions are P-type material. The symbol used to represent this device is shown in Figure 2-12. The operation of the P-channel MOSFET is similar to that for the N-channel device with the exception of the operating voltage polarities. Like the N-channel device, the P-channel MOSFET is normally off if the gate-to-source voltage is below a specific threshold value. When the gate is made negative with respect to the source and is of an amplitude beyond that of the threshold level, the transistor will conduct and current will flow between the source and the drain. In the P-channel device, the current carriers are holes instead of electrons.

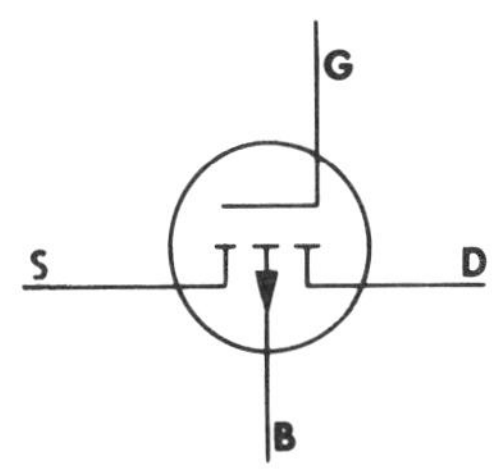

Figure 2-12
Schematic symbol of a P-channel enhancement mode MOSFET.

One of the most important characteristics of the enhancement mode MOSFET is its threshold voltage, that value of gate-source voltage required to cause the transistor to conduct. There are two basic types of threshold in common use; low threshold and high threshold. The high threshold devices are easier to make and, therefore, are more common. This high threshold value is approximately three to four volts. There is circuitry available with a low threshold in the one to two volt region. Typically, P-channel devices have high thresholds, while N-channel devices have low thresholds.

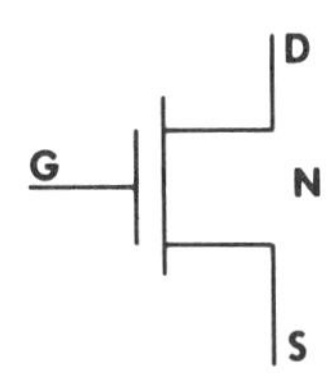

Figure 2-13
Simplified symbol of enhancement mode MOSFET.

Figure 2-13 shows a simplified schematic symbol often used to represent an enhancement mode MOSFET. Because of the complexity of the standard symbols, they are rarely used. The letter P or N is written adjacent to the symbol to designate whether it is a P-channel or an N-channel device.

Bipolars vs. MOSFETs

The primary benefits of the MOSFET over the bipolar transistor are; small size, simplicity of construction, high input impedance and low power consumption. However, these advantages are somewhat offset by the major disadvantage; slow switching speeds. Because of the high impedance nature of the device and its capacitive characteristics, switching speeds are significantly lower than those for bipolar transistors. Recent advances have improved the switching speed of MOSFETs, and switching speeds in the region below 100 nanoseconds are now possible. Nevertheless, the advantages of the MOSFET over the bipolar for many applications offset this switching speed disadvantage. Bipolar transistors are still faster and offer the further advantage of being able to handle high power applications.

MOSFET Circuits

Logic circuits are readily constructed with enhancement mode MOSFETs. Figure 2-14 shows an inverter circuit constructed with a P-channel MOSFET. With the gate input voltage (V_i) at zero, the transistor does not conduct and the output voltage (V_O) is the drain supply voltage (V_{DD}) as seen through the drain resistor (R_D). When the input voltage is made sufficiently negative beyond the threshold value, the transistor conducts and acts virtually as a short circuit. The output voltage at this time is near zero volts. An inverter using an N-channel device would be similar, but with positive logic levels and supply voltage.

In practical MOS circuits, the drain resistor (R_D) is not used. Because it occupies a substantial amount of space in an integrated circuit, it is generally eliminated and replaced by another MOSFET, biased on to act as a resistance. The transistor itself is smaller than a diffused resistor. A very popular type of MOS digital circuit combines both P-channel and N-channel devices to form what is known as a complementary MOS (CMOS) logic circuit.

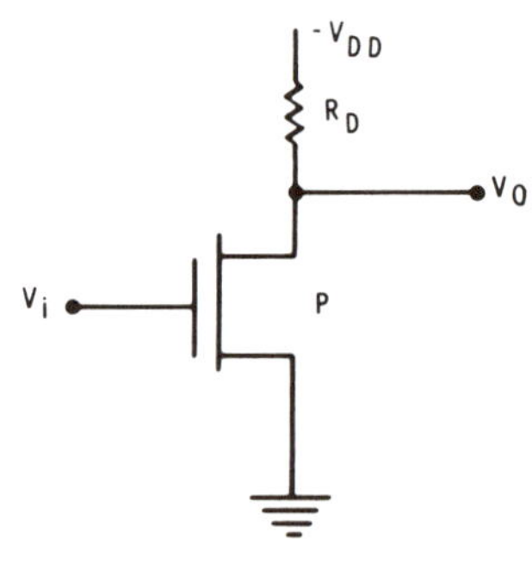

Figure 2-14
P-channel MOS inverter

Unit 3

DIGITAL LOGIC CIRCUITS

INTRODUCTION

All digital equipment, simple or complex, is constructed from just a few basic circuits. These circuits are called logic elements. A logic element performs some specific logic function on binary data.

There are two basic types of digital logic circuits: decision-making and memory. Decision making logic elements monitor binary inputs and produce outputs based on the input states and the operational characteristics of the logic element. Memory elements are used to store binary data.

Whether the digital equipment is a simple piece of test equipment or a large scale digital computer, this equipment is made up entirely of such logic circuits. Once you learn the basic logic elements and the most commonly used forms, you will be able to understand and determine the operation of any piece of digital equipment. By understanding how to apply these basic logic elements you will be capable of designing digital equipment.

In this unit you are going to study the basic types of digital logic circuits. You will learn the operation of decision making and memory circuits. You will also discover several special classes of logic circuits. We will also discuss the practical implementation of these logic circuits using switches and relays, discrete electronic components and integrated circuits. This unit provides the base upon which all of the remaining units in this program are built. The study of digital techniques is, in effect, a study of digital logic circuits and how they are applied. As a result, this is an extremely important background lesson.

TYPES OF LOGIC CIRCUITS

The two basic types of logic circuits are decision making and memory. Both types accept binary inputs and produce binary outputs. The nature of the output is a function of the state of the inputs and the characteristics of the particular logic circuit.

Decision-making logic circuits do exactly what their name implies; make decisions. The basic decision-making logic element is called a gate. A gate has two or more binary inputs and a single output. Figure 3-1 shows the generalized symbol used for representing a logic gate. More specific symbols are used to represent practical types of gates. There are several different types of logic gates, each performing a specific decision-making operation. The gate looks at its binary inputs, and based upon their states and its operation, it generates an appropriate output signal that reflects the decision it has made.

While many simple logic functions can be implemented with a gate, generally gates are combined to form more sophisticated and complex decision making logic networks called combinational circuits. A combinational circuit is made up of two or more gates and has two or more inputs and one or more outputs. Combinational circuits still perform a decision-making function but of a more sophisticated nature. Most combinational circuits perform some unique logic function such as decoding, encoding, multiplexing, comparison or an arithmetic operation with binary numbers. A generalized block diagram of a combinational logic circuit is shown in Figure 3-2.

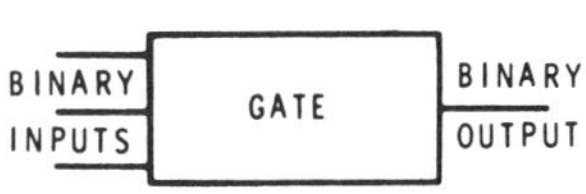

Figure 3-1
Basic symbol of a logic gate.

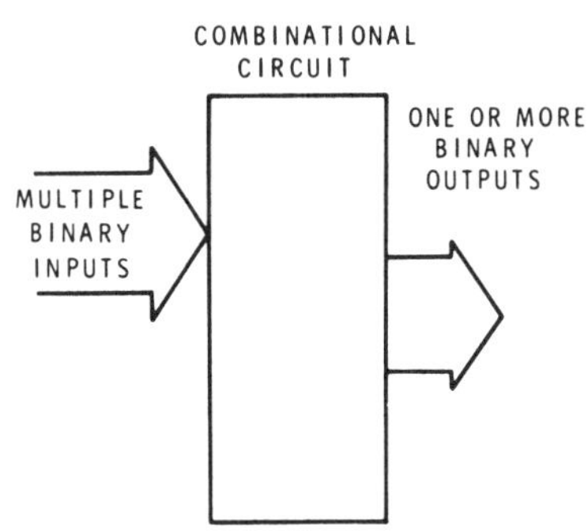

Figure 3-2
General block diagram of a combinational logic circuit.

The other type of logic element is a memory circuit. The basic memory element is a bistable storage device known as a flip-flop. This circuit has two stable states which can represent the two binary numbers 0 and 1. The circuit can be placed into either state so that it retains that state or remembers the bit stored there. Most memory circuits store a single binary bit. Many of these elements can be combined to store complete binary numbers or words.

Most memory elements are interconnected with combinational circuits to form another more sophisticated form of memory element known as a sequential circuit. A general block diagram of a sequential circuit is shown in Figure 3-3. The inputs to the sequential circuit consists of external binary data and feedback signals developed within the sequential circuit itself. The outputs of a sequential circuit are binary signals that are used to operate or control external circuits. The output of the sequential circuit is a function of the binary inputs, the binary data stored in the sequential circuit itself and the specific characteristics of this circuit.

Sequential circuits are used for a variety of operations in digital equipment. Typical sequential circuits consist of counters, shift registers, timers, sequencers and other circuits where binary data is stored and manipulated as a function of time. You will study these circuits in later units.

The three basic decision-making logic elements are the AND gate, the OR gate and the inverter. All other digital logic elements and circuits are variations or combinations of just these three basic elements. Each of these elements receive one or more binary inputs and generates a single binary output.

In order to distinguish one binary input signal from another and in order to identify both binary inputs and outputs, each signal is generally assigned a name or label. In its simplest form the label consists of a letter, a small word or mnemonic (pronounced ne-mon-ik). This label or designation indicates a specific logic signal which can assume either of the two binary states 0 or 1. In the discussions of the logic circuits presented here all inputs and outputs will be given such names and labels. In most cases a simple letter designation will be used. However, keep in mind that these designations may be words or letter and number combinations. Some typical examples are A, X, D3, CLR, JMPZ, etc.

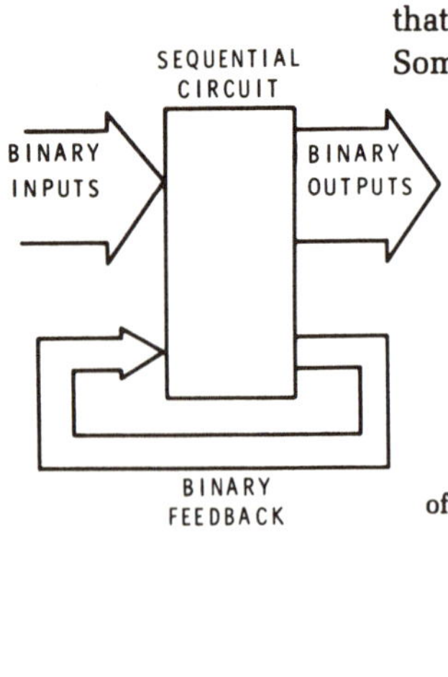

Figure 3-3
General block diagram of a sequential logic circuit.

THE INVERTER

The simplest form of digital logic circuit is the inverter or NOT circuit. The inverter is a logic element whose output state is always opposite of its input state. If the input is a binary 0 the output is a binary 1. If the input is a binary 1, the output is a binary 0. We say that the inverter has an output that is the complement of the input. The binary states 1 and 0 are considered to be complementary.

The operation of the inverter is clearly summarized by a simple chart known as a truth table. The truth table shows all possible input states and the resulting outputs. Figure 3-4 is the truth table for an inverter. The input to the inverter is designated A while the output is labeled $\overline{A}$ (pronounced A NOT or NOT A). The bar over the letter A indicates the complement of A. Note that the truth table shows all possible input combinations and the corresponding output for each. Since the inverter has a single input, there are only two possible input combinations: 0 and 1. The output in each case is the complement or opposite of the input.

The symbols used to represent a logic inverter are shown in Figure 3-5. The triangle portion of the symbol represents the circuit itself while the circle designates the inversion or complementary nature of the circuit. Either of the two symbols may be used. Note the input and output labeling. Such simplified symbols are used instead of the actual electronic schematic in order to simplify the drawing and application of a logic circuit. It is the logic function and not the circuit that is the most important.

The simplest and most widely used form of logic inverter is the transistor switching circuit shown in Figure 3-6. This inverter circuit operates with a binary input signal whose logic levels are 0 volts (or ground) and some positive voltage approximately equal to the supply voltage + Vcc. When the input B to the transistor is 0 volts or ground, the emitter-base junction of the transistor is not forward biased. Therefore, the transistor does not conduct. With the transistor cut-off, the output $\overline{B}$ with respect to ground is the supply voltage + Vcc as seen through collector resistor Rc. As you can see, with positive logic level assignments, the binary 0 input produces a binary 1 output.

INPUT	OUTPUT
A	$\overline{A}$
0	1
1	0

Figure 3-4
Truth Table for logic inverter.

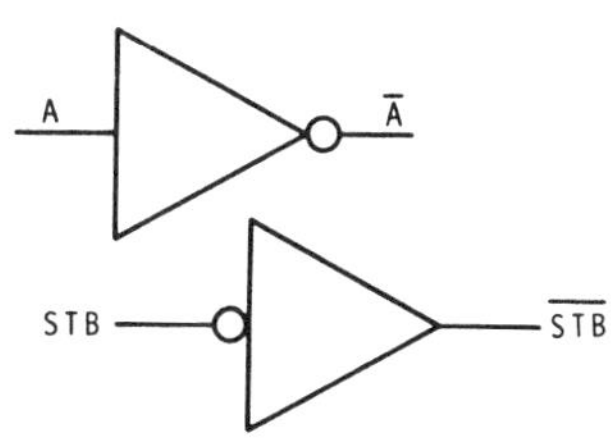

Figure 3-5
Symbols for a logic inverter.

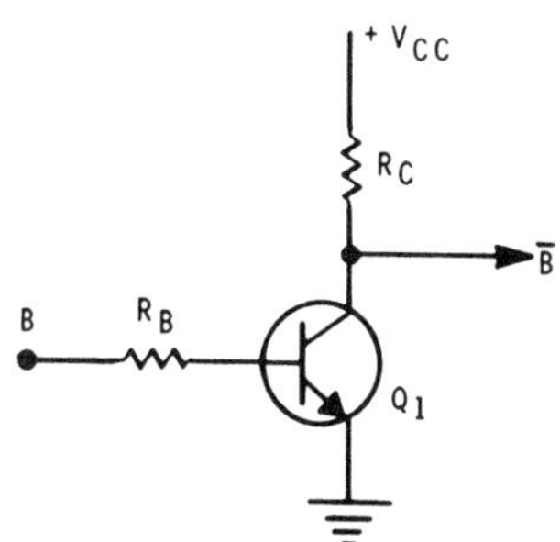

Figure 3-6
A transistor logic inverter.

Whenever a binary 1 or positive voltage level approximately equal to + Vcc is applied to the input, the emitter-base junction of the transistor becomes forward biased. Sufficient base current flows through the circuit to cause the transistor to saturate. When the transistor is saturated, both the emitter-base and base-collector junctions of the transistor are forward biased and it acts as a very low impedance. At this time, the complement output $\overline{B}$ is approximately equal to zero volts or ground as seen through the conducting transistor. The actual voltage will be equal to the saturation voltage of the transistor Vce (sat). For most good high speed switching transistors this voltage is several tenths of a volt or less and for most practical purposes can be considered to be zero. With a positive voltage binary 1 input the output is a binary 0. This circuit obviously performs logic inversion. The input and output voltage waveforms representing the operation of this circuit are shown in Figure 3-7. Most modern switching transistors are capable of turning off and on at nanosecond speeds, therefore very high frequency operation with this circuit is possible.

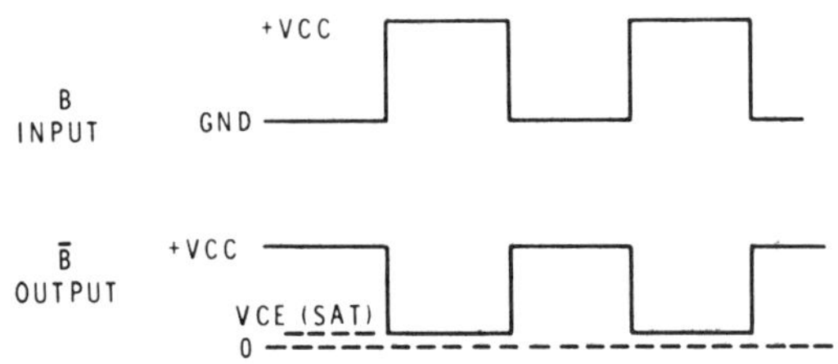

Figure 3-7
Input and output waveforms of a transistor logic inverter.

DECISION-MAKING LOGIC ELEMENTS

The two basic types of decision-making logic elements are the AND gate and the OR gate. These are logic circuits with two or more inputs and a single output. The output state is a function of the input states and how the particular gate operates. The gate makes its decision based upon the input states and its particular function, then generates the appropriate binary output. Let's consider each of these basic gates in detail.

The AND Gate

The AND gate is a logic circuit that has two or more inputs and a single output. The operation of the gate is such that the output of the gate is a binary 1 if and only if all inputs are binary 1. If any one or more inputs are a binary 0 the output will be binary 0. The AND gate is a control circuit whose output is a binary 1 only when all inputs to the gate are in the binary 1 state at the same time.

INPUTS		OUTPUT
A	B	C
0	0	0
0	1	0
1	0	0
1	1	1

Figure 3-8
Truth Table for an AND gate.

The operation of a two input AND gate is indicated by the truth table in Figure 3-8. The inputs are designated A and B. The output is designated C. The output for all possible input combinations is indicated in the truth table. The total number of input combinations is determined by raising 2 to a power equal to the number of inputs. With two inputs, each capable of assuming either of the two binary states, the total possible number of combination inputs is $2^2 = 4$. Note that the output is binary 0 for any set of inputs where either, or both, of the inputs are binary 0. The output is binary 1 only when both inputs are binary 1.

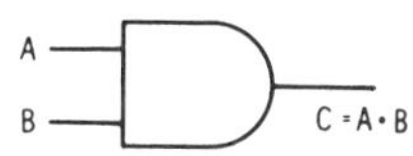

Figure 3-9
Logic symbol for AND gate.

The basic symbol used to represent an AND gate is shown in Figure 3-9. Not that the inputs and outputs are labeled to correspond to the truth table in Figure 3-8. Keep in mind that the AND gate may have any number of logical inputs.

An important point to note about the AND symbol in Figure 3-9 is the equation at the output, $C = A \cdot B$ or $C = AB$. This equation is a form of algebraic expression that is used to designate the logical function being performed. The equation expresses the output C in terms of the input variables A and B and is read, "C equals A AND B". Here the AND function is designated by the dot between the two input variables A and B. The AND function is designated by an expression similar to the product of algebraic variables.

As you will see, the operation of all logic elements and gates can be expressed in the form of an algebraic equation. These expressions permit circuits to be analyzed, designed and optimized by using standard algebraic operations and special algebraic manipulations designated by rules of Boolean algebra. Boolean algebra is a special form of two state algebra that is useful in working with binary variables. You will learn more about this in a later unit.

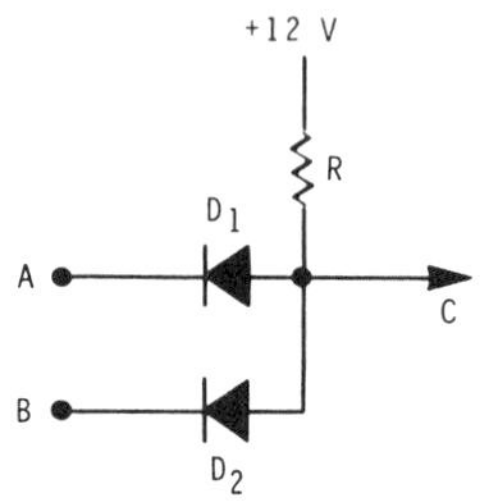

Figure 3-10
Two input diode AND gate.

The circuit in Figure 3-10 shows one electrical implementation of a logical AND gate. Here two diodes and a resistor form an electronic circuit that produces the AND function using binary signals. To analyze this circuit, assume the use of binary input signals using positive logic designations of zero volts (ground) and +5 volts. Let's also assume that perfect diodes are used. Keep in mind that in a practical circuit real diodes have a threshold voltage level before they conduct and a finite voltage drop across them during conduction.

Now let's analyze the operation of the AND circuit shown in Figure 3-10. If both inputs are a binary 0 (zero volts or ground) both diodes D1 and D2 will conduct. Assuming no forward voltage drop across the diode, the output will be zero volts. If either one of the inputs is a binary 0 while the other is at the binary 1 level (+5 volts) the diode associated with the input whose state is binary 0 will be forward biased and will conduct thereby clamping the output at the binary 0 level. The diode associated with the input at the binary 1 level will be reversed biased, therefore it will be cut-off.

If both inputs are at the binary 1 level, both diodes D1 and D2 will conduct and the output will be +5 volts. As you can see, the only time the output is at the binary 1 level is when both inputs are binary 1.

The operation of the AND gate shown in Figure 3-10 is fully illustrated by the voltage truth table in Figure 3-11 and the waveform timing diagram in Figure 3-12. Note that if positive logic is assumed, the voltage truth table corresponds to the logic truth table given in Figure 3-8. In the waveform diagram, both inputs A and B are shown switching at various times between the binary 0 and binary 1 levels of ground and +5 volts. The output C corresponding to this particular combination of inputs is also illustrated. Note that the only time that the output is binary 1 is when the inputs are both in the binary 1 state. The output is a binary 1 for a period of time during which the two inputs are coincidentally at the binary 1 level. The AND gate is sometimes referred to as a coincident gate.

INPUTS		OUTPUT
A	B	C
0V	0V	0V
0V	+5V	0V
+5V	0V	0V
+5V	+5V	+5V

Figure 3-11
Voltage truth table for diode AND gate.

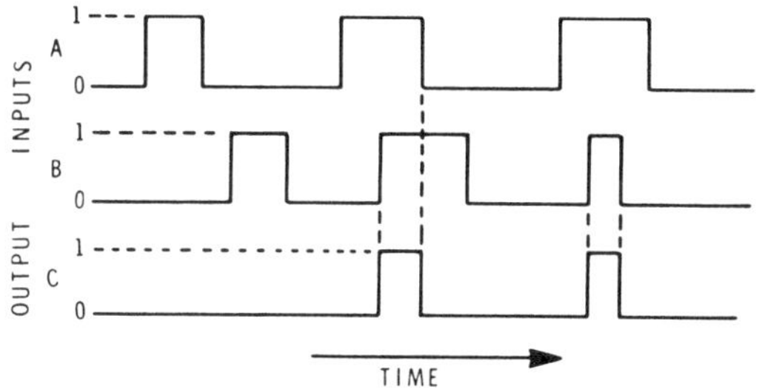

Figure 3-12
Input and output waveforms of the diode AND gate.

Figure 3-13 shows one of the most common applications of the AND gate in digital circuits. Here one input of the AND gate is used to control the passage of the other input signal to the output. The input control signal CTL enables or inhibits the passage of the other logic signal which is a train of square waves designated SQW. The output is identical to SQW during the time the CTL input is binary 1. Note the input and output waveforms as well as the logical expression for this function indicated in Figure 3-13.

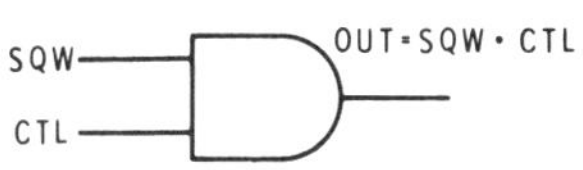

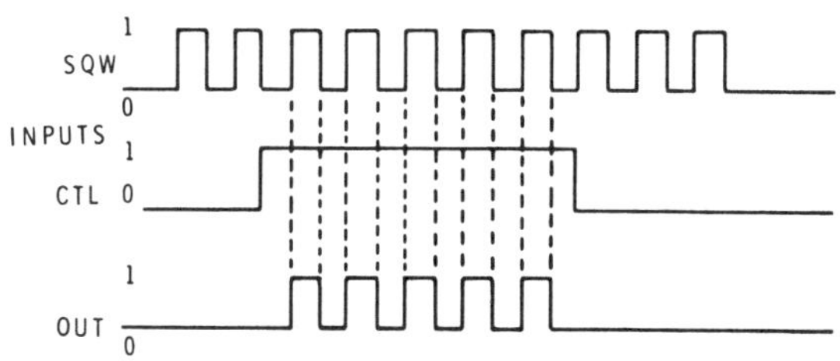

Figure 3-13
Typical AND gate application.

Remember that an AND gate may have more than two inputs, the exact number being dictated by the application. In addition, there are many other ways of implementing the logical AND function with hardware. Later in this unit and in this text you will learn about some of these circuits.

The OR Gate

The other basic logic element is called an OR gate. Like the AND gate it can have two or more inputs and a single output. It's operation is such that the output is a binary 1 if any one or all inputs are a binary 1. The output is binary 0 only when both inputs are binary 0.

The logical operation of a two input OR gate is expressed by the truth table in Figure 3-14. With two inputs there are $2^2 = 4$ possible input combinations as explained earlier. The truth table designates all four possible input combinations and the corresponding output. Note that the output is a binary 1 when either or both of the inputs are binary 1. The output is binary 1 if input D OR input E OR both are present.

INPUTS		OUTPUT
D	E	F
0	0	0
0	1	1
1	0	1
1	1	1

Figure 3-14
Truth Table for OR gate.

The logical symbol for an OR gate is shown in Figure 3-15. The inputs are labeled according to the truth table in Figure 3-14. Make note particularly of the output algebraic expression for the OR gate F = D + E. The plus sign is used to designate the logical OR function. The output F is expressed in terms of the input logic variables D and E.

The circuit in Figure 3-16 illustrates the implementation of the logical OR function using semiconductor diodes. This gate is similar to the AND circuit considered earlier except that the supply voltage and diode polarities are reversed. This circuit operates with logical input levels of zero and +5 volts as did the AND gate considered previously. Using positive logic designations, let's evaluate the operation of this circuit.

When both inputs D and E are at binary 0 (zero inputs or ground) both diodes D1 and D2 conduct. Assuming perfect diodes with no forward voltage drop, the output will be binary 0. If either one of the logical inputs is a binary 0 while the other is a binary 1 (+5 volts) the diode associated with the input at the binary 1 state will conduct causing the output to be +5 volts or a binary 1. The diode associated with the input at the binary 0 state will be reversed biased and will not affect the circuit. When both inputs are at the binary 1 or +5 volt level, both diodes D1 and D2 conduct and the output is a +5 volt or binary 1 level. The voltage truth table for this circuit is shown in Figure 3-17. Using positive logic, it corresponds to the table in Figure 3-14.

The operation of the OR gate is illustrated more fully by the waveforms shown in Figure 3-18. These diagrams show the output state for various combinations of the input voltage D and E as a function of time. Note that if one of the inputs switches to the binary 1 level the output switches to binary 1 correspondingly. The output is binary 1 when either or both inputs are binary 1. Study the waveforms in Figure 3-18 carefully at each point to be sure that you understand the circuit function.

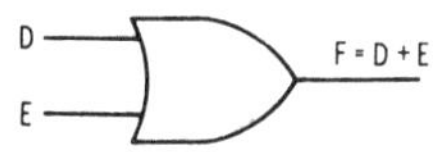

Figure 3-15
Logic symbol for OR gate.

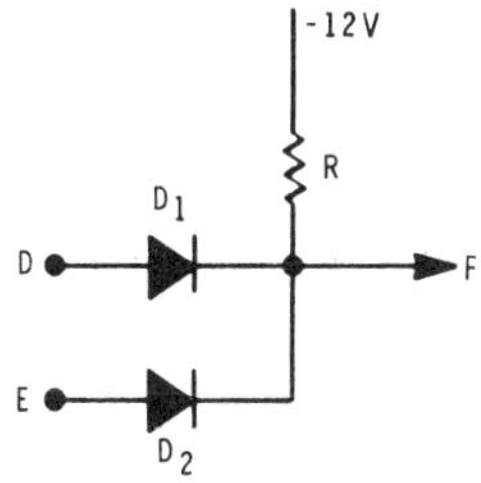

Figure 3-16
Diode OR gate.

INPUTS		OUTPUT
D	E	F
0V	0V	0V
0V	+5V	+5V
+5V	0V	+5V
+5V	+5V	+5V

Figure 3-17
Voltage Truth
Table for the diode OR gate.

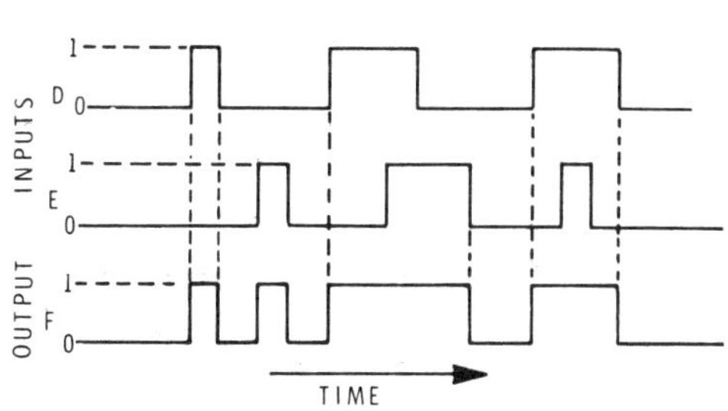

Figure 3-18
Input and output waveforms
of the diode OR gate.

Figure 3-19 illustrates a typical application for the OR gate in a digital circuit. There are two inputs to the OR gate in this application: a push-button switch SW2 and a train of pulses designated PLS. The output OPR will be a binary 1 when either SW2 of PLS is at the binary 1 (+5 volts) level. Switch SW2 is a normally closed pushbutton. The SW2 input to the OR gate is normally ground or binary 0. When the switch is depressed, the contacts open and the SW2 input becomes +5 volts or binary 1 level as seen through the resistor. The other input PLS is a series of pulse trains that switch momentarily between the binary 0 and binary 1 levels. Note the algebraic expression for the output, OPR = SW2 + PLS. The accompanying waveforms in Figure 3-19 illustrate the operation of the circuit. The OR gate permits either of the two inputs to control the output.

When considering the operation and application of an OR gate you should remember that this logic element can have two or more imputs as called for by the application. At the same time we have only illustrated one method of implementing the OR gate with electronic hardware. Many other circuit variations are used and these will be discussed later in this text.

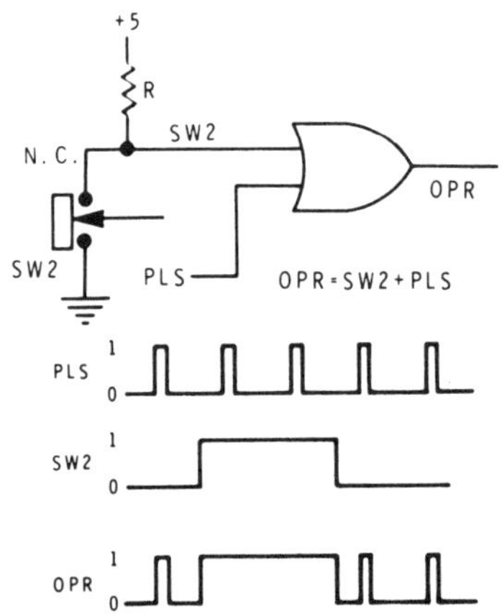

Figure 3-19
Typical OR gate application.

The Dual Nature of Logic Gates

When we explained the operation of the diode gate circuit in Figure 3-10, we indicated that it performs the logical AND function. We proved this by considering the output voltage level for each of the four possible combinations of input voltage levels. The voltage truth table for this gate is repeated here in Figure 3-20A. In considering the operation of this circuit, we assumed the use of positive logic level assignments. By doing this we were able to translate the voltage levels given in the truth table into the table shown in Figure 3-20B. Naturally this table clearly indicates that the AND function is being performed. The output is a binary 1 only when both inputs are binary 1.

INPUTS		OUTPUT
A	B	C
0V	0V	0V
0V	+5V	0V
+5V	0V	0V
+5V	+5V	+5V

A

INPUTS		OUTPUT
A	B	C
0	0	0
0	1	0
1	0	0
1	1	1

B

INPUTS		OUTPUT
A	B	C
1	1	1
1	0	1
0	1	1
0	0	0

C

Figure 3-20
Truth Table for
diode gate in Figure 3-10.

Now let's consider the function of this circuit when we assume negative logic level assignments. In this case the 0 volt level would represent a binary 1 and a +5 volt level would represent a binary 0. Using the original data as developed in Figure 3-20A and translating it into a truth table using binary 1's and 0's with negative logic level assignments, we obtain the truth table shown in Figure 3-20C. By studying this truth table, you will see that the circuit no longer appears to be performing the AND function. Close inspection of the truth table will reveal that the circuit is now performing the OR function since the output is a binary 1 when either one or both of the inputs is a binary 1. (The order or sequence of the inputs is not the same as that for the AND gate, but this is not important. It is the function that counts.) Our only conclusion can be that with positive logic the cirucit in Figure 3-10 performs AND function but with negative logic the circuit performs the OR function. This clearly indicates that the diode gate circuit is capable of performing either of the two basic logical functions, and that function is strictly dependent upon the logic level assignments given to the input and output variables.

This dual nature of logic gates applies to any logic circuit. The diode gate that you considered in Figure 3-16 is also dual in nature. With positive logic level assignments it performs the OR function as indicated previously. However, if you analyze the circuit by using negative logic level assignments you will find that the circuit performs the AND function. Keep this important fact in mind as it will help you in analyzing, troubleshooting and designing digital circuits. You must not only know how the circuit operates electrically, but also what logic level assignments are being used.

Figure 3-21 shows the logic symbols normally used to represent gates that perform the logical AND and OR functions with negative logic level assignments. The circles at the inputs and outputs represent the effect of reversing the logic level assignments from positive to negative.

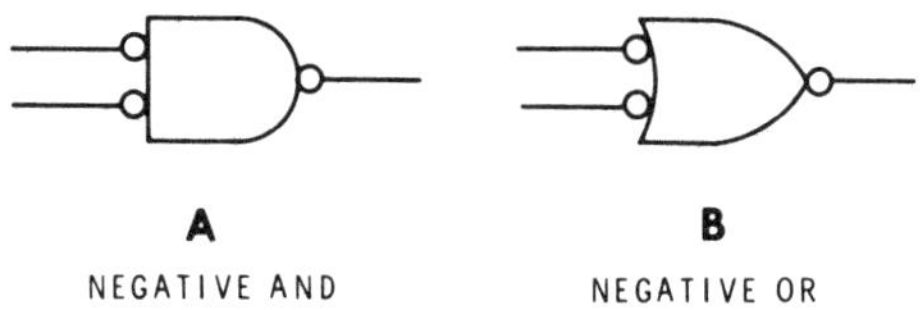

Figure 3-21
Negative logic gates.

NAND/NOR GATES

While many digital circuits can be constructed with just the three basic digital logic elements — AND, OR, and NOT — most digital equipment is implemented with special versions of these circuits known as NAND and NOR gates. Such circuits are basically AND and OR gates combined with an inverter. NAND/NOR gates are the most widely used types of digital logic elements because they offer numerous advantages over the simple diode gates considered earlier. In large complex digital logic networks, it is difficult to cascade more than just a few of the simple diode logic gates. Because there is no buffering between the gates, loading problems occur and the speed of operation suffers. For that reason, it is generally desirable to combine a simple diode logic gate with some type of transistor buffer to permit more flexible interconnection of circuits. This transistor buffer is most often an inverter.

NAND Gate

The term NAND is a contraction of the expression NOT-AND. A NAND gate, therefore, is an AND gate followed by an inverter. Figure 3-22A shows the basic diagram of a NAND gate. Note the algebraic output expression for the AND gate and the inverter. The entire AND output expression is inverted and indicated by the bar over it.

Figure 3-22 shows the standard symbol used for a NAND gate. It is similar to the AND symbol but a circle has been added at the output to represent the inversion that takes place.

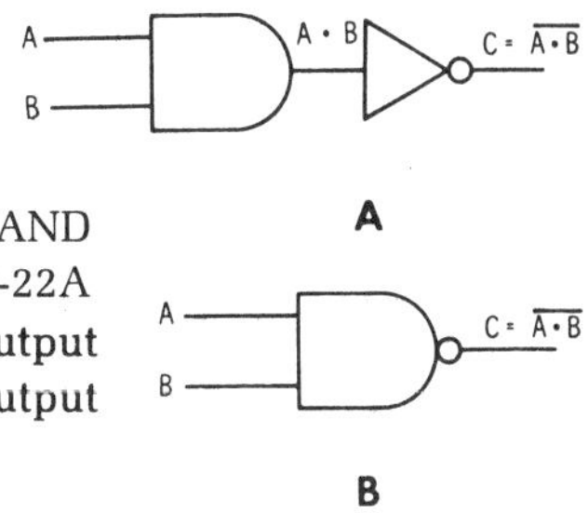

Figure 3-22
NAND gate.

The logical operation of the NAND gate is easy to infer from the circuit in Figure 3-22. This operation is indicated by the truth table in Figure 3-23. The NAND output is simply the complement of the AND output.

INPUTS		OUTPUT	
A	B	AND A•B	NAND $\overline{A \cdot B}$ = C
0	0	0	1
0	1	0	1
1	0	0	1
1	1	1	0

Figure 3-23
Truth Table of NAND gate.

NOR Gate

Like the NAND gate, the NOR gate is an improved logic element used for implementing decision-making logic functions. The term NOR is a contraction for the expression NOT-OR. Therefore, the NOR gate is essentially a circuit combining the logic functions of an OR gate and an inverter.

Figure 3-24 is a logical representation of a NOR gate. Figure 3-24 shows the standard symbol used to represent a NOR gate. Note that the output expression is the inverted OR function.

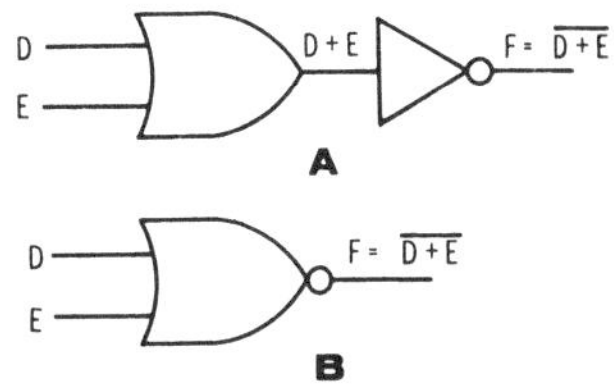

Figure 3-24
NOR gate.

The logical operation of a NOR gate is illustrated by the truth table in Figure 3-25. The NOR output is simply the complement of the OR function. Like any other logic gate, NAND and NOR gates may have two or more inputs as required by the application.

NAND and NOR gates can be used to implement any of three basic logical functions. For example, by tying all inputs together, either the NAND or NOR gate performs inversion. By combining the NAND or NOR gates with external inverters, the AND and OR operations can be performed.

INPUTS		OUTPUT	
D	E	OR $D+E$	NOR $\overline{D+E}=F$
0	0	0	1
0	1	1	0
1	0	1	0
1	1	1	0

Figure 3-25
Truth Table of NOR gate.

How NAND/NOR Gates Are Used

The AND and OR logic functions can be performed by connecting inverters on the outputs of NAND and NOR gates respectively. Since the NAND and NOR functions are simply complementary logic functions of basic AND and OR operations, then it is logical to assume that the AND and OR operations can be obtained from NAND and NOR gates simply by adding an additional inversion. As stated in the previous section on inverters, the effect of one inversion is cancelled by adding a second. Cascading an even number of inverters removes the inversion function.

Figure 3-26 shows how the AND and OR operations are performed with NAND and NOR gates. But now, how is the OR function performed with a NAND gate and the AND function performed with a NOR gate?

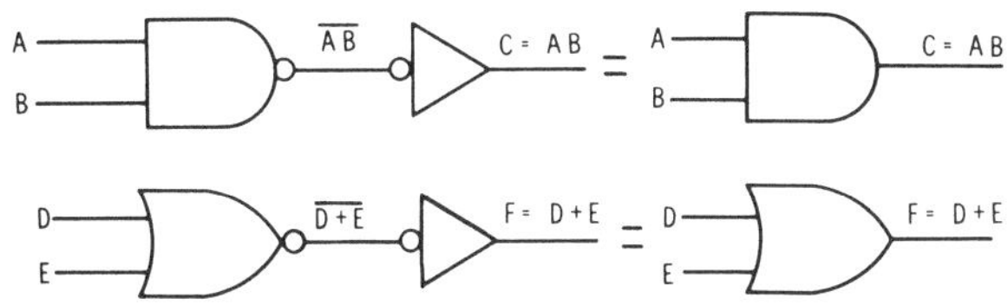

Figure 3-26
Performing AND and OR operations with NAND and NOR gates.

As mentioned earlier, any of the three basic logic functions — AND, OR, NOT — can be performed by either a NAND gate or a NOR gate. You have already seen how the AND and OR functions can be obtained with NAND and NOR gates by simply inverting the output. To obtain the other logic functions with each type of gate, inverters are used on the inputs.

Figure 3-27 shows the truth table for a NAND gate. The inputs are A and B and the output is C. Now consider the effect of adding an inverter to each input as shown in Figure 3-28. These inverters simply complement the input signals. The effect is illustrated in the truth table of Figure 3-27. The input signals to the inverters are labeled X and Y. Note that they are the complements of the signals A and B respectively. Considering the two

INVERTED INPUTS		NAND INPUTS		OUTPUT
X	Y	A	B	C
1	1	0	0	1
1	0	0	1	1
0	1	1	0	1
0	0	1	1	0

Figure 3-27
Truth Table for NAND gate.

inverters and the NAND gate as forming a single composite circuit, our inputs become X and Y instead of A and B. Output C remains the same. By observing the truth table in Figure 3-27 with this change in mind, you can now evaluate the logical function of the circuit. Disregarding the sequence of the X and Y input combinations and the A and B inputs, note the output state for each of the input states. You can see that the circuit produces a binary 1 output any time a binary 1 is applied to either one or both of the inputs. By definition this is the logical OR function. As you can see then the OR function can be performed with a NAND gate by simply placing inverters ahead of each input.

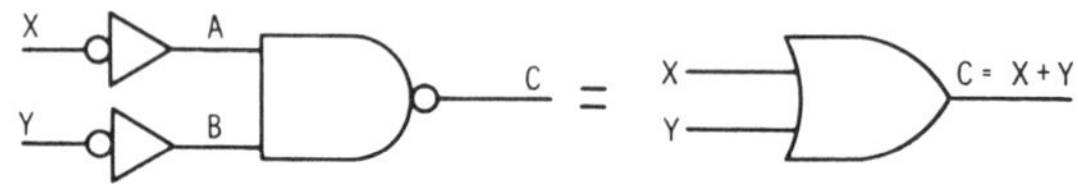

Figure 3-28
Performing the OR function with a NAND gate.

By using inverters at the input of a NOR the AND function can be performed. Figure 3-29 shows the truth table for a NOR gate. The inputs are D and E and the output is F. Now assume that inverters are connected ahead of the NOR gate and the new inputs are labeled L and M as shown in Figure 3-38. The inverted or complementary input states would be as indicated in the truth table of Figure 3-29. If we interpret the logical function of the composite circuit where the inputs are L and M and the output is F, then you can see by studying the truth table that the AND function is being performed. Note that the only time that the output F is a binary 1 is when both inputs L and M are at the binary 1 level. By definition this is the AND logical function. As Figure 3-30 indicates, a NOR gate with inverters at its input performs the AND function.

As you can see any of the three basic logic functions can be performed with either a NAND gate or a NOR gate. For that reason entire digital circuits can be constructed with just one type of gate. The choice is arbitrary and strictly up to the designer. There are some cases where both NANDs and NORs are mixed within a circuit. As you will see later, some circuit economies result when NAND and NORs are combined. However, just remember that any logic function can be implemented with NAND gates or NOR gates alone.

Figure 3-29
Truth Table for NOR gate.

INVERTED INPUTS		NOR INPUTS		OUTPUT
L	M	D	E	F
1	1	0	0	1
1	0	0	1	0
0	1	1	0	0
0	0	1	1	0

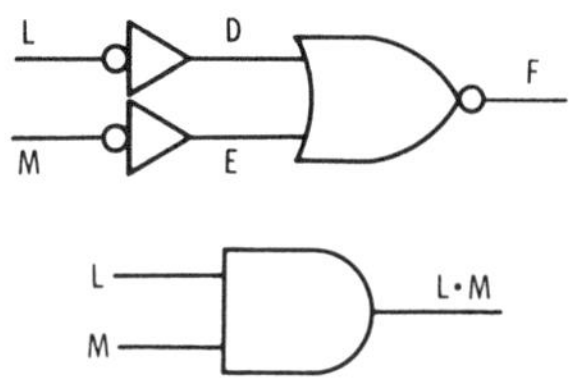

Figure 3-30
Using a NOR gate to perform the AND function.

PRACTICAL LOGIC CIRCUITS

Now that you have completed this overview of the types of logic circuits in common use in digital equipment, let's take a look at several typical ways these logic elements are implemented. There are many different ways of electrically or mechanically obtaining the particular characteristics specified by the various types of logic elements. Here we will consider several popular ways of realizing digital logic elements with hardware.

Relays and Switches

The three basic logic functions — AND, OR and NOT — can be readily implemented with relay contacts and switches. For example, Figure 3-31 shows an AND gate made with relays. Here the normally open (N. O.) contacts of two relays labeled A and B are connected in series with a battery and a lamp. In this circuit, a closed relay contact and an ON lamp represent a binary 1. An open contact and an OFF lamp represent a binary 0. A zero voltage level represents binary 0 and a positive voltage level represents a binary 1 on the relay coil. You can see, for any of the four possible input combinations there is only one where the lamp will light. If either one or both of the relays are de-energized, their contacts will be open and current will not be supplied to the lamp. However, if voltage is applied to both relay coils, both contacts will be closed and a binary 1 (ON lamp) output will occur. Contacts A and B must be closed to light the lamp. Series connected switches usually perform the AND function.

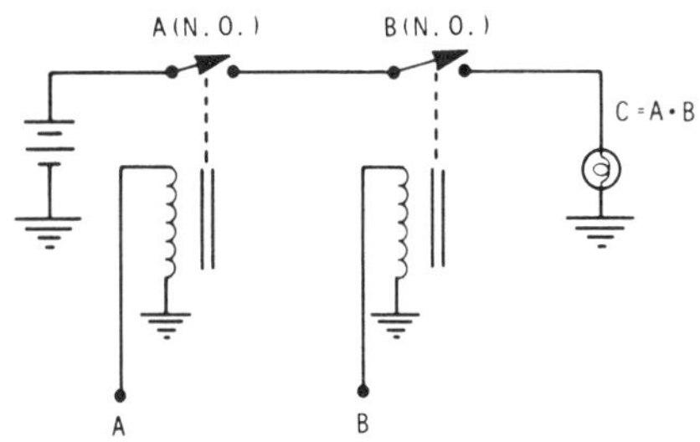

Figure 3-31
Relay AND gate.

Figure 3-32 shows an OR gate made with relays. The normally open relay contacts are connected in parallel. Here you can see that when a binary 1 voltage level is applied to either or both relay coils the D or E contacts will close, thereby supplying voltage to the lamp. An input to either relay D or E will close the circuit and turn on the lamp, representing a binary 1. Parallel connected contacts usually perform the OR operation.

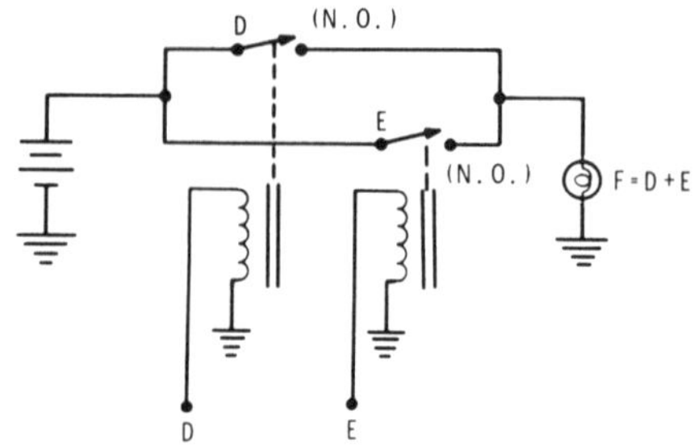

Figure 3-32
Relay OR gate.

A relay logic inverter circuit is shown in Figure 3-33. Here a relay with normally closed (N. C.) contacts is used. With a binary 0 (zero volts) applied to the relay coil, the relay is de-energized and the contact is closed representing a binary 1. This connects the battery to the lamp causing it to light. Therefore, with a binary 0 input the output is binary 1. Applying a binary 1 voltage level to the relay coil will energize the relay and open the contacts. This will cause the lamp to go off indicating a binary 0.

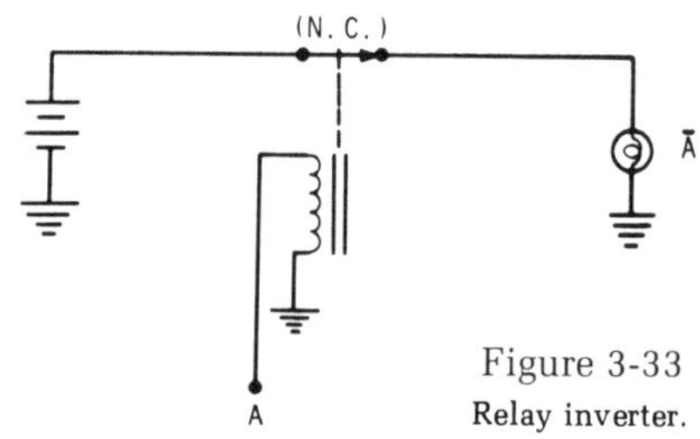

Figure 3-33
Relay inverter.

These relay switching circuits can be combined in many ways to form any logic function. In addition, manually operated switches can also be substituted for the relay contacts in some applications. Such relay or switch logic circuits are not often used today. Such circuits are large, slow in operation, and consume a significant amount of power. For most applications they are not practical. The very earliest of digital equipment including some computers were implemented with relays. However, many other different types of logic circuits are available now and these have many significant advantages over relays. There are still a few practical uses for relay and switch logic circuits. In some heavy industrial control systems where speed and power consumption is of little importance, relay logic circuits can handle high power applications and are very reliable. There are some applications where a mechanical means of operating the switches is available thereby making mechanical or manual switching logic necessary or desirable.

Discrete Component Logic Circuits

A discrete component logic circuit is a logic element made up of individual electronic components such as transistors, diodes, resistors, capacitors and other devices. These are assembled to form a complete circuit like the diode gates and inverter described earlier. For many years digital logic circuits were implemented with discrete components. They offered small sized, high performance and reasonable power consumption. However, today such discrete component circuits are rarely used. Like relay and mechanical switching logic circuits they have essentially been replaced by logic elements with greater performance, lower cost and improved features. You may still encounter discrete component logic circuits in some high power applications or in older digital equipment. Today however most digital logic functions are implemented with integrated circuits.

Integrated Circuits

An integrated circuit (IC) is a semiconductor device which combines transistors, diodes, resistors, and capacitors in ultra-miniature form on a single silicon chip. The advances in semiconductor technology have permitted the semiconductor manufacturers to design, develop and produce entire electronic circuits on a single silicon wafer that is generally less than one tenth of an inch square. These circuits are not only significantly smaller in size than discrete component logic circuits but also offer many other benefits as well. Because they can be mass produced, their cost is substantially less than discrete component circuits. Many offer significant savings in power consumption. Perhaps even more important is the elimination of the need for circuit wiring. When discrete components are used the components must be interconnected physically on a printed circuit board and then tested. With an integrated circuit the entire circuitry, all components included, are manufactured simultaneously. Manufacturing costs are reduced and reliability is improved.

Integrated circuits have been in existence for approximately 15 years. During this short time significant advances have been made. The complexity and sophistication of the circuits has increased significantly while the prices have continued to decline. Today with integrated circuit techniques it is not only possible to implement the basic logic elements, but also it is possible to fully integrate complete combinational and sequential circuits. Integrated circuits implementing the basic logic functions such as NAND, NOR, and flip-flops are known as small scale integrated circuits (SSI). Complete functional circuits of either the combinational or sequential type are generally designated as medium-scale integrated circuits (MSI). However, today technology permits even greater flexibility. Complete circuits and systems can be constructed on a single chip. For example an entire digital computer is available as a single integrated circuit. Such circuits are known as large-scale integrated circuits (LSI). Today most digital equipment is implemented with integrated circuits. There are few applications where other forms of circuitry are necessary or desirable. This text emphasizes digital integrated circuits, their operation and application.

Unit 4

DIGITAL INTEGRATED CIRCUITS

INTRODUCTION

All modern digital equipment is constructed with integrated circuits. A study of digital techniques, therefore, is a study of digital integrated circuits and their application. Because of integrated circuits, digital equipment can be analyzed and designed almost entirely at a conceptual logical or systems level as opposed to an electronics or circuits level. Digital integrated circuits are, in the true sense of the word, building blocks which are used to construct digital equipment. Previously, the designer of digital equipment had to design not only the logic involved but also the electronic circuits necessary to implement that logic. With integrated circuits, the designers job is primarily that of selecting commercially available devices and applying them to his specific application. No knowledge of electronic circuit design is necessary to the understanding and use of most digital techniques. However, by understanding the basic components and circuits used in modern digital integrated circuits your ability to work with and use them will be greater.

In this unit you will study the basic components and circuits used in the most common types of digital integrated circuits. The information in this unit will help you to analyze the operation of digital circuits and will help you to select a type of integrated circuit for a specific application.

LOGIC CIRCUIT CHARACTERISTICS

There are many different types of digital integrated circuits available to implement digital equipment. Both saturated and unsaturated bipolar transistors as well as MOSFETs are used to implement a variety of logic circuits. Each type or family of digital integrated circuits has its own special capabilities and limitations. Their characteristics vary widely and the optimum circuit to use in a given application depends upon specific needs and requirements.

Some of the most important characteristics of digital integrated circuits are logic levels, propagation delay, power dissipation, noise immunity, and fan out. By understanding the meanings of these characteristics you can quickly compare, contrast and evaluate different IC logic families.

Logic Levels

Logic levels are the voltage values assigned to the binary 1 and binary 0 states for a given type of digital integrated circuit. Nominal values for the two levels are generally given, but in practice, the actual voltage levels may vary somewhat because of internal component tolerances, power supply variations, temperature and other factors. Generally, the manufacturer will list maximum and minimum acceptable voltage values for the binary 0 and binary 1 levels.

It is important to know the logic levels for a given type of integrated circuit so that when you are working with the equipment you can readily identify input and output logic states by measuring the logic levels with a voltmeter or an oscilloscope. A knowledge of the logic levels will permit you to analyze the operation of a circuit or determine whether it is functioning properly.

Propagation Delay

The propagation delay is a measure of the speed of operation of a logic circuit. Speed of operation is one of the most important characteristics of a digital circuit. For most digital applications high speed operation is beneficial.

Propagation delay is the amount of time that it takes the output of a digital circuit to respond to the input level change. It is the accumulation of all of the rise times, delay times and storage times associated with any logic circuit. When the input voltage changes from the binary 0 to binary 1 or from the binary 1 to binary 0 levels, the output of the logic circuit will respond at some finite time later.

Figure 4-1 illustrates propagation delay. Shown here is the input to a digital circuit and the corresponding output. The circuit could be an inverter, a NAND gate or a NOR gate. A binary 0 to binary 1 transition causes a binary 1 to binary 0 transition at the output. Note that the output transition occurs a specific time after the input transition. This is the propagation delay. The propagation delay (t_p) is generally measured between the 50 percent amplitude points on the corresponding leading and trailing edges of the input and output pulses. Note also that there are also two types of propagation delay, the propagation delay occurring when the output changes from high to low (t_{pHL}) and the propagation delay that occurs on the low to high output transition (t_{pLH}). Because of the characteristics of the logic circuit, the propagation delays for the two types of level changes are generally different. They are of the same order of magnitude and close in value but nevertheless unequal.

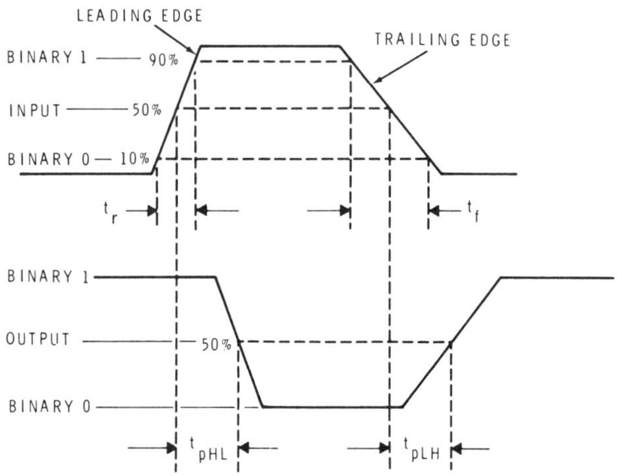

Figure 4-1
Propagation Delay.

The rise and fall times of the input and output pulses are another important consideration. The rise time (t_r) is the time it takes the pulse to rise from 10 percent to 90 percent of its maximum value. The fall time (t_f) is the time it takes for the pulse voltage to fall from 90 percent to 10 percent of its maximum value.

For most modern digital integrated circuits propagation delays are very short but finite. Propagation delays as low as one nanosecond are achievable. Some types of modern logic circuits have propagation delays as high as several hundred nanoseconds. Rise and fall times are usually less than the propagation delays. Because of manufacturing tolerances, circuit wiring and other factors, propagation delays can vary considerably from their nominal indicated value. In addition propagation delays are additive. When gates and other combinational logic circuits are cascaded, the propagation delays accumulate. If there is more than one level of logic, the total propagation delay from input to output is simply the sum of the individual gate propagation delays.

Power Dissipation

Another important characteristic of digital logic circuits is power dissipation. This is a measure of the amount of power consumed by the components in a typical logic gate or other circuit. Power dissipation, in milliwatts per logic gate, is an average value since the power consumption is usually different for the binary 1 and binary 0 output states.

The amount of power dissipated by a logic circuit is a very important consideration in the design of any digital equipment. A high power dissipation will mean high electrical energy consumption. Naturally, it is desirable to conserve as much electrical power as possible since the cost of operation of the equipment is an important consideration. This is particularly true of large scale digital systems such as computers.

The total power dissipation of the digital circuitry will also determine the size and cost of the power supply. In addition, high power dissipations mean high heat levels. In some instances special cooling requirements may be necessary to ensure proper operation of the equipment. Power dissipation is particularly important in portable or battery operated equipment. In order to reduce the cost of the battery and ensure long battery life, low power dissipation is desirable.

Gate power dissipation runs all the way from microwatts for certain types of MOS circuits to as high as 60 to 100 milliwatts per gate per certain types of high speed nonsaturated logic.

The Speed-Power Trade-Off

Two of the characteristics that we have considered so far, namely, speed and power dissipation are directly dependent upon one another in all types of digital logic circuits. The relationship between these two characteristics is such that speed is proportional to power dissipation. The faster a logic circuit switches the higher its power dissipation. In order to get high speed operation you must accept the penalty of high power dissipation. This trade-off or compromise between speed and power is one of the most important considerations that a digital designer must make in a selection of a type of logic circuit for a given application. High speed digital logic circuits use nonsaturating bipolar transistors. Because the transistors do not saturate, their emitter-collector voltage drops are higher. Combine this with the very low circuit resistance values to minimize charge and discharge times of stray capacitances and the result is high power comsumption.

MOS integrated circuits consume a very small amount of power. Their high impedance nature is partially responsible for this, however, this characteristic plus the built-in capacitances make for very slow switching speeds. The result is that the frequency of operation is severely limited. Nevertheless, the extremely low power consumption — on the order of nanowatts — makes MOS circuitry extremely desirable for portable and battery operation where high speed is not required. Other types of logic circuits fall between these two extremes. The speed-power trade off is an inherent compromise.

Noise Immunity

Noise immunity is a measure of the susceptibility of a logic circuit to noise pulses on the inputs and output of a logic circuit. Noise is considered to be any extraneous and undesired signal generated within the equipment itself or externally that is added to and appears superimposed upon the standard system logic levels. This noise can be a slowly varying dc level or very high frequency short duration voltage or current spikes. The noise may be either randomly occurring or repetitive. In any case the noise signals can cause the logic circuit to switch to an undesirable state at an improper time.

All digital logic circuits have built in noise immunity. Because of the voltage thresholds associated with the components and the circuit, most logic circuits are capable of rejecting noise spikes of a relatively high amplitude. The noise immunity of most logic circuits is from approximately 10 to 50 percent of the supply voltage. This means that a noise spike occurring on a binary 0 or binary 1 level will be rejected if its amplitude is below a level that is 10 percent to 50 percent of the supply voltage. A circuit with a noise immunity of one volt, for example, would reject noise pulses that are one volt or less different from the nominal binary 0 or binary 1 logic levels. In some cases, noise is rejected by the logic circuit by virtue of its slow response. Some noise is high frequency in nature and noise pulses are of such short duration that the logic circuits cannot respond fast enough to cause a logic state change.

Noise immunity is an important consideration of digital logic circuits since most digital systems generate a substantial amount of noise during high speed switching. In addition much digital equipment is used in noisy industrial environments where transients from the power line and other electrical equipment can cause false triggering of the logic circuitry. When selecting a particular digital integrated circuit for a logic application noise immunity is an important consideration.

Fan Out

Fan out is a characteristic that indicates how much of a load can be connected to the output of a digital circuit. Fan out is generally expressed in terms of the number of standard size loads that a logic gate output can accomodate and still maintain proper operation at the nominal logic levels, speed, temperature range and other factors. Because of the component limitations and circuit configuration naturally there is a limit to the number of loads that can be connected to a logic circuit. A typical logic gate, for example, may have a fan out of ten indicating that ten separate gate inputs can be attached to the output of this logic circuit and still maintain proper operation according to the manufacturer's specifications.

There are two basic ways that the output of a logic circuit is connected to the load. Some loads appear essentially between the output of the logic circuit and ground while others appear between the output of the logic circuit and the supply voltage.

Current Source Logic. When the loads appear between the output of the logic circuit and ground, the driving logic circuit functions as a current source. This type of logic is referred to as current sourcing logic. An illustration of this type of circuit configuration is shown in Figure 4-2A. Here the output of a logic circuit Q1 drives two similar logic circuits made up of Q2 and Q3. When transistor Q1 is on, the voltage at point X is near zero volts so no base drive is applied to Q2 and Q3, therefore they do not conduct. However, when Q1 cuts off, the supply voltage effectively furnishes current through collector resistor R_C to base resistors R_{B1} and R_{B2}. As you can see, each load consists of the emitter-base junction of a transistor and the appropriate base resistor connected between the output of the logic circuit (point X) and ground. As more of these loads are connected to the circuit, the total effective resistance of the load decreases and therefore causes increased current to be drawn from the supply through the collector resistor. The voltage divider formed by the collector resistance and the external loads causes the logic output voltage level to decrease as the number of loads increase. In order to maintain a specific minimum logic level at the collector of Q1, the number of loads must naturally be limited. In addition the size of the supply voltage and collector resistor effectively limits the maximum amount of current that the driving logic circuit can supply to the transistor loads and still ensure that they all saturate. The maximum available current is $V_{CC} \div R_C$.

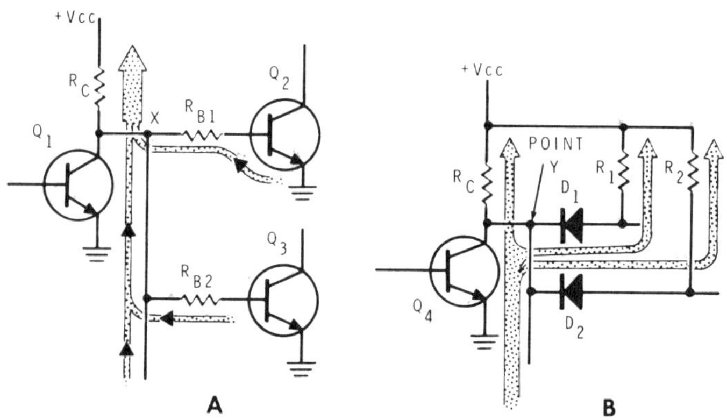

Figure 4-2 Examples of current sourcing (A) and current sinking (B) logic circuits.

Current Sink Logic. Another type of logic circuit, referred to as current sinking logic, is shown in Figure 4-2B. Here the loads appear effectively between the output of the logic gate (point Y) and the supply voltage. The logic circuit is shown driving two diode gates. When transistor Q4 is cut off its collector voltage is V_{CC}, therefore diodes D1 and D2 are not forward biased so they draw no current. However, when transistor Q4 conducts, a current path is formed through Q4, the collector resistor R_C and through each diode gate load R_1-D_1 and R_2-D_2. The greater the number of loads connected, the higher the current that will pass through the driving or sink transistor Q_4. As more loads are added, the collector current of Q_4 increases, and the binary 0 output level increases. In order to maintain saturation and a reasonably low binary 0 level, the base drive to Q4 must be sufficiently high to handle the current of the external load as well as that determined by collector resistor R_C.

INTEGRATED CIRCUITS

Since all modern digital equipment is made up of integrated circuits you should be familiar with the various types. In this section you are going to learn how integrated circuits are classified and something about their physical characteristics. You will also learn of the most popular families of integrated circuits used in digital equipment today.

Integrated circuits are classified in three basic ways: by method of manufacturing, by application and by function. Let's briefly consider each of these types of classifications.

Manufacturing Methods

There are four basic ways of making integrated circuits. The most widely used method is called monolithic. Other types of manufacturing methods include thin film, thick film, and hybrid.

Monolithic. A monolithic integrated circuit is one that is constructed entirely on a single chip of silicon semiconductor material. Semiconductor materials are diffused into the basic substrate or base material to form the various junctions making up components such as diodes, transistors and resistors. The semi-conductor materials to be diffused into the substrate are in gaseous form and are deposited on the substrate through a series of masking operations under very high temperature. The result is that the entire circuit, all components and interconnections, are on a single base, thus the term monolithic. Most integrated circuits are constructed using this monolithic technique.

There are two basic forms of monolithic integrated circuits: bipolar and MOS. Here the difference is primarily that of the type of transistors used in constructing the circuits. Bipolar circuits which can be either saturating or non-saturating are by far the most widely used. But MOSFET circuitry is becoming more popular. The MOS circuitry is easier to make and takes up less space. Therefore, much more circuitry can be placed on a silicon chip of a given size. The simplicity of the components also make the manufacturing yield much higher. The result is that MOS circuits can be constructed with higher density and at lower costs.

Thin and Thick Film Techniques. Thin and thick film integrated circuits are manufactured by depositing certain materials on a non-conducting base such as ceramic. Through a series of masking procedures, various resistive and conducting materials are deposited on the base or substrate to form resistors, capacitors, and inductors. Semiconductors are not usually manufactured in this way. Thin and thick film techniques are primarily used for manufacturing passive networks such as attenuators, filters, phase shift networks and the like. Because such networks can be made extremely small they offer the same advantages over discrete component circuits as do monolithic integrated circuits. Another advantage is that component tolerances can be closer than equivalent components made by monolithic techniques. For high quality precision circuits, thin and thick film techniques are preferred.

Hybrid Circuits. A hybrid integrated circuit is one made up of a combination of monolithic, thin film or thick film circuits. Any number of combinations are considered to be hybrid. A hybrid integrated circuit may consist of multiple monolithic chips interconnected in a single package. Another example of a hybrid is a monolithic circuit combined with a thin film or thick film passive network. Sometimes monolithic circuits and thin film or thick film circuits are also combined with individual semiconductor component chips to form a special high grade circuit for an unusual application.

Hybrids offer the advantage that a variety of different integrated circuits and components can be combined to offer special advantages not available in individual types of integrated circuits alone. For example, because of the ultra small size of a monolithic circuit, power dissipation is limited. In order to handle high power requirements it may be necessary to combine a low power monolithic circuit with a power transistor mounted on a separate chip but physically interconnected within the same package. High precision circuit requirements might be met by a combination of a monolithic circuit and a highly accurate thin film network. Because more than one type of technique is involved, and the complication of the interconnections that are necessary, hybrid circuits are more complex and expensive than other types. However they do offer the designer a wide range of capabilities while still maintaining the ultra small size and other benefits generally associated with integrated circuits.

Application

Another method of classifying integrated circuits is by their application. Primarily, this is a means of distinguishing between linear and digital circuits. Digital integrated circuits of course work with logic levels, pulses and binary data. Such switching circuits use either bipolar transistors or MOSFETs. Linear integrated circuits usually involve amplifiers of some kind and work with analog signals. They are constructed with bipolar transistors.

The chart in Figure 4-4 shows the basic integrated circuit hierarchy.

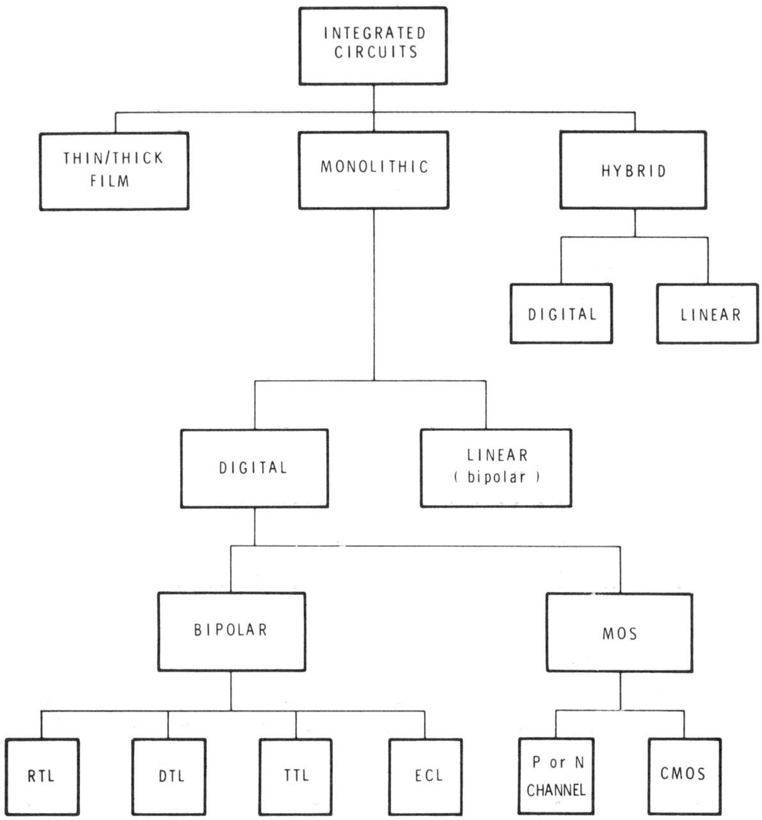

Figure 4-4 Hierarchy of integrated circuits.

Function

There are three basic classifications that identify the function of a digital integrated circuit. These are small scale integration (SSI), medium scale integration (MSI), and large scale integration (LSI).

SSI circuits are the simplest and most basic form of integrated circuits. These are amplifier or gate circuits that perform a single basic function. They must be interconnected externally in order to form complete functional or operational circuits. A typical SSI digital integrated circuit might consist of several multiple input gates or a flip-flop.

Medium scale integrated circuits are more complex. MSI circuits involve multiple gates which are interconnected to form a complete functional circuit. Most MSI circuits contain twelve or more equivalent gates or circuitry of similar complexity. An MSI circuit is usually a complete functional operating network such as a decoder, a counter or multiplexer. Such circuitry eliminates the need of having to interconnect individual gates in SSI packages to form the same function. MSI circuits greatly reduce the number of integrated circuits in a system and thereby reduce cost, assembly time, and in some cases, power consumption.

LSI circuits contain 100 or more equivalent gate circuits or networks of a similar complexity. LSI circuits are larger functional circuits or the equivalent of multiple MSI circuits. An LSI circuit often forms a complete system or instrument. The major application of LSI circuits is in semiconductor memories which store binary data. However, there are many different types of complex LSI circuits including electronic calculators, computers and certain types of test instruments.

Integrated Circuit Packaging

A primary consideration to the integrated circuit user is the packaging and physical characteristics of integrated circuits. There are three basic methods of packaging the silicon chip. These are the TO5 can, the flat pack, and the dual in-line package. These three basic types of packages are illustrated in Figure 4-3.

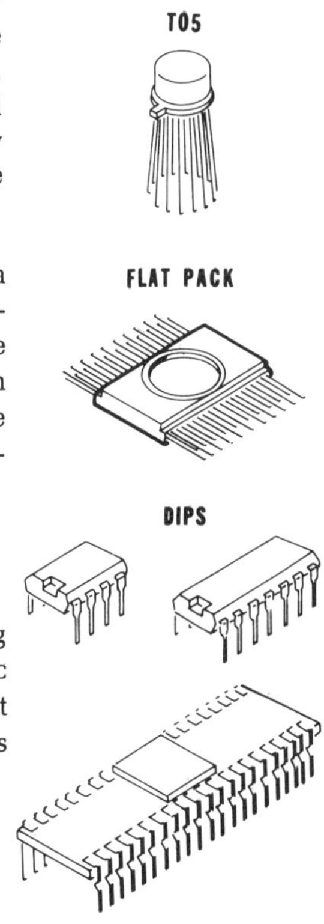

Figure 4-3 Illustrations of typical T05, flat pack and DIP ICs.

TO5. The earliest form of package used for integrated circuits was the TO5 can. This is a standard configuration for packaging transistors. This same package was modified by including additional leads. This type of package is still used with integrated circuits today but it is not the most popular form. Its advantage is that it can dissipate a substantial amount of heat. For that reason, this package is used mostly with linear integrated circuits.

Flat Pack. The flat pack was another type of housing used in the early development stages of integrated circuits. It is the smallest of all available integrated circuit packages and is designed for high density packaging. The packages are flat and are designed to be soldered or spot welded to a circuit board. Circuits can be placed close together and, therefore, a considerable amount of circuitry can be packaged in an extremely small area. Because of their ability to be packed so densely, these integrated circuits are generally made of a ceramic material that can withstand high temperatures. Closely packaged circuits cause heating and cooling problems. Therefore, the circuitry must be able to withstand such an environment. Flat pack circuits are used primarily in critical-size applications such as avionics, high reliability military systems and special industrial equipment.

DIP. The newest and most widely used form of integrated circuit packaging is the dual in-line package (DIP). It is slightly larger than the other types available but it offers many advantages. Such circuits are easy to mount and use. They are designed to be adaptable to machine insertion on printed circuit boards. They are available in various sizes, all the way from an 8-pin package (mini DIP) to a 40-pin package. Most SSI circuits are housed in 8, 14 or 16-pin dual in-line packages. MSI circuits are found in 14, 16 and 24-pin dual in-line packages. LSI circuits, because of their greater size complexity, require a greater number of input and output leads and, therefore, are usually housed in 24, 28 and 40-pin packages.

Several different types of dual in-line package materials are used. The most commonly used and least expensive is a plastic package. In this type of package, the integrated circuit chip is spot welded to a metal lead frame. The entire circuit is then encapsulated by an injection molded plastic technique.

For some critical integrated circuits, several types of ceramic packages are used. These are capable of withstanding higher temperatures and are generally hermetically sealed to provide an extra clean and safe environment for the circuit.

Temperature Ranges

Most integrated circuits are rated according to the range of temperatures over which they can operate satisfactorily. Most manufacturers generally specify both a military grade and a commercial or industrial grade circuit. The military grade circuits can be packaged in TO5 cans, ceramic flat packs, or ceramic dual in-line packages. These devices are capable of operating over a wide temperature range, usually from −55°C to +125°C. Circuits that perform properly over this wide temperature range are generally much more expensive. Such circuits are used only in high quality military equipment or in industrial equipment that is to be operated in severe environments.

For most general applications the commercial or industrial grade integrated circuits can be used. These are generally housed in plastic packages and are capable of operating over the 0°C to 70°C temperature range. Other temperature ranges are sometimes specified for different types of integrated circuits. Check the manufacturers data sheet for specific information on temperature ranges.

TRANSISTOR TRANSISTOR LOGIC

There are many different types of integrated circuit logic elements used in implementing digital equipment. All of them perform the basic logic functions but have different characteristics, capabilities and limitations. Different types of digital logic circuits have been developed to meet special needs. Over the years a variety of circuits have emerged.

One of the biggest and most important design decisions made by an engineer designing digital equipment is in the selection of a type of digital logic circuit. In this section we are going to discuss the most popular form of bipolar integrated circuit logic elements, transistor transistor logic. Non-saturating bipolar circuits and MOS digital integrated circuits will be considered in following sections.

The most popular and most widely used type of digital IC is transistor-transistor logic (TTL or T^2L pronounced T square L). Its popularity is primarily the result of its extremely low cost and the availability of a wide variety of SSI logic elements and MSI functional circuits. Its ease of use, high performance characteristics and interfacing capability are other features that make it desirable. A number of special types of TTL circuits are available to match special needs. Even though TTL integrated circuits have been available for approximately ten years, this type of logic circuit continues to remain popular for use in new equipment designs. Its versatility has made it the standard logic circuitry used in modern digital equipment.

Circuit Operation

Figure 4-5 shows the circuit of a typical TTL logic gate. It operates from a single +5-volt power supply and has typical logic levels of .4 volts for binary 0 (low) and 2.4 volts for a binary 1 (high). The circuit consists of three basic sections: a multiple emitter input transistor (Q1), a phase splitter transistor (Q2), and a totem-pole output circuit consisting of transistors Q3 and Q4. The multiple emitter-base junctions of transistor Q1, along with R1, form a diode gate. The primary advantage of this arrangement over individual diodes is that higher speed operation can be obtained.

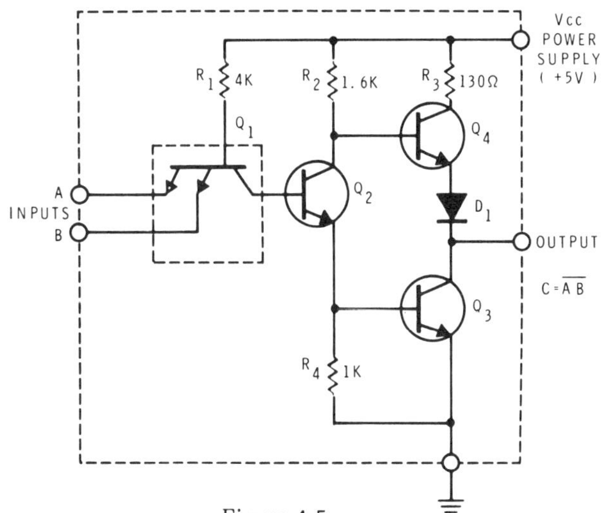

Figure 4-5
A typical transistor-transistor logic gate.

The phase splitter transistor (Q2) is a circuit that provides complementary drive signals for the two output transistors. The output circuit consists of transistors Q3 and Q4. These transistors are stacked one upon the other. Thus, this arrangement is given the name totem-pole. Transistor Q3 is simply a shunt transistor switch. Q4 in this circuit essentially serves as an active load resistor for Q3. In some logic circuits, the collector of the output transistor is returned through a collector resistor to the supply voltage. This collector resistor is known as a pull-up resistor because it causes the output to be pulled up to the supply voltage when the output transistor cuts off. In the TTL gate, Q4 serves as an active *pull-up* resistor. Current to any shunt load on the output is supplied by this transistor. This arrangement provides a much lower output impedance in the high output state and therefore higher speed operation can be obtained. In logic circuits using a pull-up resistor, any shunt output capacitance must be charged through the collector pull-up resistor. This charging time can be long depending on the amount of shunt capacitance and the value of the collector resistor. With the active pull-up arrangement in the TTL gate, any output capacitance can be charged more quickly through the very low impedance represented by Q4.

To simplify the discussion of the TTL logic circuit it is convenient to show a diode equivalent of the key parts to this circuit. This diode equivalent arrangement is shown in Figure 4-6. Diodes D1 and D2 in this circuit represent the emitter-base input junctions of transistor Q1. Diode D3 represents the base-collector junction of transistor Q1. Diode D4 represents the emitter-base junction of Q2, and D5 is the emitter-base junction of Q3. Study the diode equivalent in Figure 4-6 and relate it to Figure 4-5. Keep in mind that a PN junction silicon diode requires approximately 0.7 volts across it before it conducts. The forward voltage drop across this diode is also approximately 0.7 volts. Since diodes D3, D4 and D5 are connected in series, the voltage at point X will be the sum of the individual voltage drops or in this case approximately $3 \times 0.7 = 2.1$ volts. A voltage less than 2.1 volts at point X will mean that all three diodes will be cut off.

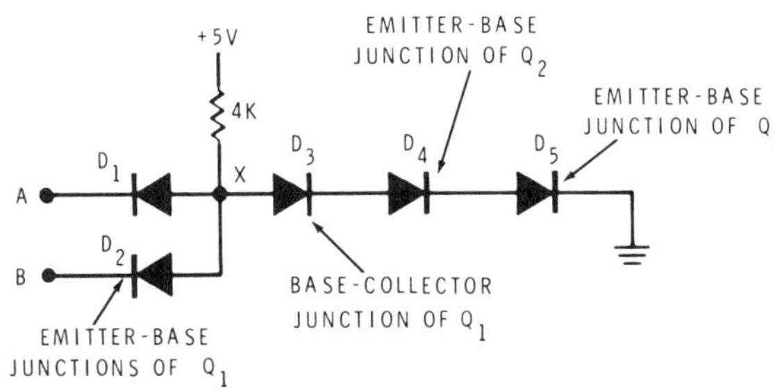

Figure 4-6 Diode equivalent circuit of TTL gate.

Now consider the operation of the circuit. If either one or both of the logic inputs are at their binary 0 level, 0.4 volts or less, the associated input emitter junction will conduct. The voltage at the base of Q1, (point X in Figure 4-6) will be the input logic level voltage plus the drop across the input emitter-base diode, in this case approximately $0.4 + 0.7 = 1.1$ volts. Current will flow through the input diode whose input is low and through the 4k pull-up resistor. Since the voltage at the base of Q1 (point X in Figure 4-6) is less than that required to cause the three diode string to conduct, the base-collector junction of Q1 will not conduct. The emitter-base junctions of Q2 and Q3 will not conduct therefore these transistors will be cut off. With Q2 off, base current will be supplied to transistor Q4 through resistor R2. Q4 will conduct if an output load is connected to ground. The output voltage at this time will be the supply voltage $+V_{CC}$ less the drop across diode D1, Q4 and the 130 ohm collector resistor. A typical TTL binary 1 output voltage level is approximately +2.4 to +3.6 volts. A binary 0 voltage level at either or both of the inputs will produce a binary 1 output.

If a binary 1 logic level is applied to both inputs, the input diode junctions will be reverse-biased. A typical binary 1 logic level input will be +2.4 volts or higher. Most of the inputs will be driven from other TTL output circuits and therefore, the output voltage will in most practical situations approach +3.6 volts. With the emitter-base input diodes reverse-biased, the diode equivalent string D3, D4, and D5 in Figure 4-6 will conduct through the 4K resistor. This means that the emitter-base junctions of Q2 and Q3 will be forward-biased as well as the base-collector junction of Q1. With Q2 conducting, its collector voltage is lower than that required to turn Q4 on. Base current normally supplied to Q4 through R2 is shunted away by the conduction of Q2. With Q2 conducting Q3 will saturate. At this time the output voltage is the emitter-collector saturation voltage of Q3 which will be +0.4 volts or less. As you can see, with binary 1's on both inputs, the output will be binary 0. From this circuit description you can see that the circuit performs the NAND function for positive logic and the NOR function for negative logic. The truth tables in Figure 4-7 sum up the operation of the basic TTL gate while the basic characteristics are summarized in Table I.

INPUTS		OUTPUT
A	B	C
+.4V	+.4V	+3.6V
+.4V	+3.6V	+3.6V
+3.6V	+.4V	+3.6V
+3.6V	+3.6V	+.4V

A

INPUTS		OUTPUT
A	B	C
0	0	1
0	1	1
1	0	1
1	1	0

B

INPUTS		OUTPUT
A	B	C
1	1	0
1	0	0
0	1	0
0	0	1

C

Figure 4-7 Truth tables for typical TTL logic gate (A) electrical, (B) positive logic NAND, (C) negative logic NOR.

Table I

TTL Characteristics

Type of logic: Current sinking
Propagation delay: 10 nanoseconds
Power dissipation: 10 milliwatts
Fan out: 10
Noise immunity: high
Logic levels: binary 0 = + .4 volts
binary 1 = +3.6 volts
Basic gate form: positive NAND/negative NOR
Supply Voltage V_{CC} +5 volts ±10 percent

TTL integrated circuits continue to be one of the most popular and widely used forms of logic elements. Many new equipment designs continue to use this type of circuit. Many manufacturers supply TTL circuits and new circuits are developed regularly. The wide range of SSI and MSI types make TTL circuitry perhaps the most versatile line of digital integrated circuits available. The most common type of TTL circuits are the 7400 series originally developed by Texas Instruments. Almost all other integrated circuit manufacturers second source this series of TTL circuits. Other TTL circuits are also available. These include the 9300 series made by Fairchild and the 8000 series manufactured by Signetics. All of these types of TTL circuits are compatible with one another. In this program you will use and demonstrate many different types of the 7400 series of TTL ICs.

Special TTL Variations

All TTL integrated circuits whether SSI or MSI, combinational or sequential use the basic TTL gate circuit shown in Figure 4-5. In addition there are several other versions of this TTL circuit made for special applications. These include gates for low power operation, higher speed operation or special logic functions.

Low Power TTL. Low power TTL circuits are similar to the basic TTL circuit described earlier. The only difference is that the resistor values in the circuit are approximately ten times higher meaning that the power consumption of the circuit is one tenth of that of the standard circuit. Low power TTL circuits are excellent for applications requiring a versatile high speed logic line with minimum power consumption. Increasing the values of the internal resistances, causes the propagation delay of the circuit to be increased. The propagation delay in a typical low power gate is approximately 30 to 40 nanoseconds. High speed is sacrificed for low power consumption.

High power TTL. The high power TTL circuit is basically the same as the standard circuit considered earlier. In this circuit resistor values are decreased significantly in order to improve operating speed. Typically, the gate propagation delay is reduced to approximately 6 nanoseconds. This increase in speed is accompanied by a power dissipation approximately twice that of a standard gate. This is approximately 22 milliwatts average power dissipation per gate. High power TTL circuits have for the most part been replaced by the newer Schottky TTL circuits which are not only faster but also consume less power.

Schottky TTL. The transistors in a TTL logic circuit operate in the saturation mode. To achieve higher speed operation than that obtainable with a standard or high power TTL gate, non-saturating transistors must be used. This is what is done to improve the speed of operation of Schottky TTL circuits.

The circuit of a Schottky TTL gate is basically the same as the standard TTL gate circuit we discussed earlier. The primary difference is that a diode is connected between the base and collector of each transistor in order to prevent those transistors from saturating. See Figure 4-8A.

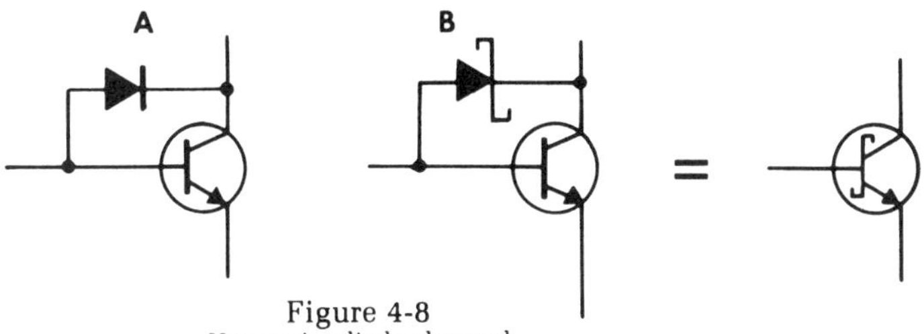

Figure 4-8
Hot carrier diode clamped transistors used in Schottky TTL circuits to prevent saturation and increase switching speed.

When the transistor begins to turn on, its collector voltage will drop quickly to some low value. When it drops beyond a certain point, the diode will conduct and shunt current away from the collector-base junction that would normally conduct during saturation. The diode effectively clamps the collector to a voltage sufficiently high to keep the base-collector junction reverse biased. This circuit permits the condition of saturation to be closely approached but still avoided.

The diode used to provide the clamping that prevents saturation is a hot carrier or Schottky diode. This type of diode, unlike other semiconductor diodes, is not a PN junction type. Instead a Schottky diode is the junction of a metal such as gold or aluminum and N-type semiconductor material. These diodes are not separate units. Instead they are part of the complex integrated circuit diffusion on the silicon chip. These diodes are high speed in operation because they do not have the normal charged storage normally associated with PN junction diodes. The forward voltage drop or voltage required for the diode to conduct is also much less than a standard PN junction.

The Schottky clamp diode as it is used on the transistors in a TTL gate is illustrated in Figure 4-8B. Note the special symbol used to represent the Schottky diode. The special symbol on the right in this figure is a Schottky diode clamped transistor. You will see this symbol used in schematic diagrams of Schottky TTL logic circuits.

The primary advantage of the Schottky TTL gate is its higher speed of operation. Since the transistors do not saturate, no charge storage problems occur. Gate propagation times as low as 3 nanoseconds are possible with this type of circuit. At the same time Schottky TTL gates achieve this rate of switching at a power dissipation of approximately 19 milliwatts, something less than the high power TTL circuit. A special low power Schottky TTL circuit with a propagation delay of 10 nanoseconds and a power dissipation of 2 milliwatts is also available. This form of TTL has one of the most favorable speed-power trade-offs of any digital integrated circuit.

Three State TTL. Three state TTL integrated circuits are a special version of TTL circuits whose output can assume three states instead of the normal two. Besides the binary 0 and binary 1 logic levels normally associated with a TTL gate output, the three state circuit has a third *open* state. This open state represents a very high impedance and is essentially equivalent to disconnecting the TTL totem pole output circuit from the output pin on the integrated circuit. This particular type of circuit is useful in digital systems using multiplexed or bussed data transmission.

A data bus is a group of wires, transmission lines or cables over which digital information or binary numbers are transferred in parallel from one point to another. There are basically two types of data busses, unidirectional and bi-directional. On a unidirectional bus data is transferred in only one direction. On a bi-directional bus data can move in either direction. Most digital busses are bi-directional in nature.

Instead of having multiple parallel paths for the transmission of digital data in two directions, a common bus is used and the information is transferred from one place to another on a time shared basis. While data from one particular source is being transferred, other data waits until the current transfer is complete. The concept of having one bus serve as a carrier for multiple signals is known as multiplexing. Circuits not currently in use are disabled while those sending and receiving data on the bus are activated.

Figure 4-9 shows a simplified diagram of a typical bi-directional digital data bus. Only one of several identical bus lines is shown. This bus line is generally responsible for transmitting one bit of data of a multiple bit binary word. The bus itself may be simply a cable several feet long or a transmission line several hundred feet long. Gates 1, 2, or 3 can transmit one bit of information down the bus line to be received by gate 8. Only one of the three gates will be enabled at a time to transmit the desired data. Note that the same bus can be used to transmit binary information from either gate 6 or gate 7 down the transmission line to be received by gates 4 or 5. Just keep in mind that only a single data transmission may take place at any given time but that it may be in either direction from any one of several sources or to several destinations.

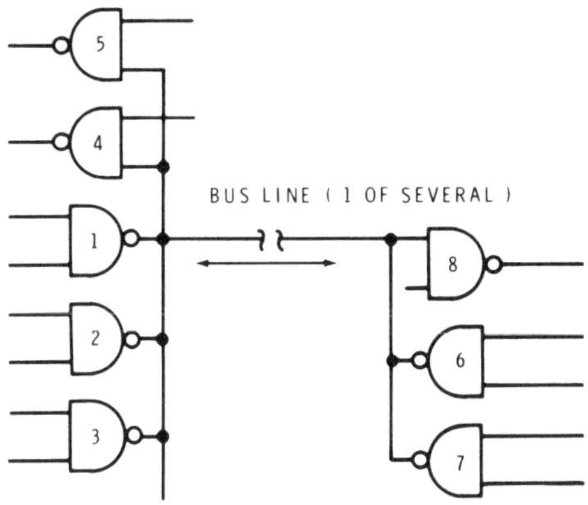

Figure 4-9
Bi-directional digital data bus.

The digital bus is relatively easy to implement with logic circuits using collector pull up resistors. The outputs of the gates from which the digital data comes are simply connected in parallel as illustrated in Figure 4-10. By connecting their outputs directly together, we effectively parallel the collector resistors, thereby reducing the total resistance to one half the value of an individual resistor. The two output transistors then share a common collector resistance. With this arrangement either transistor Q_1 or Q_2 can bring the output to the binary 0 condition. If Q_1 conducts and Q_2

is cut-off or if Q_2 conducts and Q_1 is cut-off, the output will be binary 0. The only time that the output will rise to $+V_{CC}$ is when both Q_1 and Q_2 are cut-off. The way digital data is transmitted by one gate then is to disable the gates not responsible for transmitting data. This is done by applying the appropriate input to the gate so that its output transistor is cut-off. This permits the other transistor to control the state of the output.

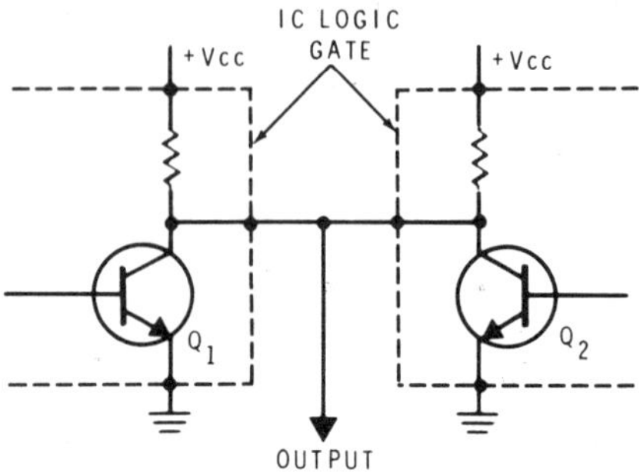

Figure 4-10
Paralleling gate outputs
to share a common output line.

Parallel gate outputs form what is known as the wired OR connection. It is given this name simply because either transistor Q_1 OR Q_2 can bring the output to the binary 0 level. This connection is frequently used to implement the logical OR function without the need of an additional gate.

Because of the active pull-up circuit in the totem pole output of a TTL gate, TTL circuits cannot be wire ORed. Improper operation or damage can occur. For this reason TTL ICs cannot be used in bussing operations. To overcome this problem, open collector TTL circuits can be used. In these circuits, the active pull-up stage is eliminated and the collector of the shunt output transistor is made available at an output pin. To this is connected an external collector pull-up resistor. The wired-OR arrangement can then be used. However, because the active pull-up transistor is removed, one of the primary advantages of a TTL gate is eliminated. The active pull-up produces higher speed operation and lower output impedance, both of which are desirable not only from a speed stand point but from one of improving noise immunity.

The disadvantage of not being able to use TTL circuits in bus applications was overcome by the development of three state logic. This type of logic was originally introduced by National Semiconductor Corporation as Tri-State Logic. Three state logic is a special form of TTL that retains the basic TTL circuit configuration including the totem pole active pull-up output circuit. However, additional circuitry has been added to produce an optional high impedance third state that can effectively remove from any common bus line those circuits not transmitting data.

A typical three state TTL circuit is shown in Figure 4-11A. The circuit arrangement is basically identical to the TTL gate circuit we discussed earlier. Q_1 is the multiple emitter input transistor, Q_2 is the phase splitter while Q_4 and Q_5 form the totem pole output circuit. Transistor Q_3 has been added in order to provide better control of output transistor Q_4. Together, Q_3 and Q_4 form a compound or Darlington transistor with high gain. Components D_1, Q_6, Q_7 and Q_8 have been added to control the third state.

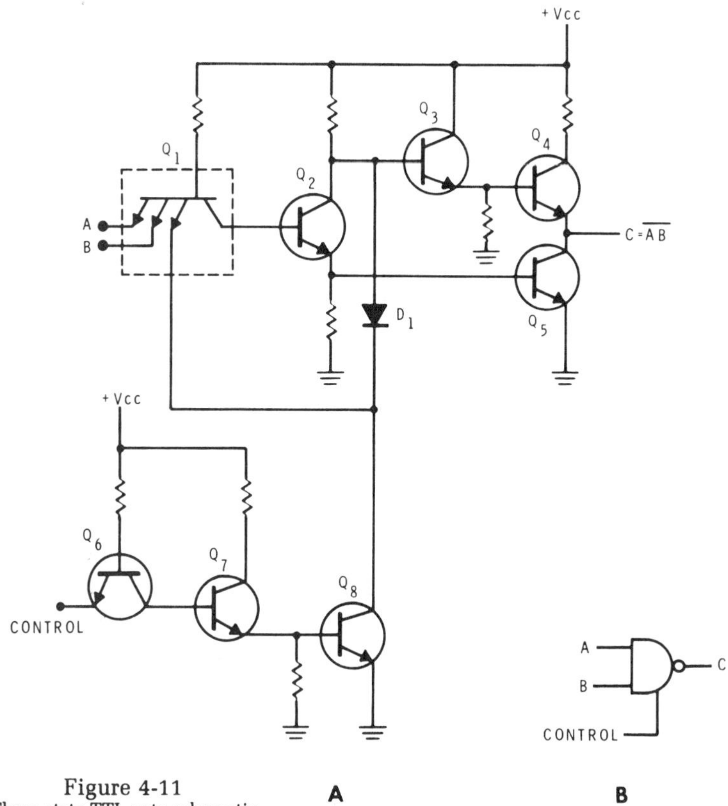

Figure 4-11
Three state TTL gate schematic (A) and the logic symbol (B).

Whenever the control input is low, transistor Q_6 saturates. When it saturates it causes the collector of Q_6 and hence the base of Q_7 to be nearly the same low input value. This means that Q_7 and Q_8 are cut-off. The TTL gate then functions normally. Here the standard TTL output logic levels for binary 0 and binary 1 are achieved according to the input states.

When the control input is made a binary 1, Q_6 cuts off. The emitter-base junctions of Q_7 and Q_8 become forward biased through the base-collector junction of Q_6 and the associated base resistor. With Q_8 saturated, the third input to Q_1 is brought to ground along with the cathode of diode D_1. As you recall, when anyone or more of the inputs of a TTL gate are brought to a binary 0 level, the output is forced high. This is done by turning on the active pull-up transistor Q_4 and turning off the shunt output transistor Q_5. But in this case, both output transistors Q_4 and Q_5 are cut-off. When Q_8 saturates, the cathode end of D_1 is grounded. All the base current for Q_3 is shunted away. This causes Q_3 and Q_4 to cut off. With both output transistors Q_4 and Q_5 cut-off, the output is effectively an open circuit. Looking from the output of the gate back into the circuit, any load sees an extremely high impedance. Because of the high quality of the circuit and the low leakage, any load sees essentially an open circuit. With this arrangement, any number of three-state TTL outputs may be paralleled to form a common bus line. When data is to be transmitted, all of those gates not transmitting data will have their control lines at binary 1 so that their outputs represent an open circuit. Only the gate designated to transmit data will be enabled.

Figure 4-11B shows the logic symbol normally used to represent a three state TTL gate.

EMITTER COUPLED LOGIC

The major switching speed limitation of bipolar logic circuits is the charge storage that occurs in the base region when both the emitter-base and base-collector junctions of a bipolar transistor are forward-biased to achieve saturation. The time required for this charge to be eliminated delays the turn-off of the transistor. Switching speeds can be greatly increased if this delay time is minimized or eliminated. In Schottky TTL circuits this storage problem is eliminated by preventing the transistors from saturating. This significantly increases the switching speed. Non-saturating transistor circuits offer the best potential for fast logic.

Another major form of logic circuits using non-saturating bipolar transistors is called emitter coupled logic (ECL). Also known as current mode logic, this circuitry is essentially a differential amplifier configuration which effectively prevents transistor saturation. ECL ICs are the highest speed logic circuits available today.

Circuit Operation

Figure 4-12 shows the schematic diagram of a typical ECL logic gate.

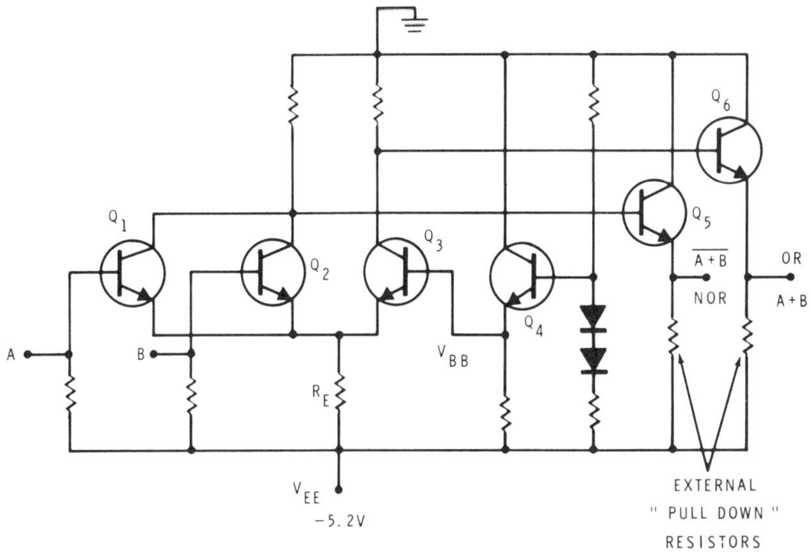

Figure 4-12

Typical emitter coupled logic gate.

Transistors Q_1 and Q_2 along with Q_3 form a differential amplifier. The inputs (A and B) are applied to transistors Q_1 and Q_2 whose emitter and collector connections are in parallel to form one side of the differential amplifier circuit. If additional inputs are required, more transistors are paralleled. Q_3 is the other side of the differential amplifier. Input logic levels are typically −1.75 volts (binary 0) and −.9 volts (binary 1). The output and input voltage swing is typically the difference between these two voltage levels or (1.75 − .9) = .85 volts. The supply voltage V_{EE} is −5.2 volts. The supply voltage V_{EE} and R_E form a current source that supplies a fixed current whose level is below the point that permits saturation.

In Figure 4-19, transistor Q_4 and the associated components form a temperature stabilized voltage source that biases the base of Q_3 to approximately −1.3 volts. The emitter of Q_3 will be approximately .8 volts more negative than its base because of its emitter base voltage drop. Therefore, the voltage at the emitters of Q_1, Q_2, and Q_3 will be approximately (−1.3 − .8) = −2.1 volts.

Assume that both logic inputs A and B are at the binary 0 logic level of −1.75 volts. With this condition transistors, Q_1 and Q_2 will not conduct because the emitter-base bias is insufficient. At this time the collectors of Q_1 and Q_2 are high while the collector of Q_3 is low. These two output levels are buffered by output emitter followers Q_5 and Q_6 and produce the proper binary 0 and binary 1 logic levels. Note that both the normal and complement outputs are available simultaneously.

If any one or both of these logic inputs rise to the binary 1 state (−.9 volts), the associated input transistor will conduct. The emitter of Q_1, Q_2, and Q_3 will then be one emitter-base junction drop more negative than the input voltage or approximately −1.7 volts. This means that when one or more of the inputs rises to the binary 1 level, the common emitter point will rise from −2.1 volts to −1.7 volts. This voltage will cause transistor Q_3 to cut-off. As you can see the current supplied by the emitter supply voltage and the common emitter resistor R_E switches from Q_3 to the conducting input transistor or transistors. With this arrangement the collectors of Q_1 and Q_2 will be low while the collector of Q_3 will be high. These logic output levels are buffered by the emitter followers Q_5 and Q_6. The circuit performs the OR and NOR functions for positive logic level assignments.

One of the major advantages of most commercial ECL IC logic circuits is the availability of both the normal and complement outputs simultaneously. This permits both OR and NOR functions to be obtained at the same time. No external inverters are required. Note that the emitter follower "pulldown" resistors are usually external to the integrated circuit itself. This permits the resistors to be located remotely at the end of a transmission line or in another desirable location depending upon the exact circuit configuration and application. It is also possible to tie together the OR or NOR outputs of ECL logic circuits to permit the wired OR function.

Circuit Characteristics

ECL logic circuits are extremely versatile, easy to use and produce high quality results. However, these circuits are normally higher in cost and of course consume significantly more power than other types of logic circuits. Their only real advantage is their high speed. Naturally, this high speed capability should be used only where absolutely necessary. Other types of logic are preferred for slower speed applications. The most widely used form of ECL is Motorola's MECL series.

Table II

ECL Characteristics	
Type of logic:	unsaturated, current sourcing
Propagation delay:	1—3 nanoseconds
Power Dissipation:	40—60 milliwatts
Fan Out:	10—25
Noise Immunity:	High
Logic levels:	binary 0 = −1.75 volts.
	binary 1 = −.9 volts.
Basic gate form:	OR/NOR
Supply voltage V_{EE}:	−5.2 volts

METAL OXIDE SEMICONDUCTOR INTEGRATED CIRCUITS

Metal oxide semiconductor field effect transistors (MOSFETs) or insulated gate field effect transistors offer numerous advantages over bipolar transistors in digital circuits. First, these devices are simpler in construction and therefore can be made much smaller. Because they occupy less space, higher density logic networks can be placed on a given silicon chip. This permits large scale digital integrated circuits to be readily constructed. Another advantage of the MOSFET is its high impedance and therefore, low power consumption. MOS digital integrated circuits consume only a fraction of the power of equivalent bipolar circuits.

The big disadvantage of MOS digital integrated circuits is their lower speed. The high impedance and capacitive nature of these circuits produce switching speeds that are several orders of magnitude slower than bipolar digital integrated circuits. Despite their low speed nature, these circuits have nevertheless found wide application in those areas not requiring high speed operation. Recent technological advances in the manufacturing of MOSFET circuitry have also helped their switching speed approach that of some bipolar circuits.

There are two basic types of MOSFETs used in MOS digital ICs: the P-channel and the N-channel. As indicated earlier, the MOSFETs used in digital ICs operate in the enhancement mode. In this mode of operation, the transistor is normally cut-off. When the appropriate voltage level is applied between the source and the gate, the transistor suddenly switches on. These devices are near perfect switches in that they have an extremely high impedance when not conducting and an extremely low impedance when conducting.

The earliest MOS devices in use in digital circuits were PMOS circuits. P-type MOSFETs are the simplest and easiest to manufacture. N-channel MOS devices are more difficult to make, however, recent technological advances have simplified their construction. N-channel MOSFETs are smaller in size and switch at higher speeds. Their switching threshold is also much lower making them more compatable with bipolar digital integrated circuits. Another class of MOS logic circuits combines both P- and N-channel devices in a single circuit. These digital circuits are known as complementary MOS or CMOS.

MOS digital integrated circuits are clearly the digital ICs of the future. Their small size, low power consumption and simplicity make them attractive for many medium and large scale applications. Entire test instruments and computers can be made on a single silicon chip with MOS techniques. Improvements in manufacturing techniques will gradually reduce the switching speeds to levels acceptable to all but the most demanding high speed applications. It is expected that MOS ICs will eventually replace many of the various types of bipolar circuits.

PMOS and NMOS Circuits

When a digital IC circuit is constructed only with P-channel MOSFETs, it is known as PMOS. If the circuitry uses only N-channel enhancement mode MOSFETs, the circuits are referred to as NMOS. The basic circuit configurations are the same for either type of transistor.

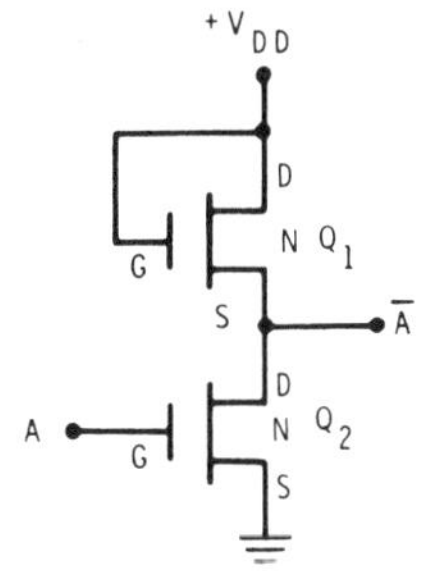

Figure 4-13
N-channel MOS logic inverter.

Figure 4-13 shows the circuit for an N-channel MOS logic inverter. Q_2 is the standard shunt inverter switch. Transistor Q_1 is connected to form a drain or load resistance. Another MOSFET (Q_1), biased into conduction, is substituted for a finite load resistor. A standard integrated resistance occupies substantially more space than the MOSFET and therefore, if used in any quantity, greatly reduces the amount of digital circuitry that can be placed on a given size silicon chip. The gate and drain of Q_1 are connected together which biases the transistor into conduction. It acts as a low resistance.

In operation, this circuit performs like any other inverter. When the logic input voltage is less than the gate-source threshold voltage, Q_2 is cut off. Since the gate and drain of Q_1 are connected together, Q_1 conducts and the output voltage is the supply voltage V_{DD} less the gate-source threshold voltage. When the input voltage exceeds the gate-source threshold of Q_2, Q_2 conducts. Its output voltage drops to a very low level. The on resistance of transistor Q_2 is made significantly less than the resistance of Q_1 by a ratio of at least 20 to 1.

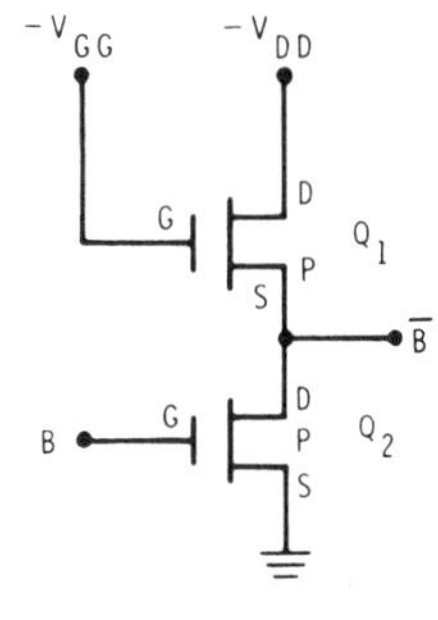

Figure 4-14
P-channel MOS logic inverter.

A logic inverter circuit using P-channel MOSFETs is shown in Figure 4-14. This circuit is similar to the N-channel inverter discussed earlier, however, a different means is used to bias the load transistor into conduction. In this circuit, the gate voltage is made more negative than the source voltage by the use of another power supply designated $-V_{GG}$. This voltage causes Q1 to conduct and act as a low value resistance. Q_2 in this circuit is the shunt inverter switch. Note the power supply polarity difference in this circuit due to the use of P-channel rather than N-channel transistors. The circuit operation is as described before. When the gate-source threshold voltage is exceeded, transistor Q_2 will conduct and the output voltage will drop to a value near ground. When the gate-source voltage is less than the threshold value, Q_2 will cut off and act as an open circuit. At this time the output voltage is some negative voltage less then $-V_{DD}$. This particular circuit arrangement produces somewhat faster switching speeds than the circuit in Figure 4-13 but the disadvantage is that it requires the additional power supply.

Figure 4-15 shows how the various logic functions are implemented using the MOSFETs. In Figure 4-15A, two N-channel devices are connected in parallel and share a common load (Q_3). If either one or both of the input devices is biased on, the output voltage will drop to a low value. When the input voltages are both less than the threshold voltage, Q_1 and Q_2 will be cut off, allowing the output to rise toward $+V_{DD}$. With this arrangement, the circuit performs a NOR function for positive logic.

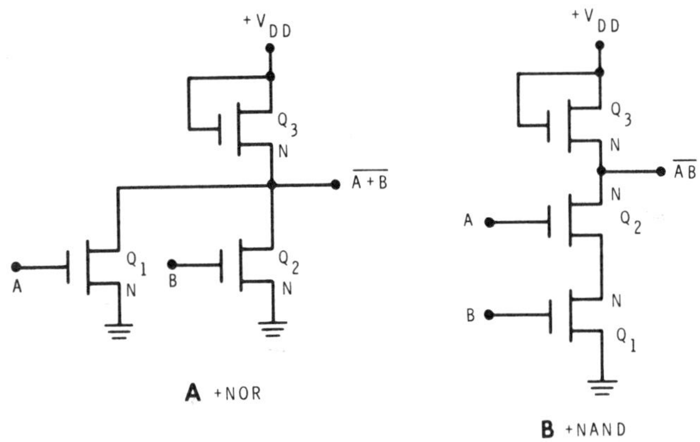

Figure 4-15
NMOS logic circuits.

The circuit in Figure 4-15B performs the NAND function for positive logic. Here transistors Q_1 and Q_2 are connected in series. In order for the output line to be brought low, input signals must appear high at both the A and B inputs simultaneously. If either one or both of the inputs are low, the output will be high. This is the NAND function.

The type of MOS logic circuits that we have discussed here are known as static logic circuits in that they perform logic functions with voltage levels. Another type of MOS logic is also widely used. Known as dynamic logic this circuitry has the same basic configuration as we have discussed here. However, the difference lies in that the circuits take advantage of the capacitive nature of the input to the MOS devices. Here the input capacitors are used to store charges or logic levels temporarily. During operation, high speed clock signals are used to transfer stored charges from one circuit to the next. The advantage of this type of MOS circuitry is still lower power consumption than the static circuits just discussed and higher operating speeds.

Complementary MOS

Complementary MOS or CMOS logic circuits use both P-channel and N-channel enhancement mode MOSFET's. A logic inverter constructed using both types of devices is shown in Figure 4-16. Here the input signal drives the gates of both devices simultaneously. When the input voltage is low or near ground, the gate-source threshold level for Q_2 is less than that required for conduction. Therefore Q_2 is cut off. However with the input low, the gate-source voltage threshold of Q_1 is exceeded. Since the gate is more negative than the source, this P-channel device conducts and connects the supply voltage V_{DD} to the output.

If the input voltage is brought to a high logic level, normally the same as the supply voltage $+V_{DD}$, the gate-source threshold voltage of Q_2 will be exceeded. Q_2 will conduct and act as a very low resistance bringing the output to a low level. At this time with the gate and source of Q_1 approximately at the same levels, the threshold is not exceeded, therefore, Q_1 is cut off. The logic inversion function is performed. It is interesting to note that in this circuit both the current sink and current source modes are used. When Q_2 conducts it sinks current from external loads connected between the outputs and the supply voltage. When Q_1 conducts it supplies current to any load connected between the output and ground. Because of the very low on resistance of a conducting device and the extremely high input impedance of other MOS devices, the logic levels are very nearly equal to ground and the supply voltage $+V_{DD}$.

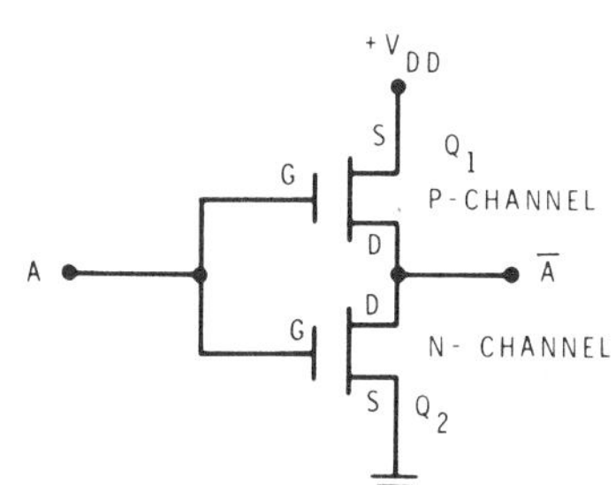

Figure 4-16
A complementary MOS logic inverter.

Figure 4-17 shows a diagram of a typical CMOS logic gate. It consists of two P-channel devices Q_1 and Q_2 connected in series and two N-channel devices Q_3 and Q_4 connected in parallel. When either one or both of the inputs go high to the positive binary 1 voltage level, the associated N-channel transistor Q_3 or Q_4 will conduct. This will cause the output to drop to a voltage level near zero volts indicating a binary 0 output. With either input high, either Q_1 and Q_2 will be cut-off. Since Q_1 and Q_2 are in series, the path between the supply voltage $+V_{DD}$ and the output is not completed unless both Q_1 and Q_2 conduct.

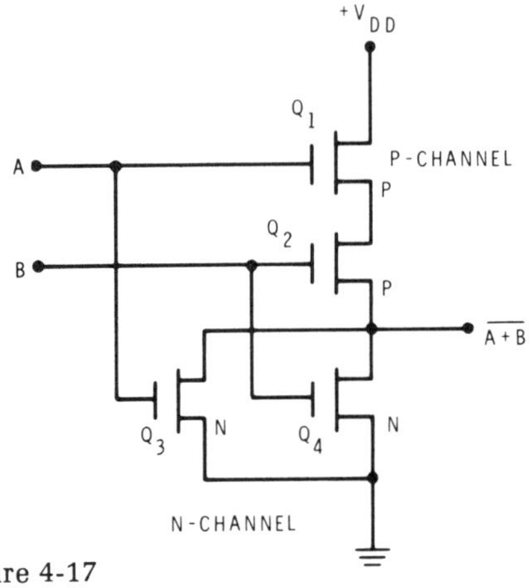

Figure 4-17
CMOS NOR logic gate.

When both inputs go low, Q_3 and Q_4 will become cut-off because their gate-source voltage is below the threshold level necessary to cause conduction. At this time the gate-source voltages of Q_1 and Q_2 will be approximately equal to $+V_{DD}$. Both devices conduct and form a low resistance path between the output and the supply voltage.

Summarizing the operation of this CMOS logic gate, you can see that the NOR logic function is being performed with positive logic. Any logic function including NAND, AND and OR can be performed by suitably arranging the P- and N-channel devices in the circuit.

If a digital designer were to specify the perfect logic circuit for all applications, this logic circuit would have characteristics very similar to those of CMOS. CMOS logic circuits offer a balanced combination of characteristics making it highly versatile and desirable. This type of logic circuit features very low power dissipation, excellent noise immunity, wide power supply voltage variations, high fan out and moderately high speed of operation.

Low power dissipation in CMOS circuits is achieved because there is never a continuous path through any of the devices in the circuit from the supply voltage to ground. A look at the inverter circuit in Figure 4-16 or the NOR gate in Figure 4-17 will indicate this. When the N-channel devices connected between the output and ground are conducting, the P-channel devices between the output and $+V_{DD}$ are cut off. Similarly, when the P-channel device between the supply voltage and the output is on, the N-channel devices from the output to ground are cut off. The current flow that does take place between the supply voltage and ground is that which flows when the output state switches. It is during this time that the P- and N-channel devices may be on together momentarily thereby causing a small current to flow. If the switching rate is high, the rate of occurrance of this current flow increases. The power consumption of a CMOS device increases with its operating frequency because of this effect.

Table III

CMOS Characteristics

Type of logic: Current sinking and current sourcing.
Propagation delay: 30—100 nanoseconds.
Power Dissipation: .01 milliwatt (static)
1 milliwatt at 1 MHz
Fan out: 50 +
Noise immunity: very high (45 percent V_{DD})
Logic levels: binary 0 = 0 volts
binary 1 = $+V_{DD}$
Basic gate form: positive NOR/negative NAND
Supply voltage V_{DD}: +3 to +15 volts

The most popular forms of CMOS logic are the 4000 series circuits made by RCA and the 74C series made by National Semiconductor. Both offer a variety of SSI, MSI and LSI circuits.

SELECTING A DIGITAL INTEGRATED CIRCUIT FOR A SPECIFIC APPLICATION

The most important decision that you will make in designing a piece of digital equipment is in selecting the type of integrated circuit. The success or failure of your design from a performance and economic standpoint will depend directly upon this choice. For this reason, you must carefully consider the requirements of your application.

The first step in selecting a digital integrated circuit is to completely define the system characteristics. Performance, economy, and reliability are the prime requirements. These are specifically stated in the terms of needed system speed, noise immunity, power dissipation and other factors. Once all of the specifications have been detailed, these can be matched against the capabilities and characteristics of the various types of integrated circuits available.

The three key factors in defining a digital system and selecting a type of integrated circuit are speed (propagation delay), power consumption and noise immunity. While these are the most important characteristics in terms of performance and economy, there are other considerations. These include cost, availability, trends and the need for complex functions. The cost and availability factors are obvious. Lowest cost circuits are best if they meet all of the requirements of your application. And the circuits you choose must be readily available. It is desirable to select a circuit with several sources of supply.

Trends

The semiconductor industry is one of the most volatile in the field of electronics. New technological developments occur frequently and therefore some devices are quickly obsoleted while others with improved characteristics are added. These changes take place so quickly that it is difficult for any designer to select an integrated circuit with long life and a stable price. It is important for you to follow the trends by reading the manufacturer's literature and the technical publications. By staying abreast of the latest technology you will be better prepared to select a circuit that not only meets your design requirements but also will have favorable life, cost and availability trends.

Complex Functions

A highly desirable characteristic of any integrated circuit type is the availability of medium scale integrated circuits (MSI). These are functional circuits either of the combinational or sequential type. Since most digital equipment is made up of only a few basic types of circuits (decoders, counters, etc.), a large portion of the equipment can be designed by simply interconnecting the functional circuits. It is much more economical to use MSI functions than to implement these same functions with SSI gate circuits. By using MSI functions the design time is reduced substantially. In addition the total package count and size is proportionally reduced. Savings in power consumption and in assembly labor can be significant. The more MSI functions available in a digital integrated circuit logic line, the more advantageous it is.

Trade Offs

Keep in mind that all of these factors are somewhat inter-related. Your choice of a type of digital IC will be a compromise based on your application and the circuits available. The speed-power trade off is one of the most critical trade-offs in selecting a circuit. Noise immunity is also a factor that may have to be traded off with some other characteristic depending upon the types of circuits available. You may have to juggle circuit specifications with cost and availability.

Figure 4-18 shows the primary characteristics of all of the types of digital logic circuits discussed in this section. It will permit you to compare the different types for your application. Only these types should be considered for new product design.

Comparison of IC Logic Family Characteristics

Characteristic	TTL	ECL	MOS	CMOS
Fan-out	10	25	20	50+
Cost	low	medium to high	medium to high	low to medium
Power dissipation per gate (mW)	12 to 22	40 to 60	0.2 to 10	0.01 static 1 at 1 MHz
Noise generation	high	low to medium	medium	low medium
Immunity to external noise	good	good	good	very good
Temperature range (°C)	−55 to +125 0 to 75	−55 to +125	−55 to +125 0 to 75	−55 to 125 −40 to 85
Typical supply voltage (V)	+5	−5.2	−27, −13 (PMOS) +5 (NMOS)	+1.5 to 18
Avg. propagation delay per gate (ns)	3 to 12	1 to 2	300 (PMOS) 50 (NMOS)	70
Avg. clock rate (MHz)	15 to 120	200 to 1000	2 (PMOS) 5 to 10 (NMOS)	5 to 10

Figure 4-18

Figure 4-19 shows a chart of speed vs. power consumption for the various types of integrated circuits. Here propagation delay is plotted as a function of power dissipation for all of the most popular types of digital logic circuits. Naturally, the best circuit is the one with the shortest propagation delay and the least amount of power dissipation. You can see from this chart the type of circuit with the most favorable speed-power rating is low power Schottky TTL.

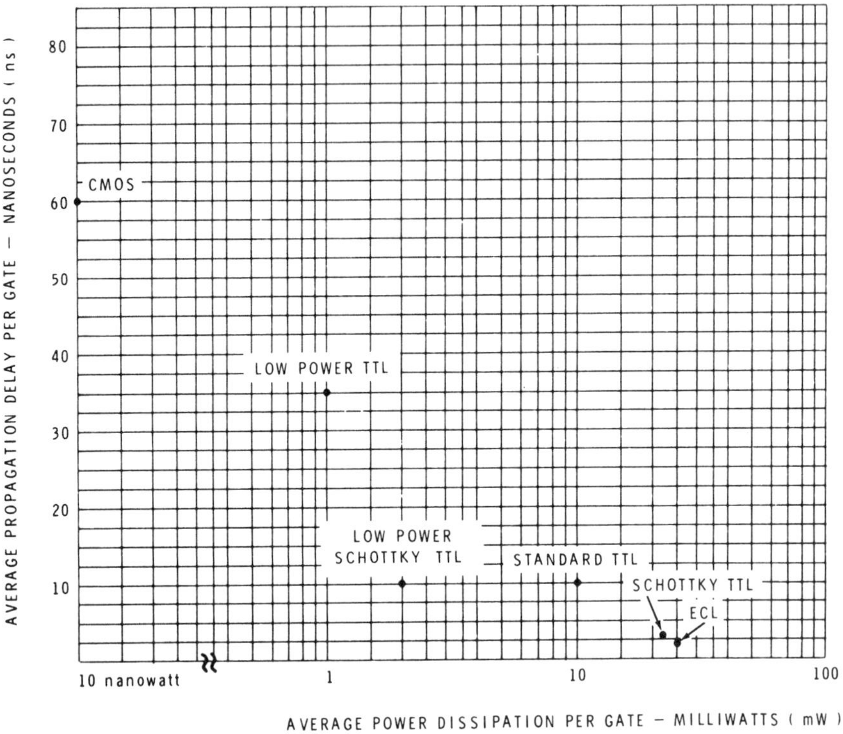

Figure 4-19
Speed (propagation delay) vs. Power dissipation for popular digital integrated circuits.

Unit 5

BOOLEAN ALGEBRA

INTRODUCTION

Boolean algebra is the special language of digital logic circuits. It is a mathematical method of expressing, analyzing, and designing logic circuits. It is similar in many ways to conventional algebra, but it is simpler and has several pertinent differences. It is easy to learn and extremely useful. It is almost essential to the proper understanding and application of digital circuits. In this unit, you will learn how to use Boolean Algebra. You will see how truth tables help you to design and understand logic circuits. And you will learn how to implement practical digital circuits using integrated circuit logic elements.

RELATING DIGITAL LOGIC CIRCUITS AND BOOLEAN EQUATIONS

1. Boolean algebra is a simplified mathematical system used to deal with binary or two value functions. It permits us to express all of the various logic functions, both simple and complex, in a convenient mathematical format. This system gives us a method of understanding and designing digital logic circuits.

In Boolean algebra, logic functions are expressed ________________.

2. (mathematically) The mathematical expression of logic functions permits a convenient means of analyzing and expressing operations in digital circuits. It also aids greatly in design. The proper application of Boolean algebra usually results in the simplest, least expensive and most efficient logic circuit design.

One of the most beneficial applications of Boolean algebra is in ________________.

3. (design) Most digital equipment in use today is made with integrated circuits. Boolean algebra is used in designing these devices. The applications of integrated circuits in the design of modern electronic equipment, also involves Boolean algebra, but only to a lesser extent. At one time, the engineer or technician designing a digital system had to design not only the logic functions but also the circuits to implement them. Boolean algebra was his primary design tool.

The modern digital designer does not need Boolean algebra today as much as he did in previous years because of the availability of ________________ ________________.

4. (integrated circuits) The engineer and technician using and designing digital circuits, finds Boolean algebra most valuable in expressing and analyzing logic circuits. His design job is basically that of choosing and using existing integrated circuits to implement the functions required by the application. Occasionally, Boolean algebra will be used to minimize a function and achieve an efficient design.

However the greatest benefit of Boolean Algebra to the modern designer is in ________________ and ________________ digital logic operations.

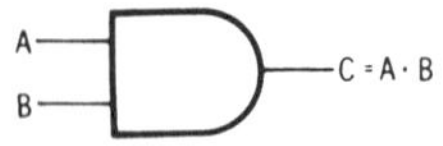

Figure 5-1

5. (analyzing or expressing) A Boolean expression is an equation that expresses the output of a logic circuit in terms of its inputs. You were introduced to Boolean expressions when you studied basic logic gates. The binary inputs and outputs were expressed as letters of the alphabet, alpha-numeric combinations, abbreviations or short words called mnemonics. For example, in the AND gate in Figure 5-1, the inputs are A and B and the output is C.

Note that the output C is expressed in terms of the inputs. The dot between the A and B indicates the AND function. The output expression $C = A \cdot B$ is read C equals A AND B. Remember that the inputs and outputs are binary signals which may assume either the binary 0 or binary 1 state.

Figure 5-2

What is the output expression of the AND gate in Figure 5-2?

______________.

6. ($X = \overline{C} \cdot D \cdot E$) In most Boolean expressions for the AND function the dot between each input variable can be eliminated and the expression written as a standard algebraic product like $X = \overline{C}\,D\,E$. The AND function is sometimes referred to as the logical product.

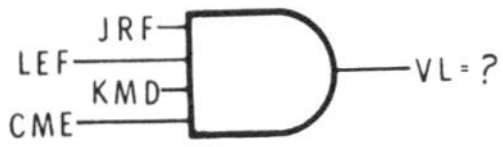

Figure 5-3

Write the output expression for the AND gate in Figure 5-3.

7. (VL = JRF • LEF • KMD • CME) In this expression each input is identified by a three letter combination called a mnemonic. The output is a two letter mnemonic. The dot between each input not only designates the AND function but also helps to separate or distinguish the inputs from one another. Occasionally you will see parenthesis used to separate the inputs and indicate the logical product or AND function.
VL = (JRF) (LEF) (KMD) (CME).

If all of the inputs in the above expression are binary 1, the output will be binary __________.

8. (binary 1) The AND function states that the output will be binary 1 only if all inputs are binary 1. If any one or all inputs are binary 0, the output will be binary 0.

Sketch the AND gate implementing the function $W = (F)(MX)(\overline{G}.)$ Label inputs and outputs.

9. (See Figure 5-4)

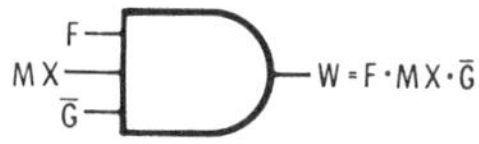

Figure 5-4

Other logic functions like inversion are also expressed with Boolean expressions.

If the input to an inverter is B and the output is designated Z, write the output expression. ______________.

10. ($Z = \overline{B}$) The output of an inverter is the complement of the input. This is expressed by putting a bar over the input variable. The term $\overline{B}$ is expressed as NOT B or B NOT. The term NOT refers to logic inversion. See Figure 5-5.

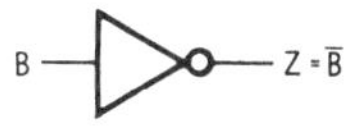

Figure 5-5

Draw the logic circuit corresponding to the expression $CTL = \overline{INV}$.

11. (See Figure 5-6) The output CTL is the complement of the input INV. $CTL = \overline{INV}$

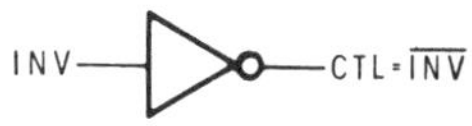

Figure 5-6

If the input is a binary 1, the output will be binary ______________.

12. (binary 0) And if the input is binary 0 the output will be binary 1.

Another common logic function is the OR. In an OR gate the output will be binary 1 if

a. only all inputs are binary 1
b. any one or more inputs are binary 1
c. all inputs are binary 0.

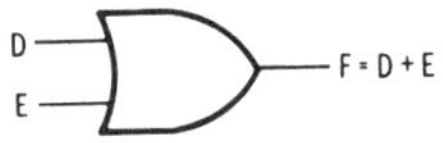

Figure 5-7

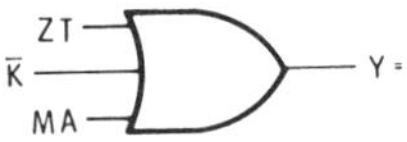

Figure 5-8

13. (b) The output of an OR gate is binary 1 if any one or more of the inputs is a binary 1. A typical OR logic gate and its related output expression is shown in Figure 5-7. The plus sign between the input variables designates the OR function. It is sometimes referred to as the logical sum.

Write the output expression of the logic gate in Figure 5-8.

14. ($Y = ZT + \overline{K} + MA$) The plus sign or OR function separates the input variables.

Draw the logic symbol for the expression $FN = MAF + \overline{DG} + BF + JS$

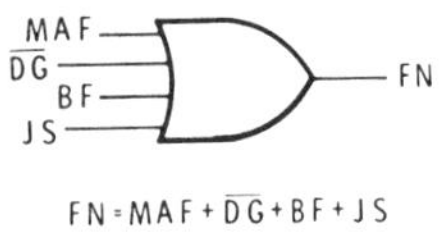

Figure 5-9

15. (See Figure 5-9)

It is important that you be able to draw the logic diagram corresponding to a given Boolean expression or write the Boolean equation from a given logic diagram.

The output of a logic circuit is a function of its __________.

16. (inputs) The output of a logic circuit is a function of the states of the inputs and, of course, the special logical characteristics of the circuit itself.

While there are some simple digital control operations that can be implemented with a single logic gate, more often it is necessary to use a number of logic gates to implement the desired decision-making function. When two or more logic elements are combined, the result is known as a combinational logic circuit. The circuit usually has multiple inputs and either a single output or multiple outputs depending upon its exact function.

Using AND and OR gates together to implement a special logic function creates a type of digital circuit called a __________ circuit.

17. (combinational) Any combination of multiple ANDs, ORs, and NOTs is called a combinational logic circuit. Such circuits are used for sophisticated decision-making functions.

There are many common combinational logic circuits used in digital equipment. They perform specific functions that tend to regularly reoccur in digital equipment. However, regardless of the type of combinational logic circuits there are two basic circuit forms. These are referred to as the sum-of-products and product-of-sums circuits. Here the term product refers to the AND function while the sum refers to the OR function.

The AND function is written in the same form as the algebraic product or the multiplication of two variables. (A and B = AB) The OR function is written as the sum of two input variables. (D OR E = D + E). The sum-of-products or product-of-sums expressions combine the AND and OR functions in a variety of ways.

Combinational logic circuits are used for ________________ ________________.

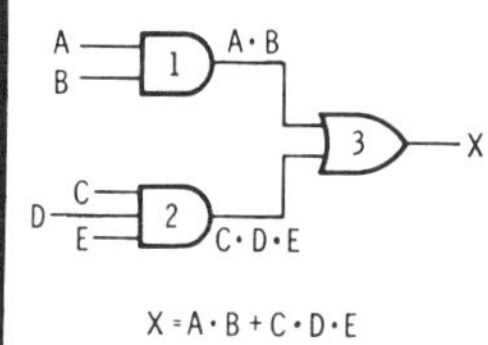

Figure 5-10

18. (decision making) Any combination of AND and OR gates is used for logical decision making purposes.

The most commonly used Boolean expression of complex decision making functions is the sum-of-products. The expression $X = A \cdot B + C \cdot D \cdot E$ is an example.

Figure 5-10 shows the logic circuit implementing this sum of products logical function. Here AND gate 1 forms the logical product AB while gate 2 forms the product CDE. These products are summed or logically ORed in gate 3 to form the output expression. This is what is meant by the sum-of-products.

Is the expression $QT = PJ \cdot \overline{W} + HV \cdot LM + GND \cdot \overline{C} \cdot F$ in sum-of-products form? ________________.

19. (yes) This expression is in the sum-of-products form. The various inputs are ANDed together in several combinations which are then ORed together. The circuit for this function is shown in Figure 5-11.

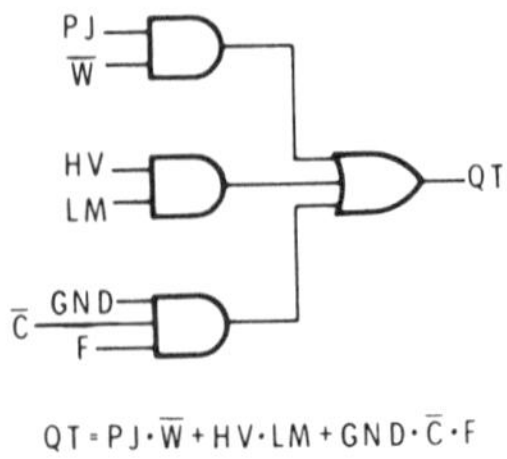

Figure 5-11

The distinguishing feature of this circuit is that the inputs feed AND gates and the output is derived from an OR gate.

The other type of Boolean expression is the product-of-sums. An example is the expression L = (M + N) (P + Q). The equivalent circuit is shown in Figure 5-12. Inputs M and N are ORed together in gate 1 while inputs P and Q are ORed in gate 2. The two logical sums are ANDed in gate 3 to produce the logical product. In this arrangement the distinguishing feature is the OR gate inputs and AND gate output.

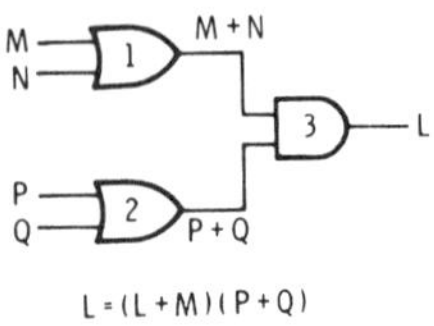

Figure 5-12

The AND function produces a logical ______________ while the OR function produces a logical ______________.

20. (product, sum) Your familiarity with this system should permit you to write the Boolean expression from any logic circuit and draw the logic circuit corresponding to a given Boolean equation.

To write the equation of a given logic circuit, you start at the inputs and write the output expression for each gate in the circuit from left to right until the output equation is developed. The unique gate symbols of course will tell you the logic function being performed. The example in Figure 5-13 illustrates this procedure.

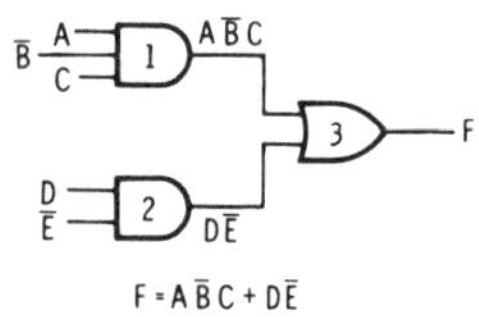

Figure 5-13

You first write the output expressions for the two input AND gates. These expressions become the inputs to the OR gate whose output equation is then written. The result is the Boolean expression for the circuit.

$$F = A\overline{B}C + D\overline{E}$$

Write the equation of the circuit given in Figure 5-14. ______________.

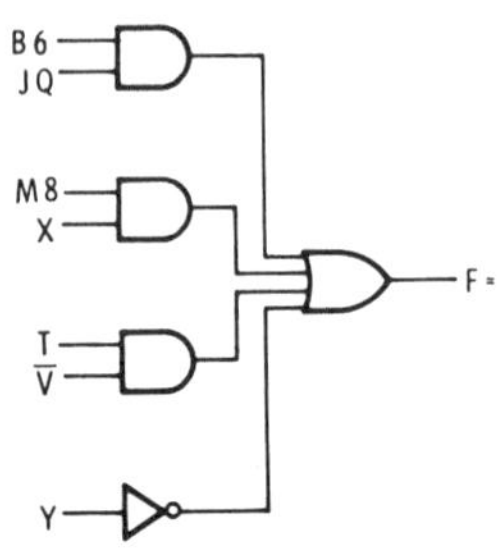

Figure 5-14

21. $(F = B6 \cdot JQ + M8 \cdot X + T \cdot \overline{V} + \overline{Y})$

You simply write the correct output for each of the input AND gates and use these as the inputs to the OR gate. Working this way from left to right quickly produces the complete output expression.

What form is this equation in? ______________.

22. (sum-of-products)

The input AND gates form products which are logically summed in the output OR gate.

This procedure of writing the equation for a given circuit also works on a product-of-sums circuit like the one in Figure 5-15.

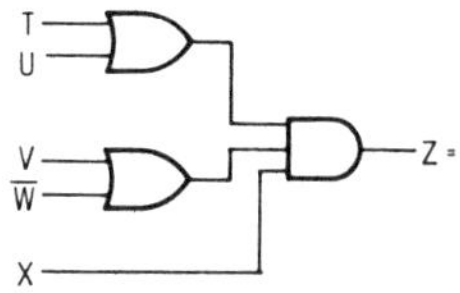

Figure 5-15

What is the output expression of this circuit? ______________.

23. $\left(Z = (T + U) \cdot (V + \overline{W}) \cdot X\right)$

The logical sums are formed by gates 1 and 2 and their outputs are combined to form the logical product in the output AND gate. See Figure 5-16.

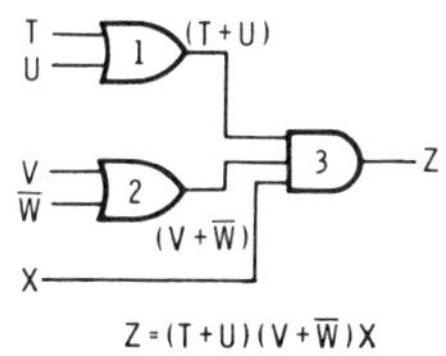

Figure 5-16

Again you work from left to right developing the output of each gate until the complete expression is formed.

What is the output of the circuit Figure 5-17? ______________.

24. $\left(OPR = (A3 + B4)\,(JMP + T\phi + CLK)\right)$

To draw the diagram corresponding to a given expression you first study the equation to determine whether it is a sum-of-products or product-of-sums. This will give you the type of output gate.

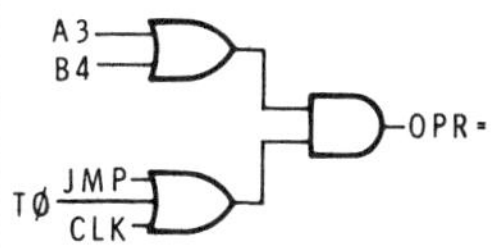

Figure 5-17

A sum-of-products circuit uses an __________ gate output.

25. (OR) A product-of-sums uses an ______________ gate output.

26. (AND) Then you work backward from the output developing the inputs and outputs from right to left.

For example, consider the expression K = LM + $\overline{C}$E. This is in sum-of-products form so the output gate is an OR. The inputs to this OR gate are LM and $\overline{C}$E. Each of these is a product that is developed by an AND gate with the appropriate inputs. See Figure 5-18.

Draw the circuit for the expression SH = (CLK + T5) ENB.

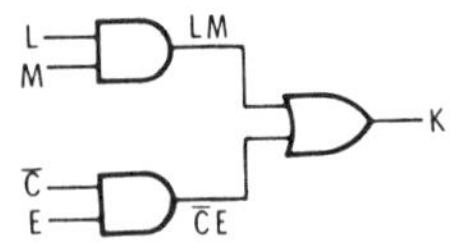

K = LM + $\overline{C}$E

Figure 5-18

27. (See Figure 5-19) This expression is in product-of-sums form but here there is only one logical sum (CLK + T5). The output is a product, however, and is produced by an AND gate. The input sum is produced by the OR gate.

Draw the circuit for the expression Y = ADD • CY + SUB • $\overline{INV}$ + MPY • ADD • SUB

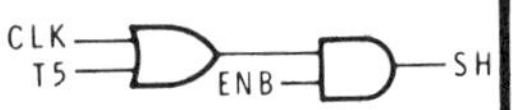

SH = ENB(CLK + T5)

Figure 5-19

28. (See Figure 5-20)

This is sum-of-products form. Note that some of the input signals (ADD, SUB) are applied to more than just one of the input gates.

So far we have only worked with two levels of logic, that is the inputs are conditioned by two sets or levels of logic gates in cascade: ANDs into OR or ORs into AND. Other more complex logic networks use three, four or even more levels of logic. The expression and circuit in Figure 5-21 is one example.

Here both the sum-of-products and product-of-sums formats are combined.

How many levels of logic are there in the circuit of Figure 5-21? ____________.

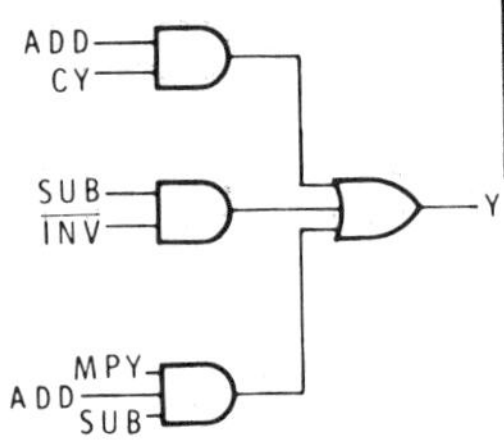

Y = ADD•CY + SUB•$\overline{INV}$ + MPY•ADD•SUB

Figure 5-20

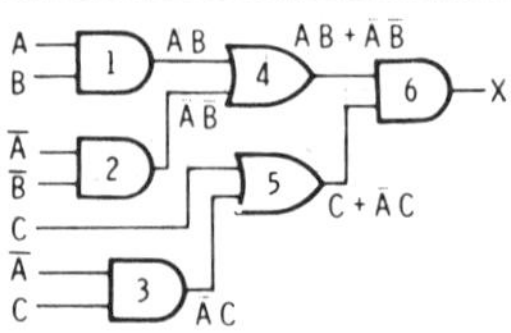

X = (AB + $\overline{A}\overline{B}$)(C + $\overline{A}$C)

Figure 5-21

29. (3) The input signals must propagate through a series of three gates before a level change occurs at the outputs. Input C on gate 5 only propagates through two levels.

Draw the circuit corresponding to the expression

$$F = (A + \overline{B})(\overline{A} + B) + (B + \overline{C})\overline{A}$$

30. (See Figure 5-22) Again you can see that the sum-of-products and product-of-sums are combined to form a three level logic circuit.

The same principles of converting from circuit to equation and equation to circuit you learned for two level logic apply to more complex multi level circuits.

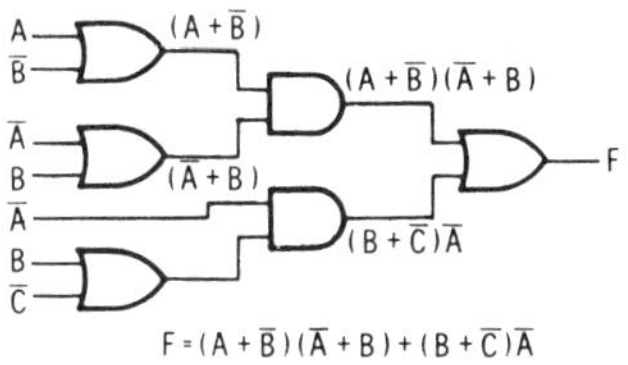

Figure 5-22

TRUTH TABLES

31. One of the most useful tools for analyzing, designing or otherwise working with digital circuits is the truth table. A truth table is a chart that lists all possible input and output signal combinations for a given logic circuit. The truth table is a tabular listing of all input and output states in binary 0 and binary 1 form. The truth table completely defines the operation of the circuit.

A truth table can be used to determine what input conditions cause a given logic circuit to produce a binary 1 output. True or False.

32. (True) You can look at a truth table and quickly identify which specific combination of input states that will produce a binary 1 (or binary 0) output.

Truth tables are used in defining the operation of simple logic circuits like inverters and gates as well as more complex combinational logic circuits. You have already seen how truth tables are used in defining the basic logic operations such as AND, OR and NOT.

In the way of a review, the truth table below defines what type of logic circuit? ______________.

INPUT A	OUTPUT $\overline{A}$
0	1
1	0

33. (inverter or NOT circuit) The truth table shows both the input and output. The output $\overline{A}$ is always the opposite or complement of the input A. Therefore, the circuit is an inverter.

The truth table for a two input AND gate is illustrated below. The inputs A and B define all possible input combinations.

INPUTS		OUTPUT
A	B	C
0	0	0
0	1	0
1	0	0
1	1	1

Column C shows the output state corresponding to each input combination. By studying the inputs and outputs you can determine the exact nature of an AND gate.

The only time the output is binary 1 is when input A is ____________ and input B is ____________.

34. (1, 1) The output is binary 1 only when both inputs are binary 1. This is true of any AND gate.

Complete the truth table below to define the operation of an OR gate where the output will be binary 1 if any or all inputs are binary 1.

INPUTS		OUTPUT
D	E	F
0	0	
0	1	
1	0	
1	1	

35. (Binary 1 outputs occur for input states of 01, 10 and 11) The total number of possible input conditons is 2^n where n is the number of inputs. In two input circuits there are $2^2 = 4$ possible input conditions. If we treat the inputs as bits in a multibit binary word, we can quickly define and record all input states by using the binary number equivalents. For example, with four possible input states we use the four numbers 0 through 3 or 00, 01, 10 and 11 in binary. These are all possible combinations for a two bit word.

A logic circuit has three inputs. How many possible logic input conditions are there? ______________.

36. ($2^3 = 8$) We represent the eight possible input conditions with the binary equivalent of the numbers 0 through 7. If the inputs are designated X, Y and Z, the input states are as shown below.

INPUTS			OUTPUT
X	Y	Z	W
0	0	0	0
0	0	1	1
0	1	0	1
0	1	1	1
1	0	0	1
1	0	1	1
1	1	0	1
1	1	1	1

This particular method of listing all possible input combinations by counting from zero through the upper limit is orderly and convenient.

The logic function defined by the table above is the ______________ function.

37. (OR) The output is binary 1 any time one or more of the inputs is a binary 1 so the OR function is defined.

There are two basic ways of using truth tables in logic work. First we can develop a truth table from a given logic circuit, and second we can develop an equation or logic circuit from a given truth table. The circuit-to-table method is useful in circuit analysis. The table-to-equation or circuit is a useful design method.

A useful circuit analysis approach is to develop a ______________ ______________ from the ______________ ______________.

38. (truth table, logic circuit) Developing the truth table for a given circuit starts with defining all of the inputs. Then the output for each gate in the circuit is developed until the final output is obtained. Consider the circuit shown in Figure 5-23.

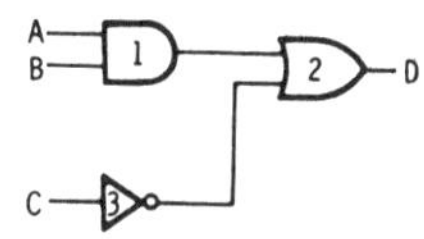

Figure 5-23

The Boolean output expression for this circuit is ______________.

39. ($D = AB + \overline{C}$) The output of gate 1 is AB and the output of inverter 3 is $\overline{C}$. These signals are ORed in gate 2 to produce $D = AB + \overline{C}$.

To develop the truth table for this circuit, we start with the inputs. Since there are three inputs there are $2^3 = 8$ possible input conditions that we define with the three bit numbers 000 through 111 as indicated in the table below. In the table we also create a column for the output of each gate or element in the circuit.

INPUTS			OUTPUTS		
A	B	C	Gate 1 AB	Inverter 3 $\overline{C}$	Gate 2 D
0	0	0			
0	0	1			
0	1	0			
0	1	1			
1	0	0			
1	0	1			
1	1	0			
1	1	1			

Observing the inputs, we can fill in the AB and $\overline{C}$ columns. The $\overline{C}$ column is simply the complement of the C input column. To complete the AB column, you consider the A and B inputs and their effect on the AND gate output.

Complete the AB and $\overline{C}$ columns above, and check your results in the next frame.

40. (See Table below)

INPUTS			OUTPUTS		
A	B	C	AB	C	D
0	0	0	0	1	1
0	0	1	0	0	0
0	1	0	0	1	1
0	1	1	0	0	0
1	0	0	0	1	1
1	0	1	0	0	0
1	1	0	1	1	1
1	1	1	1	0	1

Next, to find the D output, you note that the AB and $\overline{C}$ outputs are ORed in gate 3. You complete the D column by recording a binary 1 each time either one or both of the AB or $\overline{C}$ columns are binary 1. Check this in the table above.

Now, do the entire job yourself. Create a complete truth table for the circuit in Figure 5-24. Check your results in the next frame.

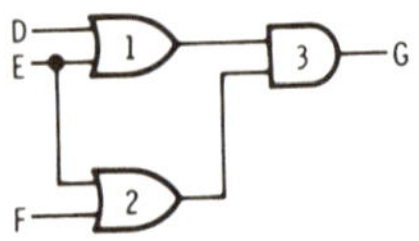

Figure 5-24

41. (See table below)

INPUTS			OUTPUTS		
D	E	F	(D + E)	(E + F)	G
0	0	0	0	0	0
0	0	1	0	1	0
0	1	0	1	1	1
0	1	1	1	1	1
1	0	0	1	0	0
1	0	1	1	1	1
1	1	0	1	1	1
1	1	1	1	1	1

To produce this table you first note the number of inputs. There are three different variables D, E, and F so there are $2^3 = 8$ possible input combinations.

Next you create a column for each gate output. In this case, gate 1 output is (D + E), gate 2 output is (E + F), and the final output at gate 3 is G = (D + E) (E + F).

To fill in the (D + E) column you OR together the D and E input columns. You record a binary 1 when either one or both column D or E are binary 1. You complete the (E + F) column in a similar manner considering all eight E and F input combinations.

Finally, you AND together the (D + E) and (E + F) columns, recording a binary 1 in the G column only when both (D + E) AND (E + F) columns are binary 1.

The truth table like the one you developed above completely defines the circuit operation for all possible conditions. And your analysis is thorough because to obtain the final output you had to derive the outputs of all other gates in the circuit.

The primary value of the circuit-to-truth table approach is in circuit ________________.

42. (analysis) In designing digital circuits we use the opposite approach. We develop a truth table as the result of our design. In designing a logic circuit we designate the number of inputs and what the output state should be for each set of input conditions. Then we write the Boolean equation from the truth table. From there the equation is readily translated into a logic diagram.

Assume that the circuit we want to design has two inputs A and B and we want a binary 1 output C to occur when A is 0 and B is 1 or when A is 1 and B is 0. Otherwise the output is binary 0.

Which truth table below defines these conditions?

a.

INPUTS		OUTPUT
A	B	C
0	1	1
1	0	1

b.

INPUTS		OUTPUT
A	B	C
0	0	1
0	1	0
1	0	0
1	1	1

c.

INPUTS		OUTPUT
A	B	C
0	0	0
0	1	1
1	0	1
1	1	0

43. (c.) The correct table is shown below.

INPUTS		OUTPUT
A	B	C
0	0	0
0	1	1
1	0	1
1	1	0

To write the output equation from this table we observe the output column C, noting those input states where binary 1s occur. Then we write a sum-of-products type Boolean expression based on these inputs. For each binary 1 state in the output we write one product term for the equation then we logically sum (OR) these terms.

The product terms are written from the input states. If the input is binary 1 we write the input letter. If the input is binary 0 we write the complement of the input letter. For example, when A is 0 and B is 1, the product term is $\overline{A}B$.

What is the product term where A is 1 and B is 0? ____________.

44. ($A\overline{B}$) The input conditions for each binary 1 output state is now defined. The product terms are then ORed. The result is C = ____________.

45. ($C = \overline{A}B + A\overline{B}$) Write the output equation from the truth table below.

INPUTS		OUTPUT
J	K	L
0	0	1
0	1	0
1	0	0
1	1	1

L = ____________.

46. ($L = \overline{J}\overline{K} + JK$) A binary 1 output occurs for two of the four possible input combinations. Therefore you know that the sum-of-product output equation will mention two product terms. The inputs are ANDed to form these products. The first output occurs when J = 0 and K = 0. Therefore the corresponding product is $\overline{J}\,\overline{K}$. The second output is binary 1 when J = 1 and K = 1. The product defining this condition is JK. To obtain the complete output expression we logically sum (OR) these products. $L = \overline{J}\,\overline{K} + JK$.

From here the logic diagram can be drawn.

Sketch the logic diagram of the equation just derived.

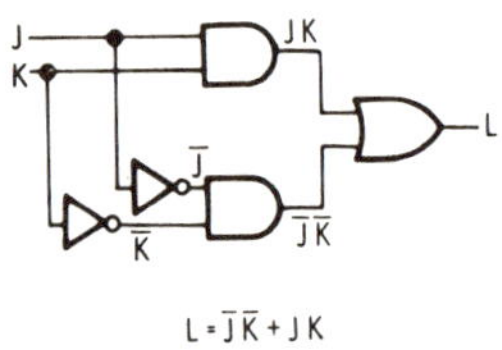

Figure 5-25

47. (See Figure 5-25)

Now try your hand at a more complex problem. Write the equation and draw the circuit corresponding to the truth table below.

INPUTS			OUTPUT
A	B	C	D
0	0	0	0
0	0	1	1
0	1	0	0
0	1	1	0
1	0	0	1
1	0	1	0
1	1	0	0
1	1	1	1

48. ($D = \overline{A}\,\overline{B}\,C + A\,\overline{B}\,\overline{C} + A\,B\,C$, See Figure 5-26)

In this problem there are three inputs, so each product term will have three factors and the number of product terms is set by the number of times a binary 1 output appears in the D output column. You develop each product by looking at the input states and writing the variable when a binary 1 input occurs and the complemented variable when a binary 0 input occurs. The product term when A = 0, B = 0 and C = 1 is $\overline{A}\,\overline{B}\,C$. Once all the product terms are developed they are ORed together to obtain the output. The logic diagram is then drawn from the equation. Note in the logic diagram that inputs A, B and C are considered to be available and signals $\overline{A}$, $\overline{B}$, and $\overline{C}$ are generated with inverters. In some circuits the complement signals may already be available from other sources in which case the inverters can be omitted.

Figure 5-26

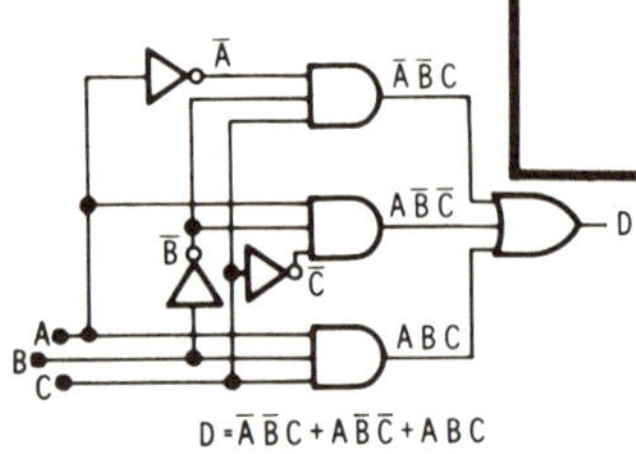

BOOLEAN RULES

49. As we mentioned earlier, the primary benefit of Boolean Algebra to a technician or engineer today is in analyzing, understanding and concisely expressing digital logic functions. The availability of a wide variety of integrated circuits has greatly minimized the use of Boolean Algebra as a design tool. However, even with modern ICs, the designer can often benefit from the use of Boolean algebra in minimizing or implementing a function.

Boolean Algebra is the algebra of two-valued functions. Many of the ordinary rules of algebra such as factoring or expanding a function, apply to Boolean expressions. However, the binary nature of the functions greatly simplifies most of the operations. There are also numerous special rules that apply to handling binary logic functions. We will explain these rules in this section and show how they are used.

Most standard algebra rules work with Boolean expressions. True or False? ______________.

50. (True) Yes, most conventional algebra manipulations will work on binary expressions. But it is the special rules that are of the most value to the digital designer.

The Laws of Intersection, for example, apply to AND gates. The two forms of this law are stated below.

$$A \cdot (1) = A$$
$$A \cdot (0) = 0$$

Remembering that A is a binary signal that can be either binary 0 or binary 1, we can prove the validity of these expressions if we remember how an AND gate works. The first expression simply says that if we apply a binary 1 to one input of an AND gate and the signal A to the other input, the output will be A. The binary 1 input simply enables the gate so that the A input state controls the output. If A = 1, the output will be 1. If A = 0, the output will be 0.

Which circuit in Figure 5-27 expresses this relationship?

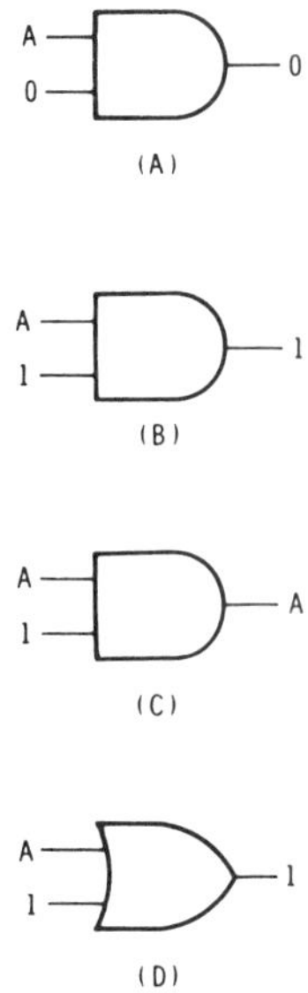

Figure 5-27

51. (c) The other form of the Laws of Intersection is just as easy to understand.

$$A\,(0) = 0$$

It says that if one input to an AND gate is 0 and the other is A, the output will always be 0. Remember that the only time the output of an AND gate can be 1 is when all inputs are binary 1. If one input is fixed at 0, the output will always be 0. The circuit in Figure 5-28 expresses this.

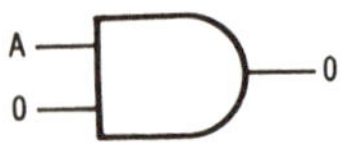

Figure 5-28

Complete the expression: D • E • (0) = ______________.

52. (D • E • (0) = 0) If one input to an AND gate is 0, the output will be zero regardless of the states of the other inputs. This proves that the Law of Intersection works for AND gates with more than two inputs. For example, D • E • (1) = D • E. You can also equate this expression to the algebraic version that says that a function (DE) multiplied by 1 is the function.

$$DE \times 1 = DE$$

Another similar set of rules exist for OR gates. These are called the Laws of Union. Expressed algebraically they are

$$B + 1 = 1$$
$$B + 0 = B$$

Sketch the logic diagrams for each of these and label the inputs and outputs.

53. (See Figure 5-29) These expressions like those for the AND gate almost perfectly define the operation of an OR gate. In the first version, if we apply a binary 1 to one input of an OR gate and a signal B to the other, the output will always be binary 1 by definition of an OR gate. If one input to an OR gate is binary 0, the other input B will control the output.

A quick way to prove the Laws of Union is to look at the truth table for an OR gate.

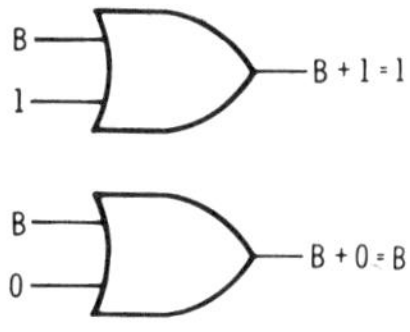

Figure 5-29

INPUTS		OUTPUT
B	C	D
0	0	0
0	1	1
1	0	1
1	1	1

To prove that B + 1 = 1 simply assume input C to be fixed at 1 and observe the output D for those cases. Do the same to prove B + 0 = B. Assume input C is 0 and observe the output D for the two corresponding B input states.

By using the truth table for an AND gate you can prove the Laws of ________________.

54. (Intersection) The procedure is the same for proving the Laws of Intersection with the truth table of an AND gate.

The Laws of Tautology apply to both AND gates and OR gates. The basic rules are given below.

$$A \cdot A = A$$
$$B + B = B$$

The related logic symbols are shown in Figure 5-30.

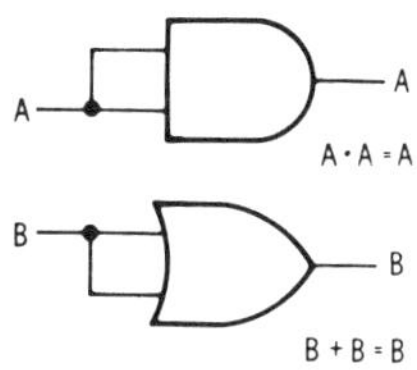

Figure 5-30

What these expressions say is that if you apply the same signal to all inputs of a logic gate, the output will be the same as the input. Again, you can prove this to yourself by looking at the truth tables for AND and OR gates.

Use the Laws of Tautology to simplify the expression QT = JMX + JMX + F9. ________________.

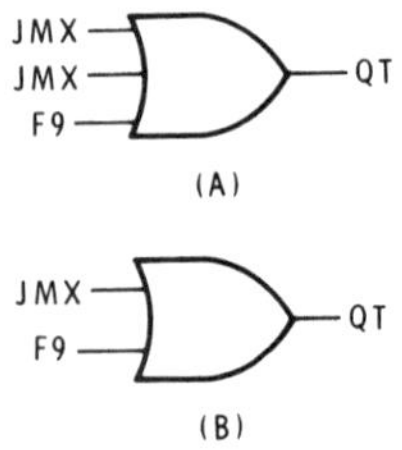

Figure 5-31

55. (QT = JMX + F9) The JMX term is redundant. In terms of circuitry you can see the simplification. The circuit in Figure 5-31A implements the original expression. Figure 5-31B shows the simplified but fully equivalent circuit.

This same simplification procedure applies to AND gates.

Use the Laws of Tautology to simplify the expression X = Q • Q • J. Draw the original and simplified logic diagrams.

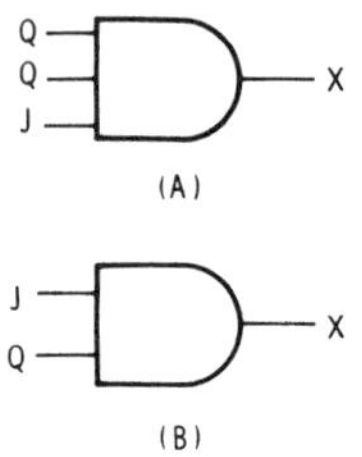

Figure 5-32

56. (See Figure 5-32)

The original circuit can be implemented as shown in Figure 5-32A. The simplified but equivalent circuit is shown in Figure 5-32B. The logical function is identical. Now you are beginning to see that the value of Boolean algebra is simplifying the design of a circuit.

Another Boolean law is the Law of Complements. These are

$$A \cdot \overline{A} = 0$$
$$B + \overline{B} = 1$$

If we apply a logic signal and its complement to a logic gate, the output becomes either a binary 0 or a binary 1 depending on type of logic gate. This is illustrated in Figure 5-33.

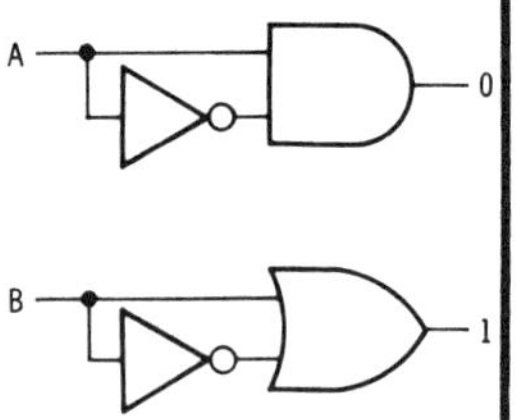

Figure 5-33

In these circuits, the output is a function of the state of the input. True or False? ____________.

57. (False) If you look closely at these circuits and analyze what happens, you will see that in either case the output is not affected by the state of the input. It can be either a 1 or a 0 and the output will be binary 0 for the AND gate and binary 1 for the OR gate. In this case the output is a function of the ____________.

58. (type of logic circuit) You can further verify these laws by referring to the AND and OR truth tables. A binary 0 on either input of an AND gate will always produce a binary 0 output. A binary 1 on either input of an OR gate will always produce a binary 1 output.

Now consider the Law of the Double Negative.

$$\overline{\overline{A}} = A$$

It says that the complement of the complement of A is equal to A. Or a signal that is complemented twice is the same as the original signal. You can see this from the circuit in Figure 5-34.

If input A is 1, the output of inverter 1 is ______________ and the output of inverter 2 is ______________.

59. (0, 1) Inverter 1 complements the 1 input to a 0. Inverter 2 complements the 0 back into a 1 so that the output of the circuit is equal to the input. Complementing a signal twice or any even number of times give us the original signal. Figure 5-35 shows this.

To simplify this two inverter circuit you could replace it with which of the circuits in Figure 5-36?

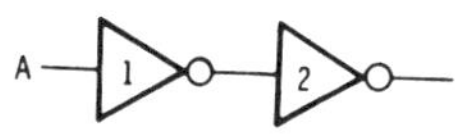

Figure 5-34

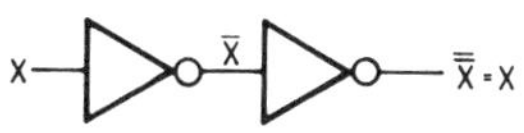

Figure 5-35

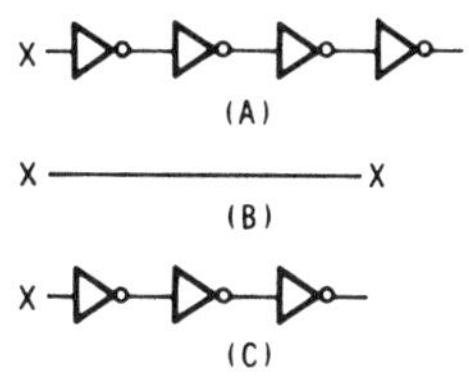

Figure 5-36

60. (B) In terms of simplification, if you encounter two inverters in cascade, you can simply remove them and substitute a piece of wire since the effect is the same. Three cascade inverters or any odd number produces the same effect as a single inverter so this is not equivalent. Four cascade inverters does produce an output that is the same state as the input, but this is certainly not simplification.

All along we have been saying that when an even number of inverters are cascaded, the output is the same as the input. This is true of course for a static (non changing) input. But when the input is switching rapidly, the output is almost, but not quite the same as the input.

Which characteristic of a logic gate do you think causes this difference when the input signal is changing rapidly?

A. Logic level
B. Power dissipation
C. Noise immunity
D. Propagation Delay.

61. (D., Propagation delay) Of course. When the input signal changes it takes a finite time for it to propagate through each circuit in a chain. The output then is delayed from the input by an amount of time of the total circuit propagation delay.

In a two inverter string where the propagation delay for each inverter is 12 nanoseconds, the output is delayed from the input by ______________ nanoseconds.

62. **(24) The total propagation delay is the sum of the individual** propagation delays. Figure 5-37 shows the input and output **waveforms of a two inverter string. The total propagation delay is t_p.**

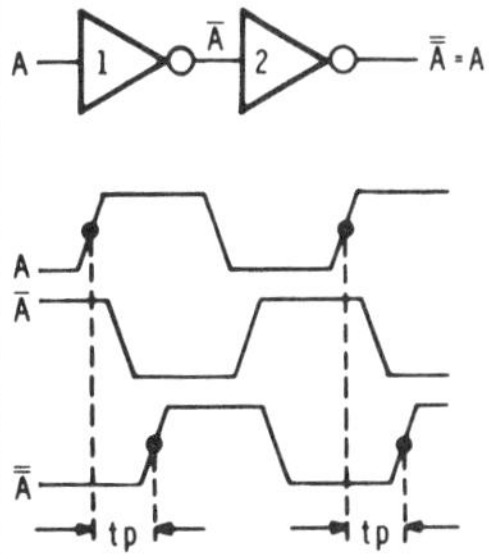

Figure 5-37

Another Boolean rule is the Law of Commutation. This is the same rule from basic algebra. The two forms of it are:

$$A \cdot B = B \cdot A$$
$$A + B = B + A$$

All it really says is that you can arrange the inputs to AND or OR gates in any order and the effect is the same. You can write the input variables in any order and they will mean the same thing.

$$W + X + Y = X + W + Y$$

True or False? ______________.

63. (True) Also JML = LJM = MLJ = JLM = LMJ = MJL. The order of the inputs is different, but the logical results is the same for each. Incidentally, don't forget that all of these Boolean rules work for 3 or more inputs although we are illustrating them with only two inputs for simplicity.

Now let's give you some practice in using these rules. What are the simplifications of the expressions below?

a. $A + \overline{B} + A =$ ______________
b. $B\,C\,\overline{B} =$ ______________
c. $C + 1 + \overline{B} =$ ______________
d. $\overline{X} + Y + X =$ ______________

Work these problems using the previously discussed rules. Check your answers in the next frame.

64. a. $A + \overline{B}$
b. 0
c. 1
d. 1

In problem (A) you first rearranged it by using the Law of Commutation $A + \overline{B} + A = A + A + \overline{B}$. By the Law of Tautalogy, you know that $A + A = A$, so the resulting simplified equation is $A + \overline{B}$.

In problem (B) you again rearrange the term using the Law of Commutation. $BC\overline{B} = B\overline{B}C$. From the Law of Complements you know that $B\overline{B} = 0$. Substituting this in the expression gives $0 \cdot C$. You know from the Law of Intersection that $0 \cdot C$ or $C \cdot 0 = 0$.

In problem (C) you should recognize the Laws of Union. $C + 1 + \overline{B} = 1$.

In problem (D), you first rearrange the equation by way of the Law of Commutation to $\overline{X} + Y + X = X + \overline{X} + Y$. The Law of Complements says that $X + \overline{X} = 1$. Therefore, the equation becomes $1 + Y$. Of course this reduces to 1 by way of the Law of Union.

Go to the next frame.

65. Now try a few more examples using the previous solutions as a guide. But this time draw the logic diagram of the original expression as well as the simplified version to get a feel for the result of Boolean simplification in terms of circuitry.

a. $L + M + \overline{M}$
b. $J K \overline{K} L$
c. $F + T + F$
d. $\overline{B} + A C \overline{A} + D$

The answers are given in the next frame.

66. a. $L + M + \overline{M} = L + 1 = 1$

The three input OR and inverter called for by the original expression reduces to a single wire with a binary 1 on it.

b. $J\,K\,\overline{K}\,L = J \cdot 0 \cdot L = 0$

The reduced expression is simply a binary 0 level.

c. $F + T + F = F + F + T = F + T$

d. $\overline{B} + A\,C\,\overline{A} + D = \overline{B} + A\,\overline{A}\,C + D = \overline{B} + 0 \cdot C + D = \overline{B} + 0 + D = \overline{B} + D$

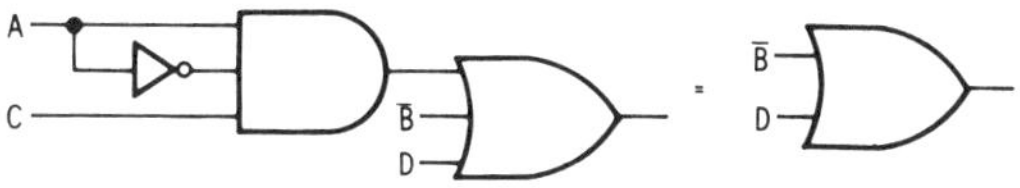

Figure 5-41

By now you should be able to see the value of Boolean algebra in simplifying a logic expression and minimizing the amount of circuitry required to implement it. Fewer parts means lower cost, smaller size and less power consumption.

Go to the next frame.

67. Another Boolean law identical to the basic algebra rule is the Law of Association. These are

$$(A \cdot B)\,C = A\,(B \cdot C) = A \cdot B \cdot C$$
$$A + (B + C) = (A + B) + C = A + B + C$$

You can see the circuitry simplification that results by applying these rules. See Figure 5-42.

Note that we trade two 2-input AND gates for a single 3-input gate.

Draw the OR gate circuit equivalents for the Law of Association.

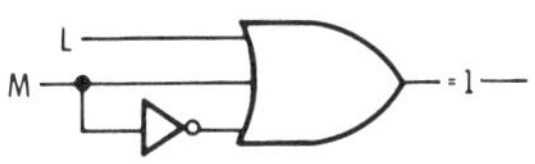

Figure 5-38

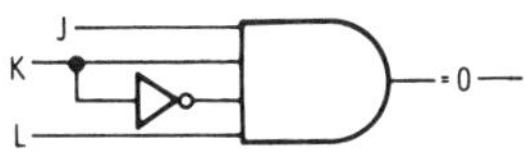

Figure 5-39

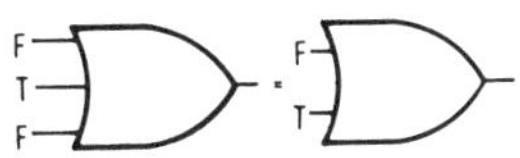

Figure 5-40

Figure 5-42

68. (See Figure 5-43)

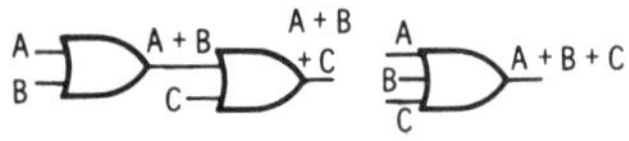

Figure 5-43

Note that the logical effect is the same for either version of the circuit. The three input gate is usually more economical than two 2 input gates.

Now let's consider the Laws of Distribution. We will use some of the rules described earlier to prove this law. The Laws of Distribution are:

$$A B + A C = A (B + C)$$

You should be able to infer this because all you are doing is factoring out one term as you would in basic algebra.

Sketch the logic diagrams for each side of the above equation.

69. (See Figure 5-44)

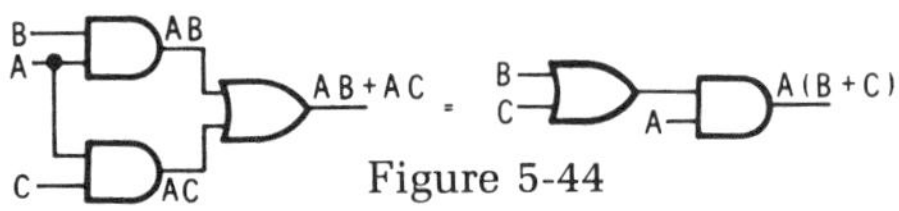

Figure 5-44

Study these two frames carefully. Not only was there a reduction in the number of circuits used, but also note that the expression changed from a sum-of-products to a product-of-sums form.

Using a truth table show that the logical effect of both circuits is identical.

70. (See tables below) Your tables should appear as shown below.

INPUTS			GATE OUTPUTS					
A	B	C	AB	AC	AB + AC	A	B + C	A (B + C)
0	0	0	0	0	0	0	0	0
0	0	1	0	0	0	0	1	0
0	1	0	0	0	0	0	1	0
0	1	1	0	0	0	0	1	0
1	0	0	0	0	0	1	0	0
1	0	1	0	1	1	1	1	1
1	1	0	1	0	1	1	1	1
1	1	1	1	1	1	1	1	1

equal

The table shows the eight possible input combinations of the three variables. It also shows the outputs of the intermediate and output gates for both the sum-of-products and product-of-sums forms. And they are equal. This shows you another use of truth tables in analyzing and understanding logic circuits.

Another version of the Laws of Distribution is

$$(A + B)(A + C) = A + BC$$

Note the format of the original and simplified versions.

The original expression is in the _______________ of _______________ form but is changed to the _______________ of _______________ form when it is simplified.

71. (product-of-sums, sum-of-products) The logic diagrams of the two versions illustrated in Figure 5-45 show this difference.

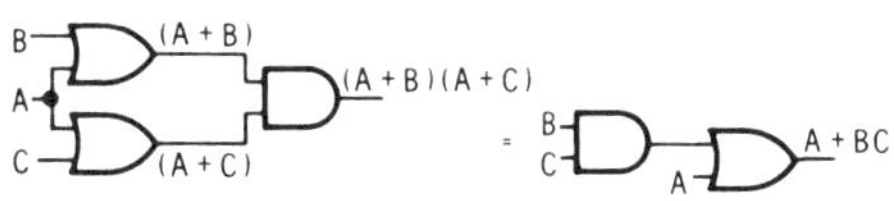

Figure 5-45

You can use truth tables to prove the equality of the two expressions or circuits. Do this now and check your results in the next frame.

72. (See table below)

INPUTS			GATE OUTPUTS					
A	B	C	(A + B)	(A + C)	(A + B) (A + C)	A	BC	A + BC
0	0	0	0	0	0	0	0	0
0	0	1	0	1	0	0	0	0
0	1	0	1	0	0	0	0	0
0	1	1	1	1	1	0	1	1
1	0	0	1	1	1	1	0	1
1	0	1	1	1	1	1	0	1
1	1	0	1	1	1	1	0	1
1	1	1	1	1	1	1	1	1

equal

Another way to prove the equality of these expressions is to use some of the Boolean rules learned earlier.

$$(A + B)(A + C) = A + BC$$

First expand the expression on the left side of the equation as you would any algebraic expression. Multiply each term by the others and sum the products as shown.

$$(A + B)(A + C) = AA + AC + AB + BC$$

Next, this can be reduced by substituting A for AA and factoring an A out of the first three terms.

$$A + AC + AB + BC = A(1 + C + B) + BC$$

Then, the Laws of Union will reduce (1 + C + B) to 1 and the expression becomes

$$A(1) + BC$$

Finally, the Laws of Intersection makes this

$$A + BC$$

Go to the next frame.

73. Next there are the Laws of Absorption. There are four versions.

$$A(A + B) = A$$
$$A(\overline{A} + B) = AB$$
$$AB + \overline{B} = A + \overline{B}$$
$$A\overline{B} + B = A + B$$

Now use the previously explained Boolean rules to prove the first expression

$$A(A + B) = A$$

74. (See procedure below) The related Boolean rule is indicated in parenthesis.

A (A + B)
Expand by multiplying (Distribution)

AA + AB
Replace AA by A (Tautology)

A + AB
Factor out A

A (1 + B)
Replace (1 + B) by 1 (Union)

A (1)
Replace A (1) by A (Intersection)

A
Next prove $A(\overline{A} + B) = AB$

75. (See procedure below)

$A(\overline{A} + B)$
Expand by multiplying (Distribution)

$A\overline{A} + AB$
Replace $A\overline{A}$ by 0 (Complements)

$0 + AB$
Replace $AB + 0$ with AB (Union)

AB

The other Laws of Absorption are more difficult to prove. You can use truth tables as we did before. But it can be done with Boolean Algebra by using a trick. Let's prove that

$$AB + \overline{B} = A + \overline{B}$$

There isn't anything we can do with the expression on the right side of the equation. So let's multiply the $\overline{B}$ term by $(A + 1)$. Since $A + 1 = 1$ and $\overline{B}(1) = \overline{B}$ we will not change the original meaning of the expression. It is equivalent to multiplying a term by 1. Therefore

$$AB + \overline{B} = AB + \overline{B}(A + 1)$$

Now, complete the proof yourself using the previously explained Boolean rules.

76. (See proof below) The related Boolean rules are indicated.

$AB + \overline{B}(A + 1)$
Expand by multiplying (Distribution)

$AB + A\overline{B} + \overline{B}$
Factor out from first two terms (Distribution)

$A(B + \overline{B}) + \overline{B}$
Replace $(B + \overline{B})$ with 1 (Complements)

$A(1) + \overline{B}$
Replace $A(1)$ by A (Intersection)
$A + \overline{B}$

You can use the same trick to prove the expression.

$$A\overline{B} + B = A + B$$

Do it for practice then check your solution in the next frame.

77. (See proof below)

$A\overline{B} + B$
Multiply B by (A + 1) (Distribution)

$A\overline{B} + B(A + 1)$
Expand by multiplying (Distribution)

$A\overline{B} + AB + B$
Factor A out of first two terms (Distribution)

$A(\overline{B} + B) + B$
Replace $(\overline{B} + B)$ by 1 (Complements)

$A(1) + B$
Replace A (1) by A (Intersection)

$A + B$

The equality of the expression $A\overline{B} + B = A + B$ can also be proven by using a ____________________.

78. (truth table) Don't overlook truth tables as a means of verifying the operation of a logic circuit or in proving the equivalence of two equations or circuits.

Another important Boolean rule is DeMorgan's theorem. There are two forms of it as indicated below.

$$\overline{AB} = \overline{A} + \overline{B}$$
$$\overline{A + B} = \overline{A}\,\overline{B}$$

The best way to prove the equality of these expressions is by a truth table. The first expression above is proven this way as indicated below.

INPUTS		OUTPUTS				
A	B	$\overline{A}$	$\overline{B}$	AB	$\overline{AB}$	$\overline{A} + \overline{B}$
0	0	1	1	0	1	1
0	1	1	0	0	1	1
1	0	0	1	0	1	1
1	1	0	0	1	0	0

equal

There are two variables A and B so there are four possible input combinations. These are indicated in columns A and B. Columns are also provided for the other terms called for by the expressions

$$\overline{A}, \overline{B}, \overline{AB} \text{ and } \overline{A} + \overline{B}$$

Using the inputs as a guide these other columns are completed. Go through each column yourself to be sure you understand how each state is obtained. Note the equality of the $\overline{AB}$ and $\overline{A} + \overline{B}$ columns.

Now, prove the expression $\overline{A + B} = \overline{A}\,\overline{B}$ yourself using the same procedure. Check your results in the next frame.

79. (See Table below)

INPUTS		OUTPUTS				
A	B	$\overline{A}$	$\overline{B}$	A + B	$\overline{A + B}$	$\overline{A}\,\overline{B}$
0	0	1	1	0	1	1
0	1	1	0	1	0	0
1	0	0	1	1	0	0
1	1	0	0	1	0	0

equal

The correct proof of the expression $\overline{A + B} = \overline{A}\,\overline{B}$ is shown above.

Like other Boolean rules, DeMorgan's theorem is useful in minimizing logic equations, particularly those that have whole terms or expressions negated with a complement bar over them. For example, the expression

$$X = \overline{A\,\overline{B}\,C} + \overline{A + \overline{C}}$$

cannot be simplified by using the previously given Boolean rules. It can only be reduced by using DeMorgan's theorem.

If you look closely at the two forms of DeMorgan's theorems you will see that they essentially show how an AND expression can be changed to an OR expression and vice versa.

$$\overline{AB} = \overline{A} + \overline{B}$$
$$\overline{A + B} = \overline{A}\,\overline{B}$$

Logic expressions of either form can be readily converted to another form more desirable in simplifying the expression.

Like other Boolean rules, DeMorgan's theorem is used to ________________ logic expressions.

80. (Simplify, minimize, or reduce) DeMorgan's theorem provides one more tool for you to use in minimizing certain types of logic expressions. In addition, it can be used to change the form of an expression for AND to OR or OR to AND.

The following procedure can be used in making the conversions.

To change a Boolean expression from one form to another, use the following procedure:

1. Change all AND (•) expressions to OR (+) expressions.
2. Complement the individual terms that were ANDed or ORed.
3. Complement the entire expression.

Let's try it out on the expression $\overline{AB}$.

1. Change AND to OR or vise versa.

$$\overline{AB} \quad \text{becomes} \quad \overline{A + B}$$

2. Complement each term.

$$\overline{A + B} \quad \text{becomes} \quad \overline{\overline{A} + \overline{B}}$$

3. Complement entire expression

$$\overline{\overline{A} + \overline{B}} \quad \text{becomes} \quad \overline{\overline{\overline{A} + \overline{B}}} = \overline{A} + \overline{B}$$

The result $\overline{AB} = \overline{A} + \overline{B}$ is one of our DeMorgan's relationships. Now try converting the expression $\overline{A}\,\overline{B}$ using this procedure.

81. ($\bar{A}\,\bar{B} = \overline{A + B}$) (See procedure below)

1. $\bar{A}\,\bar{B}$ Change AND to OR $\bar{A} + \bar{B}$
2. $\bar{A} + \bar{B}$ Complement each term $\bar{\bar{A}} + \bar{\bar{B}} = A + B$
3. $A + B$ Complement entire expression $\overline{A + B}$

The procedure is easily remembered and can be applied to other more complex expressions. The examples below illustrate some typical applications.

$$\overline{A\bar{B}} \quad \text{becomes}$$

$$\overline{A + \bar{B}} \quad \text{then}$$

$$\overline{\bar{A} + B} \quad \text{and finally}$$

$$\overline{\overline{\overline{\bar{A} + B}}} = \bar{A} + B = \overline{A\,\bar{B}}.$$

Note that while only two terms are shown in the basic DeMorgan's theorems, the same rules apply to logic expressions with three or more terms.

$$X + Y + Z \quad \text{becomes}$$

$$X \cdot Y \cdot Z \quad \text{then}$$

$$\bar{X}\,\bar{Y}\,\bar{Z} \quad \text{and finally}$$

$$\overline{\bar{X}\,\bar{Y}\,\bar{Z}} = X + Y + Z$$

Use DeMorgan's rules to change the form of the expression:

$$\overline{\bar{J} + K + \bar{L}}$$

82. (J $\overline{K}$ L) You use the same procedure for converting any Boolean expression with DeMorgan's theorem.

The following example will show you how DeMorgan's theorems can significantly reduce a Boolean expression.

$\overline{(A\ \overline{B})\ (\overline{B}\ C)\ (C\ \overline{D})\ (A\ C)}$	Original expression
$\overline{(A\ \overline{B})} + \overline{(\overline{B}\ C)} + \overline{(C\ \overline{D})} + \overline{(A\ C)}$	Expanded by DeMorgan's
$(\overline{A} + B) + (B + \overline{C}) + (\overline{C} + D) + (\overline{A} + \overline{C})$	Each term in expression expanded by DeMorgan's
$\overline{A} + B + B + \overline{C} + \overline{C} + D + \overline{A} + \overline{C}$	Expression expanded by Law of Association
$\overline{A} + \overline{A} + B + B + \overline{C} + \overline{C} + \overline{C} + D$	Expression rearranged by Laws of Commutation.
$\overline{A} + B + \overline{C} + D$	Expression simplified by Laws of Union.

A significant simplification takes place. Note that DeMorgan's theorems were applied twice, once to the original expression and again to each term in the expanded form resulting from the first application. In simplifying an expression like this you use DeMorgan's theorem as many times as necessary for any terms to achieve the greatest minimization.

You do not need to use DeMorgan's theorem on all Boolean expressions. As a rule, the only place you will use it is when a NOT bar appears over all or part of the logic equation.

Now, take a close look at the first expression for DeMorgan's theorem.

$$\overline{A\ B} = \overline{A} + \overline{B}$$

What logic function is indicated by the term on the left? _______________.

83. (NAND) This expression relates the positive NAND function ($\overline{AB}$) to the negated NOR or negative NOR function ($\overline{A} + \overline{B}$).

What about the other form of DeMorgan's $\overline{A + B} = \overline{A}\,\overline{B}$? Which logic function is indicated by $\overline{A + B}$? ______________.

84. (NOR) Yes. The positive NOR function ($\overline{A + B}$) is indicated. It is equivalent to the negated NAND or negative NAND function $\overline{A}\,\overline{B}$. You saw earlier how NAND and NOR gates can perform NAND or NOR functions in positive or negative logic. These operations are clearly expressed by the two forms of DeMorgan's theorem. You will find DeMorgan's theorems useful in dealing with NAND and NOR gates.

MINIMIZING LOGIC EXPRESSIONS

85. You have already seen how the Boolean rules are used to simplify logic expression and you have had a little practice in minimizing some simple logic equations yourself. But now we want you to polish your skill with Boolean so that you can deal with any practical logic function that you might encounter. The problems presented here will give you the necessary practice.

Let's start with the expression $F = (\overline{A} + B)(B + \overline{C})$

Draw the logic diagram for this expression.

86. (See Figure 5-46)

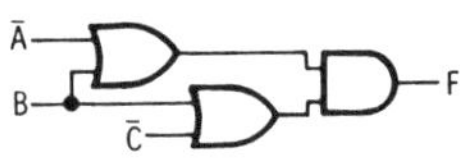

$F = (\overline{A} + B)(B + \overline{C})$

Figure 5-46

Now, using Boolean algebra, minimize the equation. The first step is to expand the expression by multiplying as you would in algebra.

$F = (\overline{A} + B)(B + \overline{C}) = \overline{A}\,B + \overline{A}\,\overline{C} + B\,B + B\,\overline{C}$

Basic Boolean rules can then be used as indicated below to reduce the equation.

$\overline{A}\,B + \overline{A}\,\overline{C} + B\,B + B\,\overline{C}$
Laws of Tautology

$\overline{A}\,B + \overline{A}\,\overline{C} + B + B\,\overline{C}$
Laws of Commutation

$\overline{A}B + B + B\overline{C} + \overline{A}\,\overline{C}$
Factor a B out of the first three terms

$B(\overline{A} + 1 + \overline{C}) + \overline{A}\,\overline{C}$
Laws of Union

$(\overline{A} + 1 + \overline{C}) = 1$

$B(1) + \overline{A}\,\overline{C}$
Laws of Intersection

$B + \overline{A}\,\overline{C}$

Now draw the logic diagram of the minimized version of the original expression.

87. (See Figure 5-47)

As you can see, there is a significant simplification in that the minimized circuit uses one less gate.

Just to be sure that the two circuits and expressions do indeed produce the same logical function, prove their equality with a truth table.

Check your results in the next frame.

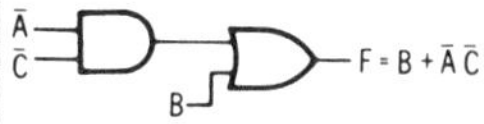

Figure 5-47

88. (See table below) The truth table proves conclusively that

$$F = (\bar{A} + B)(B + \bar{C}) = B + \overline{A}\,\overline{C}$$

the two expressions are equivalent.

INPUTS			GATE OUTPUTS							
A	B	C	$\bar{A}$	$\bar{C}$	$(\bar{A} + B)$	$(B + \bar{C})$	$(\bar{A} + B)(B + \bar{C})$	$\bar{A}\,\bar{C}$	$B + \bar{A}\,\bar{C}$
0	0	0	1	1	1	1	1	1	1
0	0	1	1	0	1	0	0	0	0
0	1	0	1	1	1	1	1	1	1
0	1	1	1	0	1	1	1	0	1
1	0	0	0	1	0	1	0	0	0
1	0	1	0	0	0	0	0	0	0
1	1	0	0	1	1	1	1	0	1
1	1	1	0	0	1	1	1	0	1

equal

Now let's try the expression below.

$$X = A\,\bar{B}\,\bar{C} + A\,B\,\bar{C} + A\,\bar{B}\,C + \bar{A}\,B\,\bar{C}$$

First draw the logic diagram.

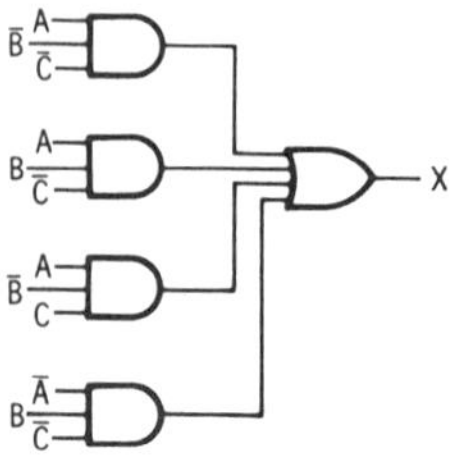

Figure 5-48

89. (See Figure 5-48)

Observing the equation, try to spot common factors in each of the terms, then regroup the terms and rearrange the factors using the Laws of Commutation.

$$X = A\,\overline{B}\,\overline{C} + A\,B\,\overline{C} + A\,\overline{B}\,C + \overline{A}\,B\,\overline{C}$$
$$X = A\,\overline{B}\,\overline{C} + A\,\overline{B}\,C + B\,\overline{C}\,A + B\,\overline{C}\,\overline{A}$$

Next we factor out the common expressions, $A\overline{B}$ in the first two terms and $B\overline{C}$ in the last two terms.

$$X = A\,\overline{B}\,(\overline{C} + C) + B\,\overline{C}\,(A + \overline{A})$$

By the Laws of Complements this becomes:

$$X = A\,\overline{B} + B\,\overline{C}.$$

Draw the corresponding logic diagram.

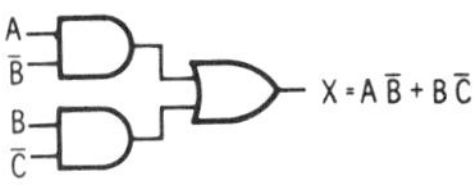

Figure 5-49

90. (See Figure 5-49)

Comparing this circuit with the one in Figure 5-53 (Frame 89), you can see the result of minimization. The three gate circuit performs exactly the same logic function as the original five gate circuit at a considerable savings.

Next, let's minimize a logic equation with DeMorgan's theorem.

Reduce the equation

$$G = \overline{\overline{\overline{(A\,\overline{B}\,\overline{C}} + B\,C)}\,(A\,\overline{B})}$$

Work the problem yourself keeping in mind that DeMorgan's theorem must be used several times to expand the equation into a form that will permit it to be minimized with the other Boolean rules.

Check the solution in the next frame.

91. The correct solution is shown below.

$G = \overline{\overline{(A\,\overline{B}\,\overline{C}} + B\,C)}\,(A\,\overline{B})$
DeMorgan's

$G = \overline{(A\,\overline{B}\,\overline{C}} + BC) + \overline{(A\,\overline{\overline{B}})}$
DeMorgan's

$G = \overline{A} + B + C + B\,C + \overline{A} + B$
Law of Commutation

$G = \overline{A} + \overline{A} + B + B + C + B\,C$
Law of Tautology

$G = \overline{A} + B + C + B\,C$
Law of Distribution

$G = \overline{A} + B + C\,(1 + B)$
Law of Union

$G = \overline{A} + B + C\,(1)$
Law of Intersection

$G = \overline{A} + B + C$

USING NAND/NOR GATES

92. Throughout this unit we have shown logic equations implemented with AND and OR gates. However, most modern digital system's and circuits are made with NAND or NOR gates. As you saw in an earlier unit, any of the three basic logic operations can be realized with either NAND or NOR circuits.

The three basic logic functions are ______________, ______________, and ______________.

93. (AND, OR, and NOT) A NAND gate or a NOR gate can be used as an inverter by tying all of the inputs together as shown in Figure 5-50.

Both circuits produce an output that is the complement of the input.

When used as inverters NAND and NOR gates are represented by the proper symbol. Sketch this symbol.

94. (See Figure 5-51) These are the commonly used symbols for an inverter.

On schematic and logic diagrams, it is desirable to use a symbol that represents the logic function being performed despite the nature of the circuit used to implement it.

A NAND gate can be used for AND operations by inverting its output as shown in Figure 5-52.

The inversion of the NAND is removed by the added inverter producing the pure AND function.

The inversion is indicated by the ______________ in the inverter logic symbol.

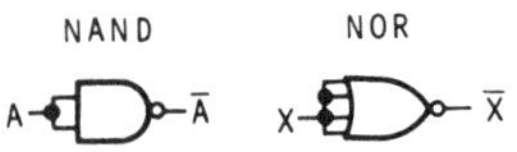

Figure 5-50

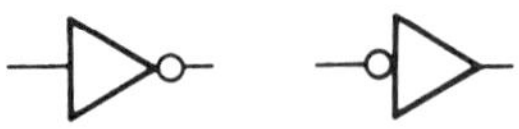

Figure 5-51

Figure 5-52

95. (circle) The small circle associated with a logic symbol indicates the complement function.

A NAND can also perform the OR function as you saw in a previous unit. This is accomplished by inverting the inputs as shown in Figure 5-53.

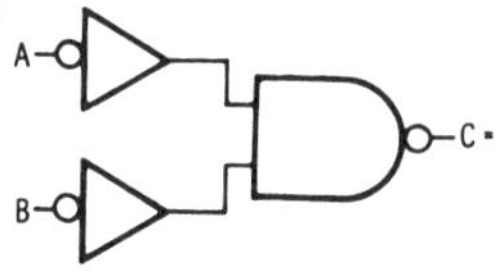

Figure 5-53

Write the output equation for this circuit. ______________.

96. ($C = \overline{\bar{A}\,\bar{B}}$) Now using DeMorgan's theorems, reduce this expression.

$$C = \overline{\bar{A}\,\bar{B}} = \underline{\qquad\qquad}$$

97. ($C = A + B$) Correct. When you invert the inputs on a NAND the OR function is performed.

DeMorgan's theorem tells us that a NAND can also perform the NOR function, that is the negated OR function.

$$\overline{A\,B} = \bar{A} + \bar{B}$$

When the NAND is used for the negated OR function the symbol in Figure 5-54 is used.

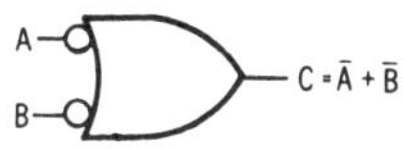

Figure 5-54

When this symbol is used, it is clear what function is being performed.

DeMorgan's theorem tells us that a positive NAND is also a ______________ ______________.

98. (negative or negated OR) Putting inverters ahead of the inputs creates the pure OR function as indicated in Figure 5-55.

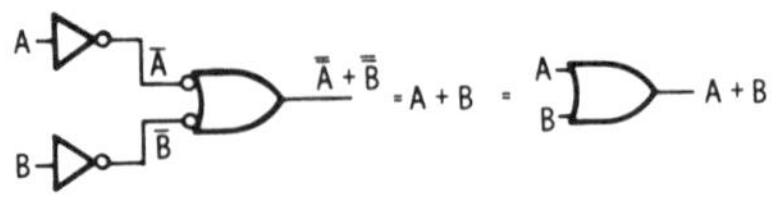

Figure 5-55

You should realize that the symbol used for the negated NOR shows circles or inverters at the inputs. In a true physical sense there are no inverters there, but the logic effect produced is as if there were.

In writing the Boolean equation from a logic diagram simply treat the symbol literally and interpret each circle as an inversion and each gate as the designated logic function.

Write the equation of the circuit in Figure 5-56.

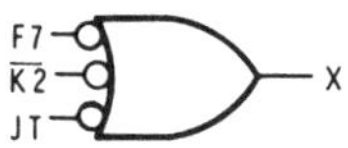

Figure 5-56

X = ______________.

99. ($X = \overline{F7} + K2 + \overline{JT}$) If you had drawn this circuit with the equivalent NAND symbol as indicated in Figure 5-57, you might expect a different logic operation to be performed.

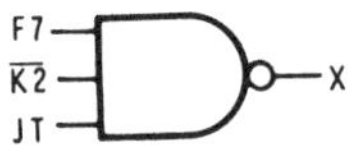

Figure 5-57

Write the equation for this circuit. X = ______________.

100. ($X = \overline{F7 \cdot \overline{K2} \cdot JT}$) **Use DeMorgan's theorem on this expression and you will get X = ______________.**

101. ($X = \overline{F7} + K2 + \overline{JT}$) This little exercise proves that a positive NAND is also a negated NOR.

The NOR gate can also be used to perform any logic function. By inverting its output the OR operation is performed as shown in Figure 5-58.

In an earlier unit you saw how the NOR could also be used to express the AND function. Sketch the logic symbols illustrating this.

102. (See Figure 5-59)

Write the output equation of this circuit. F = _______________.

103. ($F = \overline{\overline{D} + \overline{E}}$) Now reduce it with DeMorgan's theorem. F = _______________.

104. ($F = \overline{\overline{D} + \overline{E}} = D \cdot E$) This is of course the AND function.

DeMorgan's theorem tells us that the positive NOR can also perform as a _______________ _______________.

105. (negative or negated AND) The DeMorgan's expression for this relationship is

$$\overline{D + E} = \overline{D} \cdot \overline{E}$$

When the negated AND function is used the symbol in Figure 5-60 indicates the function.

When a negated AND is to be used for the AND equation, a(n) _______________ must be connected ahead of each input.

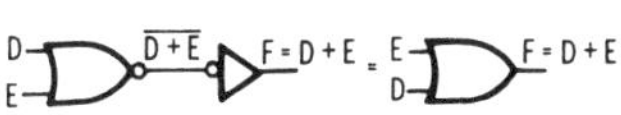

Figure 5-58

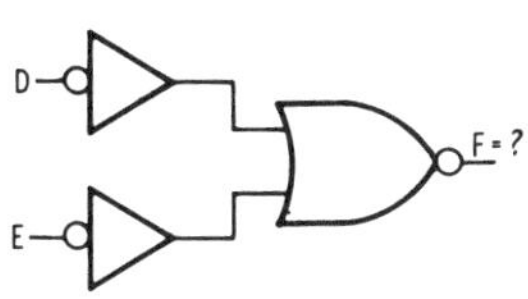

Figure 5-59

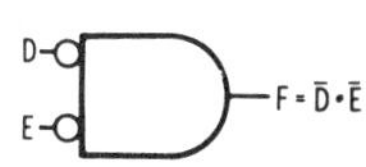

Figure 5-60

106. (inverter) The positive NOR or negated AND connected as an AND is shown in Figure 5-61 below.

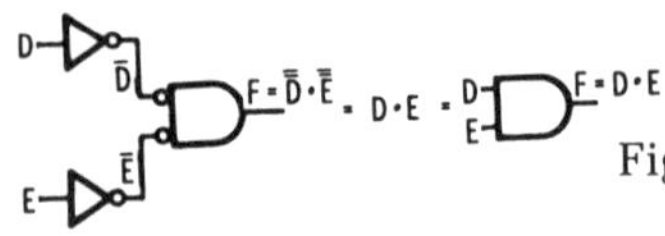

Figure 5-61

Interpret the negated AND logic symbol literally for the purpose of writing equations, even though in reality no inverters are physically present at the inputs where the circles are indicated.

Figure 5-62 summarizes the function and symbols of NAND and NOR gates.

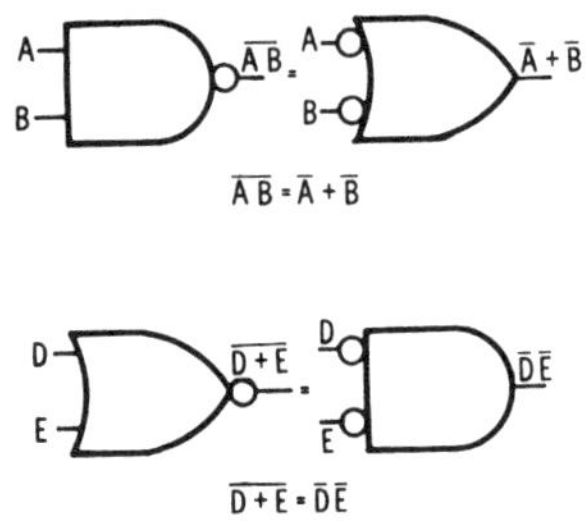

Figure 5-62

The AND and OR equivalent circuits with NANDs and NORs are summarized in Figure 5-63.

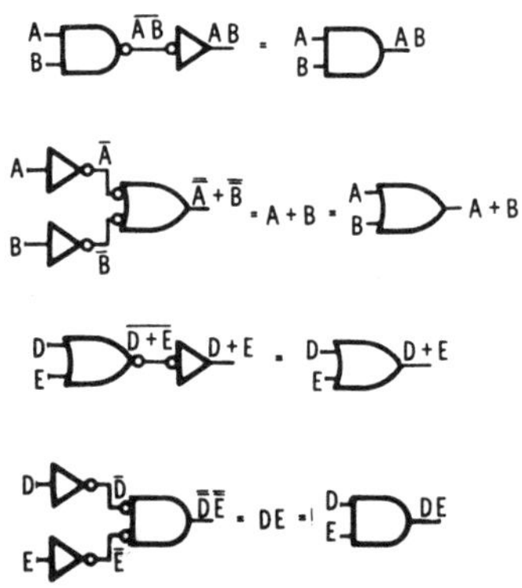

Figure 5-63

Now let's show how any Boolean equation can be implemented with NAND or NOR gates. Such logic circuits are generally known as ______________ circuits.

107. (combinational) A combination circuit is any logic function implemented with gates.

Here is a straight forward procedure for implementing any Boolean expression with NAND or NOR gates.

1. Minimize the equation using Boolean algebra and DeMorgan's theorem.
2. Draw the logic circuit of the reduced equation using standard AND and OR symbols.
3. Select a type of logic gate either NAND or NOR.
4. Replace each AND or OR symbol with the circuit for the NAND or NOR equivalent of the specified logic function (see Figure 5-68).
5. Eliminate any redundancy in the NAND/NOR circuit.

Let's take several examples to illustrate this procedure. Go to the next frame.

108. Consider the equation $F = X + \overline{Y}Z$. Following the procedure given above we will arrive at an implementation for NAND or NOR gates.

1. The equation is already in its minimum form.
2. The circuit is shown in Figure 5-64.

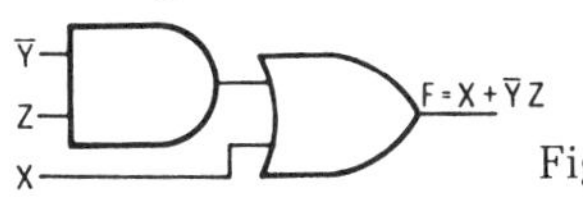

Figure 5-64

3. Let's use NAND gates.
4. The NAND circuit is given in Figure 5-65.

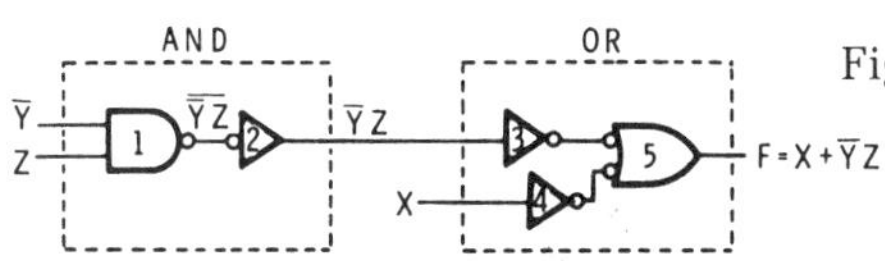

Figure 5-65

5. The redundancy in Figure 5-65 is the combination of inverters 2 and 3. Two cascaded inverters produces no logic inversion so they can be replaced by a direct connection as indicated in Figure 5-66.

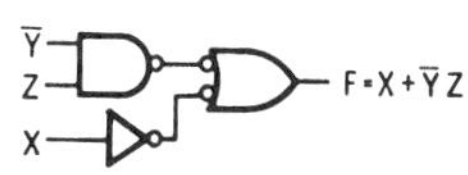

Figure 5-66

Using the same equation and procedure, develop the circuit implementation with NOR gates.

109. (See Figure 5-67) By comparing the circuits in Figure 5-66 and 5-67 you can conclude that the circuit using ______________ gates is the most economical.

110. (NAND) When NAND gates are used to implement this function fewer logic elements are used so the circuit is more economical. It will be less expensive and will consume less power. There are fewer levels of logic (2) in the NAND circuit (4 in the NOR circuit) so its propagation delay will be less. Nevertheless, either circuit performs the same logic operation.

Once you become familiar with DeMorgan's theorem and its relationship with NAND/NOR gates, the procedure described here will come naturally to you. A little practice and experience will make you competent in dealing with modern NAND and NOR integrated circuits to implement logic functions. Keep in mind that your goal is to arrive at the minimum circuit for your application.

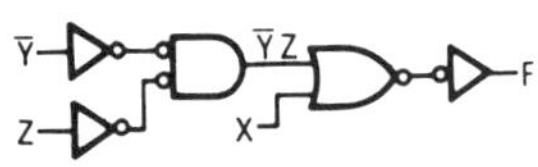

$F = X + \overline{Y}Z$

Figure 5-67

Unit 6

FLIP-FLOPS AND REGISTERS

INTRODUCTION

In this unit you are going to learn about flip-flops. A flip-flop is a digital logic element used for storing one bit of binary data. It has two stable states, one representing a binary 1 and the other a binary 0.

The flip-flop is the basic logic element used in sequential logic circuits. The primary characteristic of a sequential circuit is memory. Such circuits are used for a variety of storage, counting, sequencing and timing operations.

A major use of the flip-flop is in storage registers where a multibit binary word is stored. A register is made up of a number of flip-flops, each storing one bit of the number.

FLIP-FLOPS

1. A flip-flop is a digital logic circuit whose basic function is memory or storage. A flip-flop is capable of storing a single bit of binary data. It can assume either of two stable states, one representing a binary 1 and the other a binary 0. If the flip-flop is put into one of its two stable states, it will remain there as long as power is applied or until it is changed.

A flip-flop is used to ____________ binary data.

2. (remember, memorize, or store) A flip-flop remembers to which state it was previously set. It effectively memorizes the data we give it. We give it this data by applying appropriate logic inputs to it. To determine the value of the bit stored in the flip-flop, we look at its outputs.

How many bits of binary data can a single flip-flop store? ____________.

3. (one) The flip-flop has two states, one used to represent a binary 1, the other a binary 0.

There are three basic types of flip-flops: the latch, the D type, and the JK. Let's start with the simplest, the latch. The latch or set-reset flip-flop is the simplest form of binary storage element. The symbol shown in Figure 6-1 is used to represent this type of flip-flop.

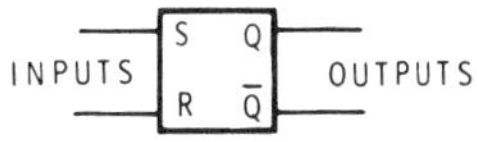

Figure 6-1

The flip-flop has two inputs, S and R, and two outputs Q and $\overline{Q}$. Applying the appropriate logic signal to either the S or R input will put the latch into one state or the other. The S input is used to *set* the flip-flop. When a flip-flop is set, it is said to be storing a binary 1. The R input is used to reset the flip-flop. A reset flip-flop is said to be storing a binary 0.

A latch is storing a binary 1. Therefore it is in the ____________ state.

4. (set) If a latch is set it is storing a binary 1. If it is reset it is storing a binary 0. The latch, also called a set-reset or RS flip-flop, is put into one state or the other by applying a logic signal to either the S or R input.

As for knowing which state the flip-flop is in, we look at the outputs. The latch has two outputs labeled Q and $\overline{Q}$. These are called the *normal* and *complement* outputs respectively. As in other logic circuits, any letter or alphanumeric mnemonic can be used to designate logic signals. See Figure 6-2.

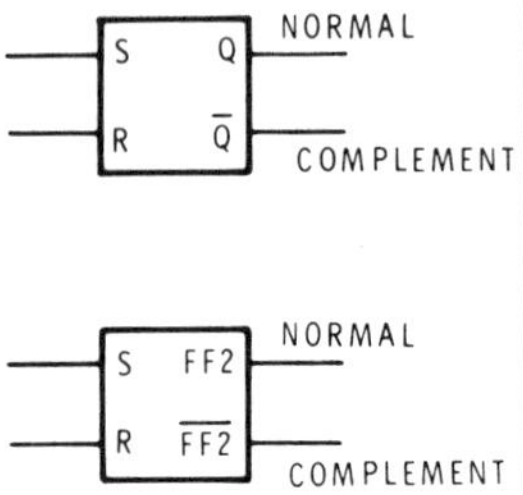

Figure 6-2

For example, the designation FF2 (meaning flip-flop number 2) could be used as shown.

To tell what state the flip-flop is in, you look at the *normal* output. The logic level present there tells you which bit, 0 or 1, is being stored.

For example, suppose that you measure the voltage at the normal output of a flip-flop with a voltmeter and find it to be the level normally associated with a binary 0. You know immediately then that the flip-flop is storing a binary __________.

5. (binary 0) Yes. If the normal output is a binary 0 level then the flip-flop is reset or storing a binary 0. The normal output always tells the state of the flip-flop. At the same time, the complement output has the state opposite that of the normal output.

If a flip-flop is reset, its complement output is binary __________.

6. (binary 1) Correct. If a flip-flop is reset, its normal output is a binary 0. The complement is a binary 1. The complement output is just as useful in determining the output state of the flip-flop as long as you remember the above relationship. The simple table below sums up the whole thing.

FLIP-FLOP STATE	OUTPUTS Q	$\overline{Q}$
SET	1	0
RESET	0	1

This relationship is true for the latch and all other types of flip-flops.

A latch is readily constructed with logic gates as shown in Figure 6-3. Here two NAND gates are wired back-to-back so that the output of one feeds the input to the other.

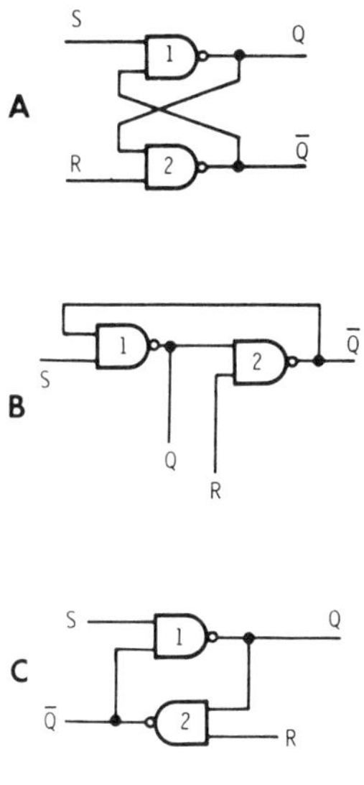

Figure 6-3

Three methods of drawing the latch are illustrated in Figure 6-3. They are all electrically the same, but the version in Figure 6-3A is the most widely used. The other versions are used occasionally so it is a good idea to be familiar with the various configurations so that you will recognize them on a logic diagram when you see them.

The latch is also sometimes drawn using the negative logic NOR symbols. Sketch the latch in its most popular configuration using this symbol.

7. (See Figure 6-4)

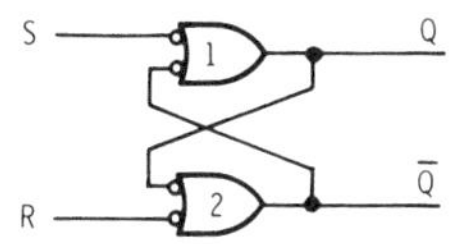

Figure 6-4

Either the NAND or NOR logic symbols can be used to illustrate a latch, but the operation of the circuit is a function of how a positive NAND/negative NOR gate works.

Let's see how much you remember about this circuit.

If both inputs to a 2 input TTL NAND gate are binary 1, the output will be ____________.

8. (binary 0 or low) In order for the output to go low, both (all) inputs must be binary 1 or high. Otherwise, any other combination of inputs will produce a binary 1 (high) output. If both inputs are open, the output will also go low. For that reason, an open input has the same effect on the gate as a ________ input.

9. (binary 1 or high) The operation of the NAND gate is summarized by the truth table below.

INPUTS		OUTPUT
A	B	C
low	low	high
low	high	high
high	low	high
high	high	low

High usually refers to the most positive logic voltage level while low refers to the least positive logic level.

Observing the truth table above, what input state (high or low) seems to have the most predominant affect on the state of the output? Specifically, which input state produces the greatest number of common output states? ________.

10. (low) Good thinking! A look at the NAND gate truth table shows that of the four output states, three of them are high. The high output state is created by one or more low inputs. For that reason, we say that the predominant input state is a low or binary 0 for this type of gate. Now, let's consider the operation of the latch. Refer to Figure 6-5.

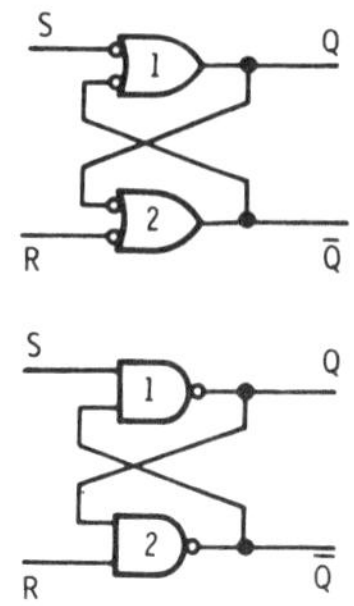

Figure 6-5

If the S and R inputs are both binary 1 (or open), which is the normal condition for this type of latch, the circuit is simply storing a bit put there by an earlier manipulation of the inputs. For example, if the flip-flop is set, the normal (Q) output from gate 1 (the set gate) will be high (binary 1). This output is fed back around to the upper input on gate 2 (the reset gate). The lower input to gate 2 is a binary 1 (or open) so its output $\overline{Q}$ is low. The output from gate 2 is fed to the lower input of gate 1. This input holds the Q output high. You can see now why they call this circuit a latch. Because of this feedback arrangement, the flip-flop is latched into this state. It will stay this way until you change it. And the way you change it is by applying a low level to either the set or reset inputs.

If the latch is set and a low level is applied to the R input (S remains high), the $\overline{Q}$ output will become ____________.

11. (high or binary 1) Right. If we apply a low to the R input of the latch, it will reset. The low level on the R input will force the output of gate 2 high. This will cause both inputs to gate 1 to be high so its output will be low or binary 0 thereby indicating the reset state.

Assume that the latch is set. A low input is applied to the S input. What will happen?

A. The flip-flop will change states.
B. Nothing will happen.
C. The flip-flop will reset.

12. (B. Nothing will happen) If the flip-flop is set, applying a low to the S input will not do anything. The low level from the $\overline{Q}$ output fed back to the set gate keeps the Q output high. In the same way, applying a low level to the R input while the latch is reset will not produce a state change.

So, summing it all up we can say that to set the latch you must apply a binary 0 to the S input. To reset it you must apply a binary 0 to the R input. The waveform timing diagrams in Figure 6-6 shows the effect of various inputs on the outputs.

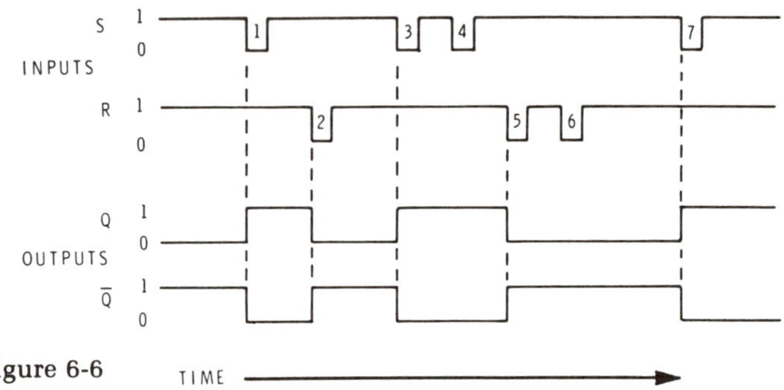

Figure 6-6

Go through these waveforms from left to right observing the effect of each input pulse on the outputs.

Looking at the waveforms above, what state is the flip-flop in *prior* to the application of pulse 1? ____________.

13. (reset) Yes. Since the Q output is low and the $\overline{Q}$ output is high, before the occurrence of (to the left of) pulse 1, the latch is in the reset state storing a binary 0. Now look at these waveforms again.

When pulse 1 occurs on the S input, the latch sets with the outputs going to their proper levels. Pulse 2 comes along next on the R input so the latch resets. Pulse 3 again sets the flip-flop. Note that pulse 4 also occurs on the S input. Since the flip-flop is already set naturally nothing will happen. Pulse 5 then resets the latch. Pulse 6, also occuring on the R input, has no affect on the state of the latch. Finally, pulse 7 again sets the latch.

When the state of the latch is not being changed, the normal state of the inputs is binary ____________.

14. (binary 1) Both inputs should normally be high on a NAND gate latch unless you are changing its state. The high inputs do not disturb the state of the latch so either a binary 1 or binary 0 may be stored there. Short duration input pulses that switch from high to low should be used when the latch is to be set or reset.

Now look at the latch circuit in Figure 6-7.

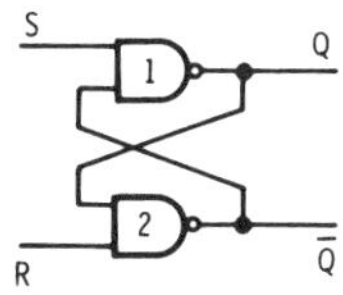

Figure 6-7

What happens if both inputs go low at the same time?

A. The latch becomes set.
B. The latch becomes reset.
C. The state of the latch cannot be determined.

15. (C. The state of the latch cannot be determined.) With both S and R inputs low, the Q and $\overline{Q}$ outputs will *both* be high. No longer are the outputs complementary, therefore, we really don't know what state the flip-flop is in. It is in some ambiguous state and is neither set or reset. This condition is one of the peculiarities of a latch. When you are using it you have to be careful to avoid simultaneous low inputs on the S and R terminals. This ambiguous state is generally undesirable because it can produce undesirable operation of a logic circuit if it is not avoided or accounted for. The ambiguous condition actually represents a third state in which the latch can exist. This state is sometimes referred to as the "limbo" state.

One way to avoid this condition is to modify the latch as shown in Figure 6-8.

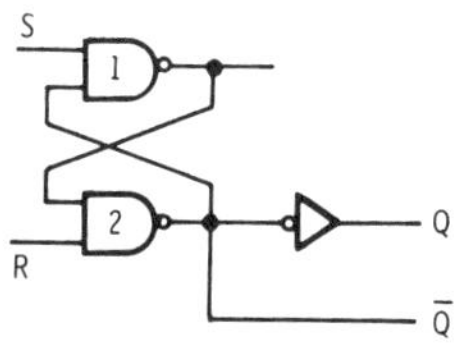

Figure 6-8

The normal and complement outputs are both derived from gate 2. The inverter ensures that the outputs are always complementary even if both inputs do go low simultaneously.

What happens if both inputs are high?

A. The flip-flop will be in the ambiguous state.
B. Nothing, this is the normal state of the inputs.
C. The state of the flip-flop will reverse.

16. (B. Nothing, this is the normal state of the inputs). Unless the latch is being set or reset, both inputs should be high (or open).

The operation of a NAND gate latch can be summarized by the truth table below.

INPUTS		OUTPUTS		STATE
S	R	Q	$\overline{Q}$	
0	0	1	1	Limbo
0	1	1	0	Set
1	0	0	1	Reset
1	1	X	$\overline{X}$	Either set or reset

The truth table accounts for all possible input and output states. Note that when both S and R inputs are binary 1, the output state of the flip-flop is designated X, where X can be either a 0 or a 1 as determined by previous input conditions.

If the flip-flop is set, you know that the last low input level was applied to the ________ input.

17. (S) The only way the flip-flop can become set is by the application of a binary 0 or low input to the S input.

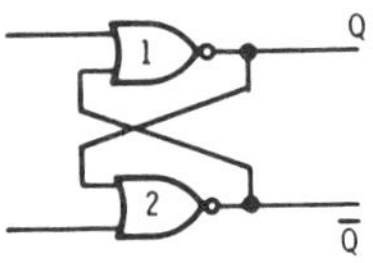

Figure 6-9

The latches we've discussed so far use positive logic NAND gates. We can also make latches out of positive logic NOR gates. Such a latch is shown in Figure 6-9.

It is identical to the other latches just discussed in that the two gates are wired back-to-back. Even the logic symbol is the same. But does this latch function the same way?

Recalling your knowledge of positive logic NOR gates, determine how the latch works and select the correct answer below.

A. The NOR latch is set or reset by binary 1's at the inputs instead of binary 0's.
B. The NOR latch requires an inverter on each input to make it function like an NAND latch.
C. The NOR latch functions exactly like the NAND latch.

18. (A. The NOR latch is set or reset by a binary 1 at the inputs instead of a binary 0.)

First, refresh your memory on NOR gate operation by referring to the truth table below.

INPUTS		OUTPUT
A	B	C
low	low	high
low	high	low
high	low	low
high	high	low

A high or binary 1 on either or both inputs produces a low or binary 0 output. This is considerably different from the NAND gate so the effect is to make the operation of a NOR latch completely different from the NAND latch.

While NOR and NAND latches have exactly the same function, they achieve it in a slightly different way. To set the NOR latch you apply a binary 1 to the S input. To reset it you apply a binary 1 to the R input. Normally, both inputs should rest at binary 0. And, if both inputs are made binary 1 simultaneously, the "limbo" or ambiguous state occurs. This is the exact opposite set of input conditions that exist in the NAND latch. Take a look at the circuit in Figure 6-10.

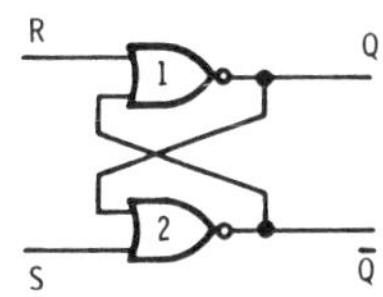

Figure 6-10

Note a subtle difference. The R and S inputs are reversed from those on the NAND latch. The reason for this has to do with the characteristics of the NOR gate. Applying a binary 1 to the R input forces the output of gate 1 low. This makes the upper and lower inputs to gate 2 low or binary 0 so its output is a binary 1. With this arrangement ($Q = 0, \overline{Q} = 1$) the flip-flop is clearly reset. As you can see, the interpretation of the outputs is the same. In fact, it is the same for *any* flip-flop.

When the NOR latch is in the "limbo" condition, both outputs will be ____________.

19. (low) Yes, both outputs will go to the binary 0 state if a binary 1 level is applied to the R and S inputs simultaneously. This is exactly the opposite of what happens in the NAND latch. Nevertheless, this ambiguous condition is generally avoided unless there is some specific application for it.

To sum up the operation of a NOR latch, complete the truth table below. Then check your answer in the next frame.

INPUTS		OUTPUTS		
S	R	Q	$\bar{Q}$	STATE
0	0			
0	1			
1	0			
1	1			

20. (See correct truth table below).

INPUTS		OUTPUTS		
S	R	Q	$\bar{Q}$	STATE
0	0	X	$\bar{X}$	Either set or reset
0	1	0	1	Reset
1	0	1	0	Set
1	1	0	0	Ambiguous

As before in the NAND latch, the X output indicates either set or reset.

One of the most common and useful applications for a latch flip-flop is in switch buffering. Pushbutton switches are used in digital equipment to control various aspects of its operation. However, most pushbutton switches produce contact bounce. When the button is depressed or released, the switch contacts do not make an immediate solid electrical or mechanical connection. The contacts "bounce" open and closed for a brief period of time.

The waveform in Figure 6-11 indicates this effect.

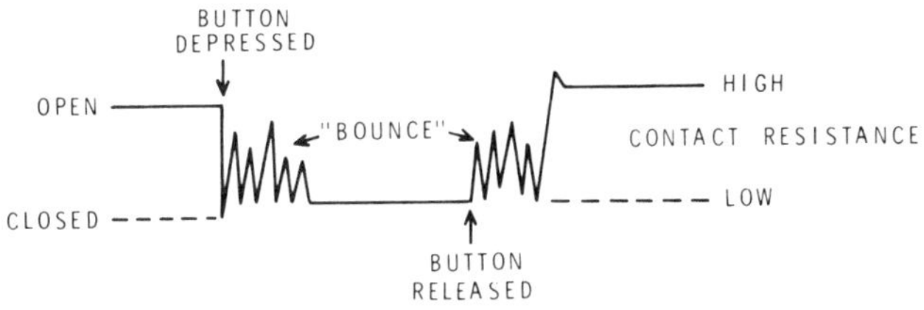

Figure 6-11

This waveform could represent contact resistance. Naturally, if current is being switched, this waveform would represent the voltage across the switch. Instead of getting solid off-on switching, you get pulses. Such pulses can repeatedly trigger digital circuits. In pressing the button once you would expect to get a single pulse or level change. Instead, the contact bounce gives you several. This effect is usually detrimental to the performance of digital circuits.

The circuits in Figure 6-12 show two ways of using a pushbutton switch to supply a logic pulse or level change. Such circuits are usually producing a considerable amount of contact bounce.

Instead of the switch supplying a single logic level change when depressed or released, ____________ are produced because of the contact bounce.

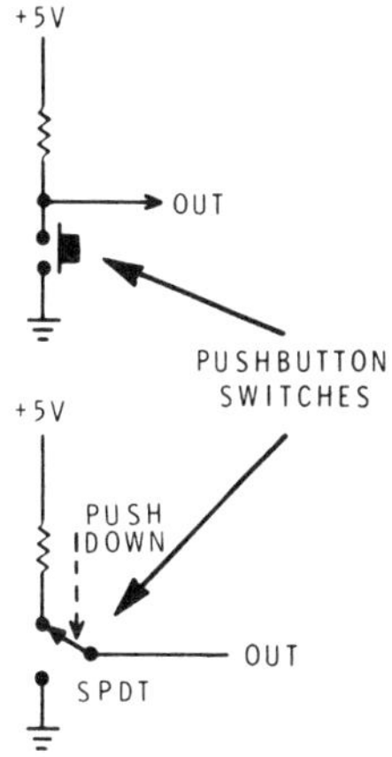

Figure 6-12

21. (pulses) Pushing a pushbutton will usually generate a series of pulses instead of a single, clean logic level change because of contact bounce. To overcome this problem, the switch can be combined with a latch as shown in Figure 6-13.

An SPDT break-before-make (non-shorting) momentary contact pushbutton switch is normally used. With the switch in position A (not depressed or normally closed N.C.), the output of gate 1 is held high. Depressing the switch so that the grounded arm contacts position B forces the output of gate 2 high and the output of gate 1 low. So how is the "bounce" removed? Well, as the button is depressed the arm of the switch breaks contact with point A. Even though it may bounce several times between open and ground, it has no effect on the state of the latch. The effect is the same as trying to repeatedly set a latch that is already set. Nothing happens. As the contact arm is in transit between points A and B, both inputs to the latch are open so the latch simply remains set. As point B is contacted, the latch resets. The slightest disturbance will trigger the state change of the latch. The feedback in the latch carries the state change through quickly. Even if switch bounce occurs, the latch is insensitive to it. The result is a single clean logic level change at the output. Releasing the pushbutton causes the latch to change back to its original state.

A NOR latch can also be used to buffer contact bounce. Which circuit in Figure 6-14 accomplishes this?

A. Circuit A
B. Circuit B
C. Circuit C

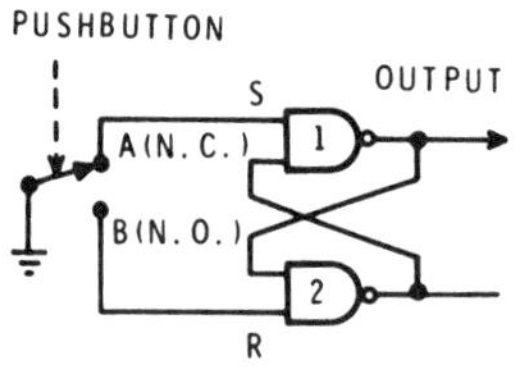

Figure 6-13

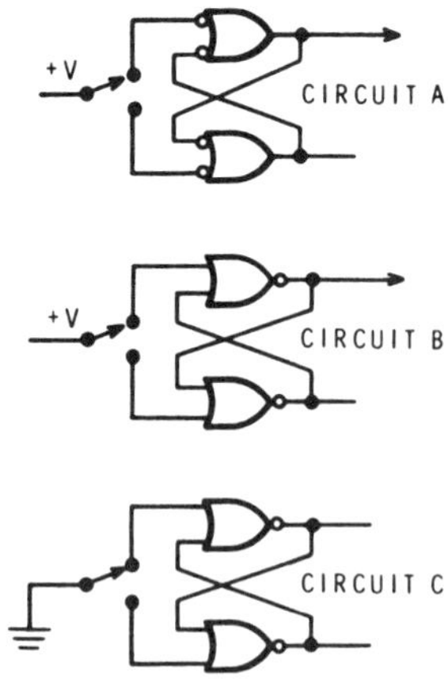

Figure 6-14

22. (Circuit B) The NOR latch removes contact bounce just as well as the NAND latch, but the arm of the switch must have a plus voltage (the supply voltage or binary 1 logic level) on it. The NAND latch requires a ground or binary 0 level on the switch arm for proper operation.

Another switch buffer latch circuit is shown in Figure 6-15.

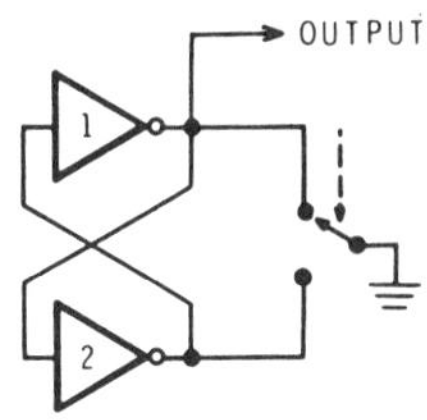

Figure 6-15

The latch is made of inverters so the outputs and inputs are common. The switch normally holds the output of inverter 1 low so that inverter 2 output is high. Pressing the switch reverses this state. The output is a "bounceless" level change.

The most important requirement for a switch used with a latch buffer is that it must be —

A. An SPST type.
B. Momentary contact only.
C. An SPDT type.

23. (C. An SPDT type.) The switch must have two terminals so an SPDT unit is required. And it is important that it be a break-before-make type so that the A and B contact points are not shorted momentarily in switching from one position to the other. This would put the latch into its "limbo" state briefly and false triggering will occur.

Another point is that the switch doesn't necessarily have to be a momentary contact type pushbutton. Any SPDT switch, slide or toggle, can be used. If it supplies logic level changes to a digital circuit it will probably need buffering.

24. Now let's consider the D type flip-flop. Its symbol is shown in Figure 6-16.

Like any other flip-flop the D flip-flop has two outputs that are used to determine its contents. That is, the outputs indicate what bit is stored there.

If we are using positive logic and the $\overline{Q}$ output is high, then the contents of the flip-flop is —

A. Binary 0

B. Binary 1

C. Can't tell, insufficient information given.

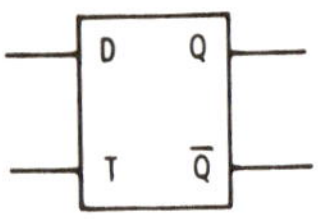

Figure 6-16

25. (A. binary 0) The $\overline{Q}$ output is high indicating a binary 1. The Q or normal output therefore is the complement or a binary 0. The Q output tells you the state of the flip-flop directly. Since it is a binary 0, the flip-flop is reset. The point here is that you read the outputs of a D flip-flop as you would a latch or any other flip-flop for that matter.

Now look at the inputs. Like on the latch there are two. But they work differently. The D input is where you apply the data or bit to be stored. Of course, it can be either a binary 1 or a binary 0. The T input line controls the flip-flop. It is used to determine whether the input data is recognized or ignored. If the T input line is high or binary 1, the data on the D line is stored in the flip-flop. If the T line is low or binary 0, the D input line is not recognized. The bit stored in the flip-flop previously is retained. The D line can essentially do anything and it will just be ignored if T is low.

If both the D and T inputs are binary 1, the normal output will be ____________.

26. (binary 1) Right. The T line is high so the flip-flop stores whatever is on the D input line. Since a binary 1 is present, it is stored, and the normal output indicates this condition. As long as the T line is high, the normal output will simply follow or track the D input.

You can get a better idea about how the D flip-flop works by taking a look at its insides. The logic diagram of one type of D flip-flop is shown in Figure 6-17.

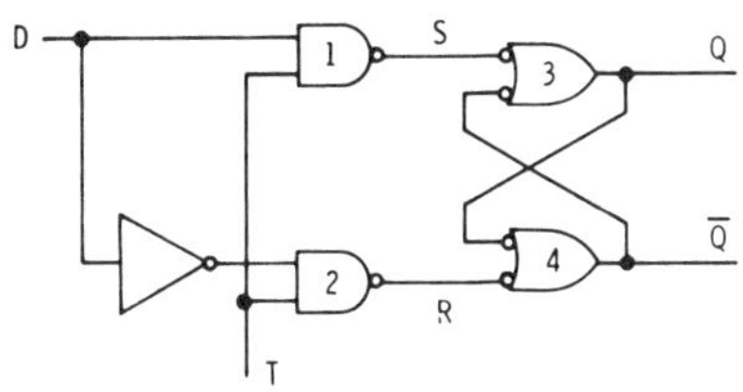

Figure 6-17

Gates 3 and 4 form a latch where the bit is stored. Gates 1 and 2 are enabling gates that pass or inhibit the input. The inverter makes sure that the S and R inputs to the latch are always complementary to avoid any possibility of the ambiguous state occurring.

With the T input low, the S and R latch inputs are ____________.

27. (high or binary 1) Yes. The low input on the T line holds the outputs of gates 1 and 2 high. This is the normal state for the inputs of a NAND latch to assume. In this state the latch is undisturbed.

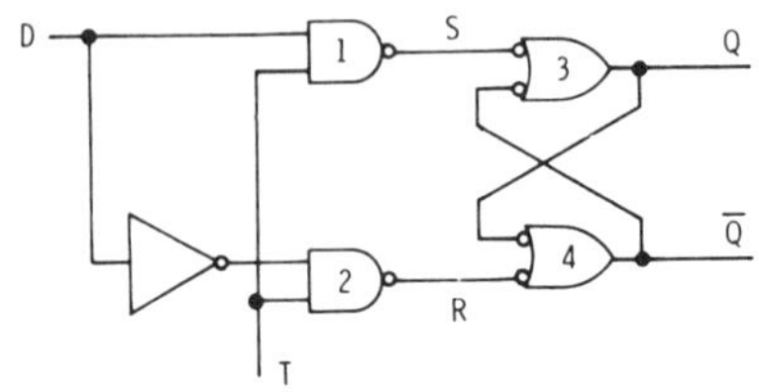

Figure 6-17

Refer to Figure 6-17. Suppose a binary 1 is applied to the D input. Of course, nothing happens if the T input is still low. Now, make the T input go high. This enables both gates 1 and 2. The binary 1 on the D input makes gate 1 output go low. The inverter puts a low on the input to gate 2 so its output stays high. The low output of gate 1 sets the latch causing it to store the binary 1. Returning the T input low disables the input, but the binary 1 is retained.

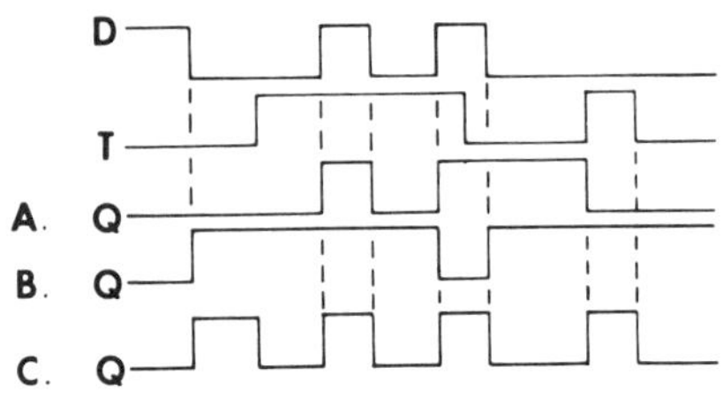

Figure 6-18

Now, look at the waveforms in Figure 6-18. These represent the D and T inputs to a flip-flop. Which of the three waveforms given represents the Q output produced by these inputs?

28. (A) With the indicated input waveforms, you will get the Q output illustrated in Figure 6-19. It takes a bit of concentration to arrive at this, but it's not difficult. All you have to remember is that the Q output is identical to the D input as long as the T input is high. And, when the T line goes low, the flip-flop stores the *last* state it sees on the D input.

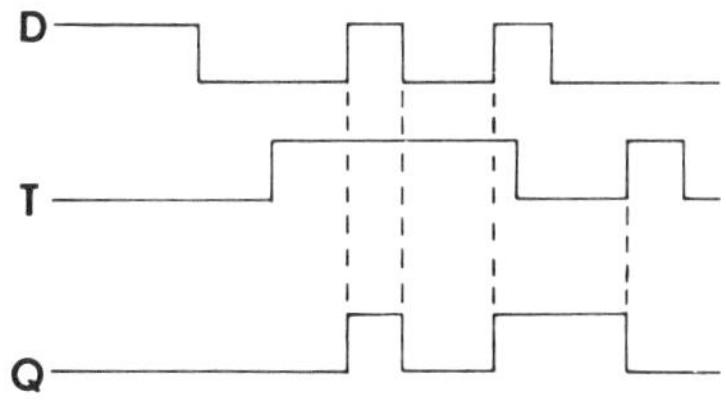

Figure 6-19

A D flip-flop is reset. The state of the D input prior to T going low was binary __________.

29. (binary 0) The circuit in Figure 6-20 shows another method of implementing the D flip-flop. As in the other circuit, gates 3 and 4 form the storage latch while gates 1 and 2 handle the input control.

Note that no separate inverter is needed. This arrangement functions exactly like the other circuit but is more economical of logic circuits. This circuit is quickly and easily made from a common quad 2 input NAND IC.

The operation of a D type flip-flop can be completely described by a truth table. From your present knowledge of D flip-flop operation, complete the truth table below.

INPUTS		OUTPUTS	
D	T	Q	$\bar{Q}$
0	0		
0	1		
1	0		
1	1		

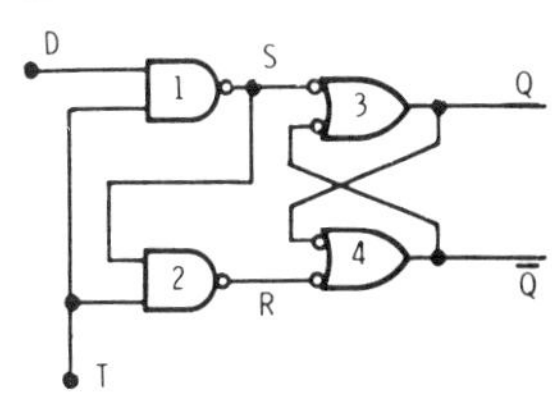

Figure 6-20

30. (See truth table below)

INPUTS		OUTPUTS	
D	T	Q	$\overline{Q}$
0	0	X	$\overline{X}$
0	1	0	1
1	0	X	$\overline{X}$
1	1	1	0

Note that when T is binary 1, the Q output is the same as the D input. When T is binary 0, the Q output can be either binary 0 or 1 depending upon a previous input. This is indicated by the X state in the table. Note that a D flip-flop does not have an ambiguous state.

A D flip-flop can also be constructed with positive NOR gates as indicated in Figure 6-21.

Figure 6-21

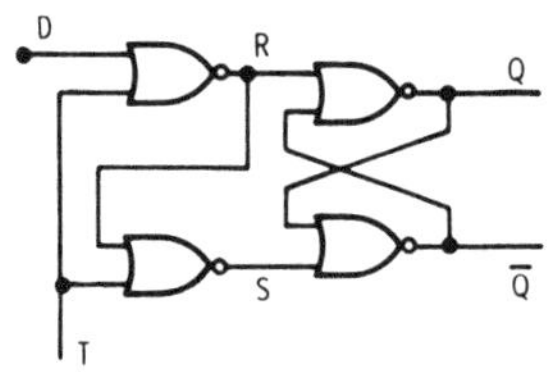

To explain the NOR latch better, however, it is desirable to redraw it so that the gates are shown as they are used. See Figure 6-22.

Figure 6-22

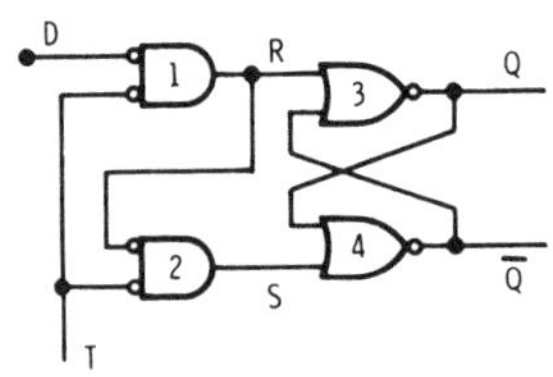

Gates 3 and 4 make up the latch that can be set or reset by the inputs from gates 1 and 2. Gates 1 and 2 control the input in that they determine whether the D input will be transferred to the latch. Functionally they perform the *AND* operation.

Now, analyze the circuit shown above and determine how it operates.

The NOR latch operates exactly like the NAND latch.

A. True

B. False

31. (False) The NOR flip-flop does not perform exactly like the NAND flip-flop, but it is similar. Both circuits store one bit of information. The recognition of the D input is determined by the state of the T input. And, here lies the difference. In the NAND flip-flop, the T line has to be high in order for the flip-flop to store the D input state. In the NOR flip-flop, the T line must go *low* in order to recognize the D input. Bringing the T line high disables the D input. The last D input state prior to the T input going high is stored.

The most common application of the D flip-flop is as an element in a storage register. A register is a group of flip-flops used to store a binary word. Each flip-flop stores one bit of the data word.

How many flip-flops would be needed to store one BCD digit of data? ______________.

32. (4) Correct. A single BCD digit consists of four bits. To store this word, we need one D flip-flop for each bit. The result is a 4 bit storage register.

Figure 6-23 below illustrates a 4 bit register.

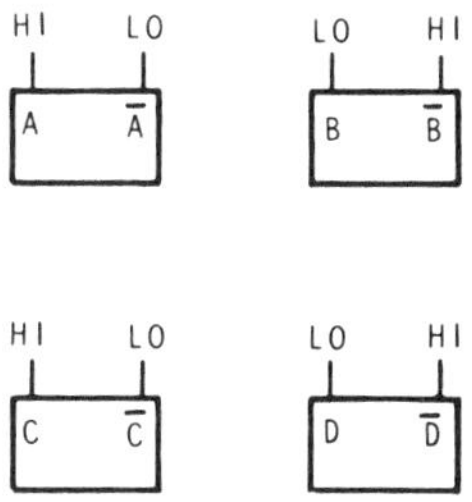

Figure 6-23

Each flip-flop is labeled with its own designation, A through D, so that it can be identified. Also shown here are the states of the flip-flop.

Given this information, what number is stored in this register? (assume positive logic)

A. 1010
B. 0101
C. Can't tell, not enough information

33. (C. Can't tell, not enough information) With the information given you really can't tell what the number is. True, you can look at normal outputs of the flip-flop and write down the corresponding bit values. This will give you two possible bit patterns depending upon whether you read from right to left or left to right. These bit patterns are 1010 (left to right) and 0101 (right to left).

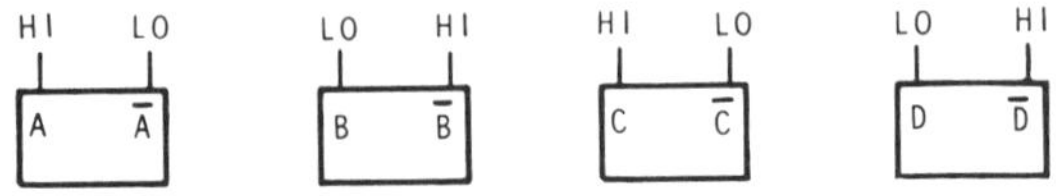

Repeat Figure 6-23

There is one missing ingredient. Which bit is the most significant (MSB)? The answer is, it could be A or it could be D. Usually the LSB is designated as the earliest letter of the alphabet or the lowest number if the number designations are used. If A is the LSB above then the number stored there is 0101 or a decimal 5. If D is the LSB the number is 1010 or a decimal 10. It's always necessary to identify the LSB and/or MSB positions on a register in a logic diagram.

A five bit register with flip-flops A, B, C, D and E is storing a number. The flip-flop states are: A-reset, B-set, C-set, D-reset and E-set. If A is the LSB, what is the decimal value of the binary number stored in the register? ______.

34. (22) Good work. The binary number is written from the flip-flop states, 0 = reset and 1 = set. The A flip-flop is the LSB so the binary number is EDCBA or 10110. This converts to a decimal 22.

Another type of register you should be familiar with is the switch register. This is exactly what its name implies, a register made of switches. The group of switches form a register for storing a single word or bit pattern using one switch per bit. The position of the switch (up/down, on/off, open/closed, etc.) determines the bit value.

A four bit switch register is shown in Figure 6-24. Four single pole double throw (SPDT) switches store the bit value as a physical position. The A, B, C, and D outputs are ground (0) or +5 (1) depending upon their position.

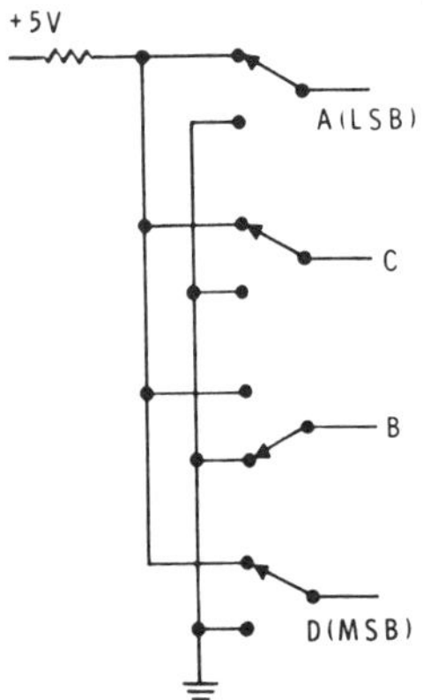

Figure 6-24

What *decimal* number is stored in this register? ______.

35. (11) Right. The binary number stored in the switch register is 1011 or decimal 11. You can determine the switch register contents by monitoring its electrical outputs. Or in most switch registers, the switches are mounted adjacent to one another horizontally with the LSB on the right and their position (usually up or down) is readily observable. Up usually means 1, down means 0. Therefore, a visual identification of the switch register contents is possible.

A frequent operation in digital equipment is the transfer of data from one register to another. Figure 6-25 illustrates how the data in a switch register can be transferred to a register made of D flip-flops. The switch outputs are fed to the D inputs. The control of the transfer is by the common T lines on the flip-flops.

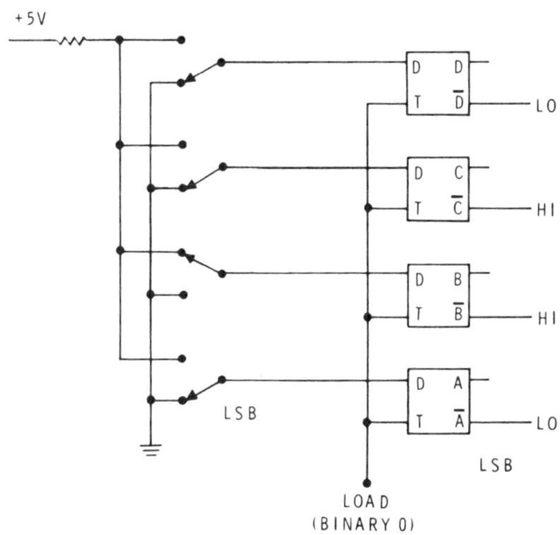

Figure 6-25

Assuming the use of positive logic and NAND flip-flops, what is the content of the flip-flop register above?

A. 1001

B. 0010

C. 0110

36. (A. 1001) The data provided to you was the flip-flop complement outputs. From this you could determine the normal outputs and hence the contents of the register. You could also determine the output of the switch register, 0010, by inspection. With the T inputs to the flip-flop at binary 0, the data input from the switch register is not recognized by the flip-flops. But, if the LOAD control line goes high momentarily, the flip-flop register contents will become the same as the switch register, 0010.

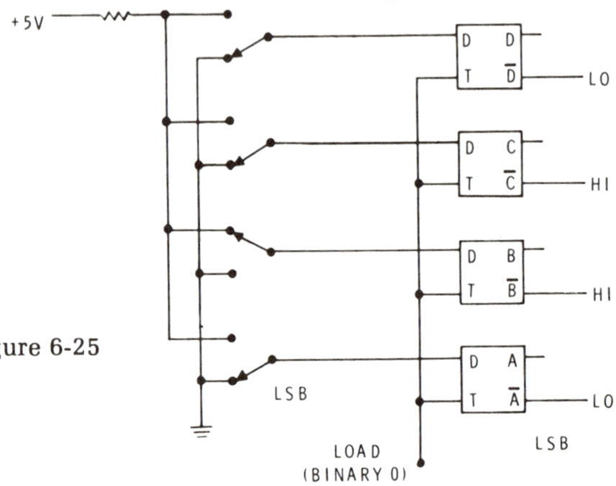

Repeat Figure 6-25

There are two important points to note here. First, the LOAD input controls the transfer of the data from the switch register to the flip-flop register. This LOAD input is the parallel or simultaneous control of all the flip-flop T inputs. This line is also sometimes called the STROBE input since it is usually only enabled or "strobed" momentarily with a binary 1 pulse to transfer the data.

Second, the data transfer is a parallel one. That is all bits from the switch register are loaded into the flip-flop register simultaneously.

Instead of drawing the individual flip-flops, most registers are shown as only a single box with the inputs and outputs identified as shown in Figure 6-26.

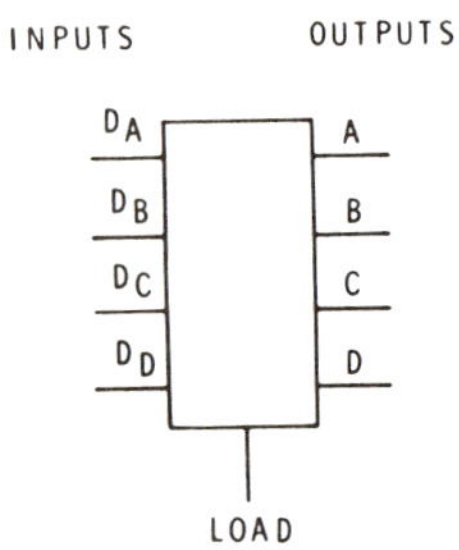

Figure 6-26

This is particularly true of MSI integrated circuit registers. Many IC registers also do not have the complement outputs available.

How is a D flip-flop register like the one in Figure 6-26 reset?

A. Apply a binary 0 to the LOAD line.
B. Load the input 0000

37. (B. Load the input 0000) Correct. To reset the register you simply load 0000. You apply 0000 to the inputs and momentarily bring the load line high. The register will then be cleared or reset since each flip-flop will be storing a binary 0.

Some commercially available D flip-flop IC registers have a single reset line which will clear all flip-flops in the register at once without having to load 0000.

JK FLIP-FLOPS

38. The JK flip-flop is the most versatile type of binary storage element in common use. It can perform all of the functions of the RS and D type flip-flops described earlier plus it can do several other things that these simple flip-flops cannot. Naturally, it is more complex and expensive than the other types so for that reason it isn't always used where simpler and less expensive circuits will do.

An integrated circuit JK flip-flop is really two flip-flops in one. It usually consists of two latches, one feeding the other, with appropriate input gating on each. See Figure 6-27.

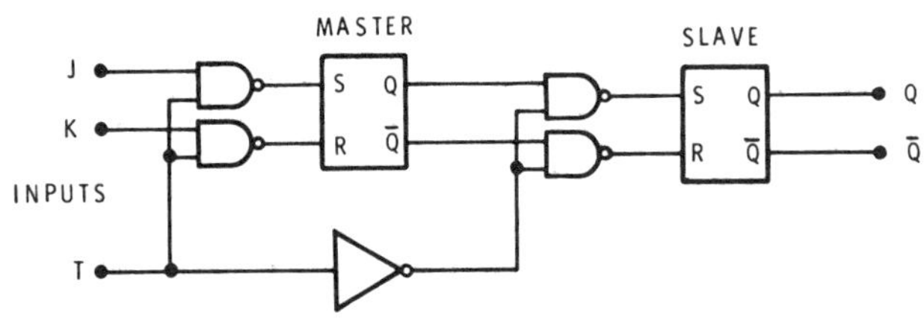

Figure 6-27

The arrangement is called a master-slave JK flip-flop. The master flip-flop is the input circuit. Logic signals applied to the JK flip-flop set or reset this master latch. The slave flip-flop is the latch from which the outputs are taken. The slave latch gets its input from the master latch. Both latches are controlled by a clock pulse. Since there are two *places* to store bits in a JK flip-flop, there can be times when both master and slave latches are identical or times they are complementary. But only one of these latches is responsible for indicating the state of the JK flip-flop.

Which latch, master or slave, determines the state of a JK flip-flop?

____

39. (slave) Yes, the slave latch designates the state being stored. If it is set, the JK flip-flop is storing a binary 1. The slave latch is the output. And to determine the state of a flip-flop, look at its outputs. Even though it's called the slave, it tells the value of the bit stored. The master flip-flop controls the state that the slave latch assumes. You can see this better by studying the logic diagram of a typical JK flip-flop in Figure 6-28.

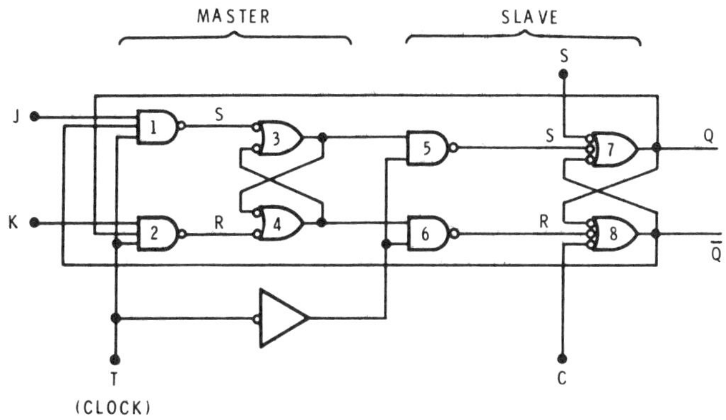

Figure 6-28

The logic gates are the positive NAND type. Gates 3 and 4 make up the master latch while its input is controlled by gates 1 and 2. The slave latch is made up of gates 7 and 8. Gates 5 and 6 control the transfer of the master latch state to the slave latch. Note that clock signal T controls the input gating circuits. The inverter keeps the clock to the master and slave input gates complementary. The clock pulse controls the JK flip-flop while the J and K inputs determine exactly how it will be controlled. (Note: You will also see the T input referred to as CP or CK, designating clock pulse or clock.)

Refer to the logic diagram of the JK flip-flop in Figure 6-28. What other inputs control the JK flip-flop? ______________

40. (S and C inputs) The set (S) and clear (C) inputs also control the JK flip-flop. These are inputs to the slave latch that can be used to set or reset (clear) the flip-flop. These inputs override all other circuitry in the JK flip-flop. These inputs are used to preset the state of the flip-flop prior to any other operation involving the JK inputs and the clock. They work just like the inputs on any latch.

To set the JK flip-flop, which of the following conditions must exist?

A. S input high, C input low
B. S input low, C input high
C. S and C inputs low
D. S and C input high

41. (B. S input low, C input high) To set the slave latch *and* the JK flip-flop, the S input should be low, the C input high. This forces the normal Q output high indicating that a binary 1 is being stored. To reset the JK flip-flop, the C input is made low while S is high. Normally, the S and C inputs will be high when they are not being used to preset the flip-flop. This arrangement is identical to that for the NAND latch.

Now let's consider how the J, K and T (clock) inputs affect the flip-flop. Refer to Figure 6-28.

Consider the time when the clock input is low. Gates 1 and 2 will be inhibited so the J and K inputs cannot control the state of the master latch. The master latch can be in either state.

What is the state of the slave latch when the clock line is low and assuming the S and C inputs are open?

A. Binary 1
B. Binary 0
C. The same as the master latch

42. (C. Same as the master latch) Yes, the slave latch will have the same state as the master latch when the clock input is low. The output of the inverter in the clock line is binary 1, causing gates 5 and 6 to be enabled during this time. Therefore, the state of the master latch is simply transferred to the slave latch. For example, if binary 1 is stored in the master latch, the output of gate 3 will be high and the output of gate 4 will be low. This will make the output of gate 5 low and the output of gate 6 high. This low on the input to gate 7 will force its output high, thereby setting the slave latch and storing a binary 1.

Now if the clock T goes high, gates 1 and 2 will be enabled. The output of the inverter will inhibit gates 5 and 6. The master latch cannot further change the slave latch. But now with gates 1 and 2 enabled, the J and K inputs can affect the state of the master latch.

What other signals affect the master latch at this time? _________.

43. (Q and $\overline{Q}$) The JK flip-flop outputs Q and $\overline{Q}$ are fed back around to gates 1 and 2 where, along with the J and K inputs, they will determine the state of the master latch.

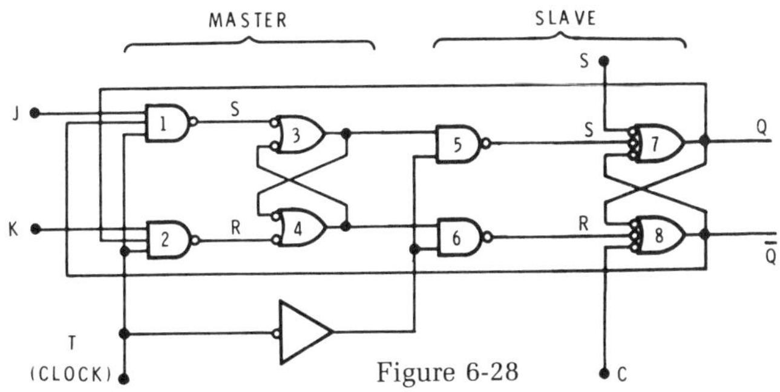

Figure 6-28

If both J and K inputs are low, the outputs of gates 1 and 2 will be held high, so no change takes place in the master latch.

If the J and K inputs are both high (or open), then the state of the master latch will be determined by the Q and $\overline{Q}$ outputs. For example, if the slave latch is set, the master latch will be reset. If the slave is reset, the master will be set. The reason for this is the way the outputs are crisscrossed back to gates 1 and 2. Remember that with the J, K and T inputs high, the state of the master latch will be determined by the Q and $\overline{Q}$ outputs.

If T, J, and K are high, and the slave latch is storing a binary 0, the master latch will be storing a binary _________.

44. (binary 1) Now let's consider the effect of the J and K inputs. These inputs are analogous to the set and reset inputs on a latch. If J is 1 and K is 0, we will set the master latch. If J is 0 and K is 1, the master latch will be reset. Remember, the T input line must be high for this to happen.

Does the JK flip-flop get set or reset as soon as the desired states are applied to the J and K inputs when T is high? ________.

45. (No) The state of the JK flip-flop is the state of the slave latch. The state of the slave latch is determined by that of the master latch. The state of the master is, in turn, determined by the J and K inputs. And to top it off, the clock input determines *when* each of these latches will be affected. With the clock input high, only the master latch will be affected. The inverter on the clock line blocks gates 5 and 6 so the slave latch is not disturbed. The states of the JK inputs will ultimately determine the output state but only at a specific time. When the clock line switches from high to low (trailing edge), the state of the master latch is transferred to the slave latch.

Which of the following conditions occurs when the clock input is high?

A. The state of the master latch is determined by the JK inputs.
B. The state of the slave latch assumes the state of the master latch.
C. Neither the master or slave latches are affected.

46. (A. The state of the master latch is determined by the JK inputs). When the clock (T) is high gates 1 and 2 will be enabled therefore the master latch will be changed by either the Q and $\overline{Q}$ outputs or by the J and K inputs. When the clock goes low, the state of the master latch is transferred to the slave latch through gates 5 and 6 which are enabled at this time. Gates 1 and 2 are inhibited so the J and K inputs have no effect.

If the J and K inputs are both left open or high, the flip-flop will change state each time the clock input switches from high to low. Consider the following: Clock high, JK inputs high and slave latch set. The master latch then is ________.

47. (reset) Right. The Q and $\overline{Q}$ lines fed back to gates 1 and 2 cause the master latch to be reset. Then when the trailing edge of the clock pulse occurs, (clock switches from 1 to 0), the reset state in the master is transferred to the slave. The JK flip-flop is now reset. When the clock again goes high, the slave latch then sets the master latch. As the clock goes low, the set state in the master is transferred to the slave. As you can see then, with the JK inputs high, the flip-flop complements itself each time the clock switches from high to low (trailing edge). We call this operation toggling.

Assume the J and K inputs are open or high. Which waveform in Figure 6-29 represents the normal flip-flop output?

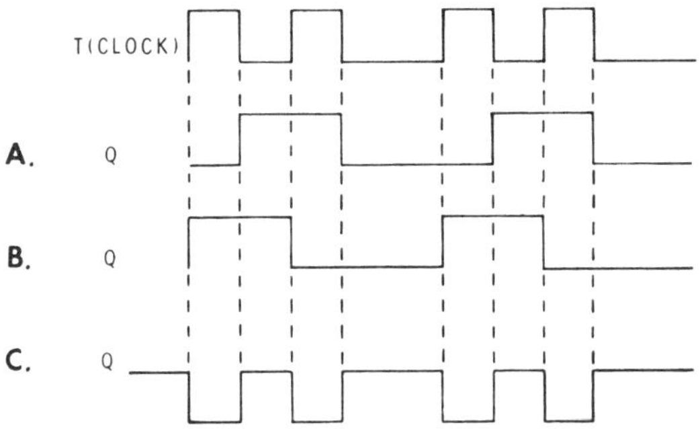

Figure 6-29

48. (A.) When the clock switches from 1 to 0, the state of the flip-flop will change. The output is not affected during the 0 to 1 transition (leading edge).

If you will look at the clock and output waveforms in Figure 6-30, you will see a definite relationship. What is it?

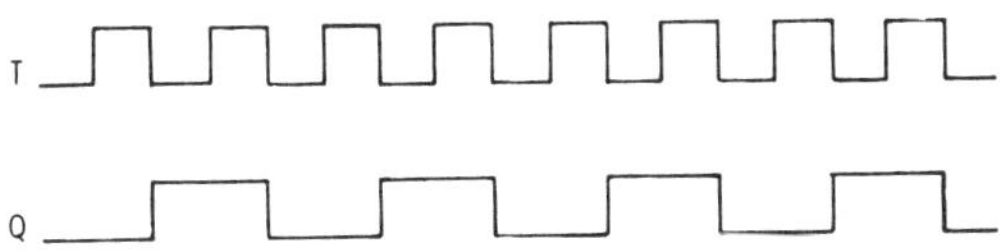

Figure 6-30

A. The output is the complement of the clock input.
B. The output has a frequency twice the input.
C. The output has a frequency one half the input.

49. (C. The output has a frequency that is one half the input.) The reason for this is simply that the flip-flop changes state on only the trailing edge or every other transition of the clock. Therefore, the JK flip-flop in its toggling mode is a two to one frequency divider. It halves any input frequency applied to the clock input. If an input of 50 KHz is applied, the output will be one half or 25 KHz. Cascading JK flip-flops permits frequency division by any factor of 2 (2, 4, 8, 16, 32, 64, etc.). The frequency division ratio is 2^n where n is the number of flip-flops cascaded.

We have now considered all of the modes of operation of the JK flip-flop, but let's review them each briefly.

First, there are the S and C inputs. The effect of these can be summed up by the truth table below.

INPUTS		OUTPUTS	
S	C	Q	$\overline{Q}$
1	1	X	$\overline{X}$
0	1	1	0
1	0	0	1
0	0	1	1

This is exactly the same truth table we established for the NAND latch.

The JK flip-flop has an ambiguous state. True or False?

50. (True) Yes, the JK flip-flop does have an ambiguous state. If both S and C inputs are low, both the Q and $\overline{Q}$ outputs will be high. Therefore, care should be taken to see that this condition cannot occur.

The set (S) and clear (C) inputs are used to preset the flip-flop to some desirable condition prior to another operation. The most common operation is to reset it. For that reason many IC JK flip-flops have only a C input line. Use of the S and C inputs is referred to as asynchronous operation. The state of the flip-flop changes immediately upon the application of the appropriate input level. No other conditions are necessary. This is not true of the J and K inputs. Their effect is dependent upon the state of the clock signal. Therefore, we call the J and K input effects synchronous because they cause state changes only on the occurrence of a specific clock transition, that is in synchronism with the clock.

In the JK flip-flop discussed here, state changes take place on the ____________ to ____________ transition.

51. (high to low, 1 to 0) When the clock input switches from 1 to 0, state changes occur. It is on this transition that the contents of the master latch is transferred to the slave latch. For some types of JK flip-flops just the opposite is true. Be sure to check the manufacturer's data sheet for details on any device you are using.

The synchronous operation of the JK flip-flop is summed up in the truth table below. Note that only the normal (Q) output condition is shown, but it is given twice, once prior to a clock pulse (t) and then after one clock pulse (t + 1). The output state X can represent either set (1) or reset (0).

INPUTS		OUTPUTS	
J	K	Q (t)	Q (t + 1)
0	0	X	X
0	1	X	0
1	0	X	1
1	1	X	$\overline{X}$

Refer to the truth table.

What happens when the J and K inputs are both low and a repetitive clock signal is applied?

A. The flip-flop complements.
B. The flip-flop state doesn't change.
C. The flip-flop is reset.

52. (B. The flip-flop state doesn't change) The clock can change all it wants to but it will not affect the state of the flip-flop when both J and K inputs are low. The flip-flop simply retains its previous condition which can be either set or reset. This is an inhibit mode.

INPUTS		OUTPUTS	
J	K	Q (t)	Q (t + 1)
0	0	X	X
0	1	X	0
1	0	X	1
1	1	X	$\overline{X}$

According to the truth table above, how does the flip-flop get reset?

A. J input = 0, K input = 1.
B. Apply binary 0 to the C input.
C. Let the clock toggle it.

53. (A. J input = 0, K input = 1) To reset the JK flip-flop, apply a 0 to the J input and a 1 to the K input then apply a clock pulse. The flip-flop will reset. To set the JK flip-flop you apply a 1 to J and 0 to K and again apply a clock pulse. The flip-flop will set on the trailing edge of the clock.

What happens when both J and K inputs are high and a clock pulse is applied?

A. The flip-flop is set.
B. The flip-flop complements.
C. The flip-flop is inhibited.

54. (B. The flip-flop complements) With the J and K inputs at binary 1, the flip-flop toggles or complements each time the clock switches from 1 to 0. The flip-flop acts as a 2 to 1 frequency divider. Modern integrated circuit flip-flops are available in a variety of configurations. Some ECL flip-flops can toggle at rates as high as 1 GHz.

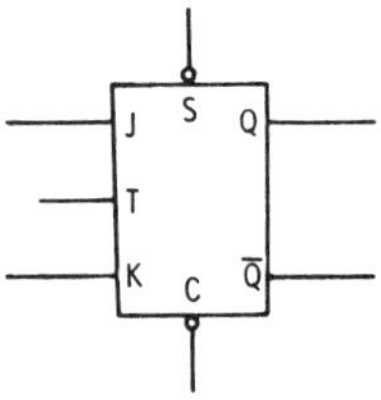

Figure 6-31

That completes the basic operation of a JK flip-flop. The symbol used to represent it is shown in Figure 6-31.

As a final check of your understanding of this important device, consider the input waveforms shown in Figure 6-32.

Which output waveform is produced by this set of inputs?

____

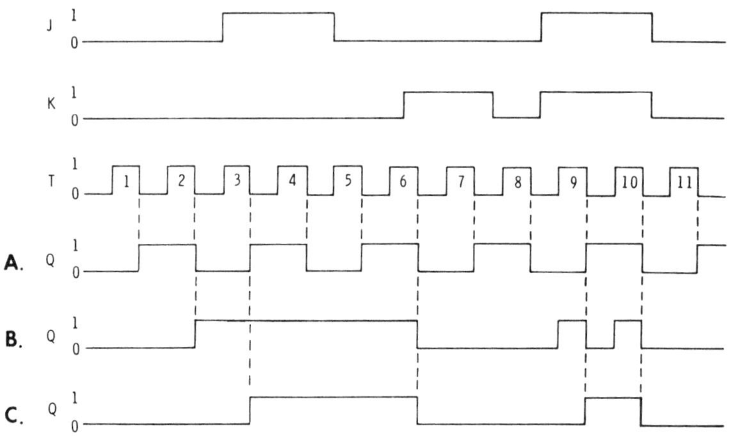

Figure 6-32

55. (C) Good work. You seem to have the idea. The J and K inputs affect the flip-flop state, but state changes occur only on the 1 to 0 transition of the clock pulse. This is synchronous operation.

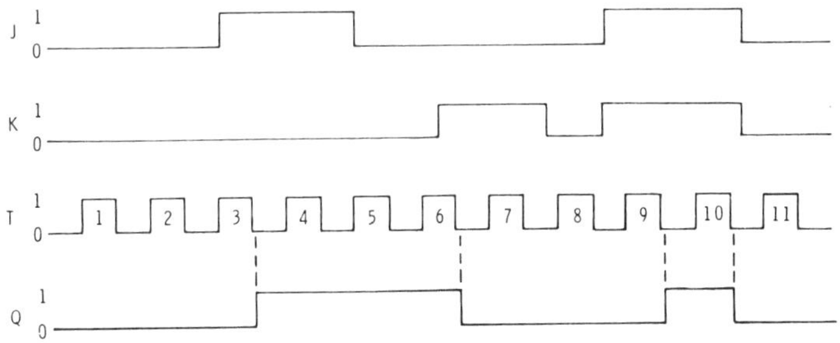

Figure 6-33

In Figure 6-33, the normal output of the flip-flop is low prior to the occurrence of the first clock (T) pulse. Pulses 1 and 2 occur but since both the J and K inputs are low, the flip-flop is inhibited and no state change takes place. Then, the J input goes high. On the trailing edge of the next clock pulse (3), the flip-flop sets. When pulse 4 occurs, J is still high so the flip-flop would be set again. However, it is already set. The J input goes low, then clock pulse 5 occurs. With both J and K low, the trailing edge of pulse 5 has no

effect. The flip-flop remains set. Next, the K input goes high and J remains low. On the occurrence of the 1 to 0 transition of pulse 6, the flip-flop resets. Pulse 7 tries to reset it again. The K input goes low. Pulse 8 occurs and since J and K are low, the flip-flop remains reset. The J and K inputs go high simultaneously. Pulses 9 and 10 then toggle or complement the flip-flop, 9 setting it and 10 resetting it. After this J and K go low inhibiting the flip-flop. Pulse 11 has no effect.

The JK flip-flop is highly versatile. They are widely used in storage registers, shift registers, frequency dividers and counters. You will learn more about each of these circuits in a later unit.

Unit 7

SEQUENTIAL LOGIC CIRCUITS: COUNTERS, SHIFT REGISTERS, and CLOCKS

INTRODUCTION

In this unit you are going to learn about sequential logic circuits. These are logic circuits that are used for a variety of timing, sequencing and storage functions. The key characteristic of sequential logic circuits is memory. Sequential logic circuits are capable of storing binary data. The output of a sequential logic circuit is a function not only of the various input states applied to the circuit but also the result of previous operations which are stored in the circuit itself.

The main circuit elements of a sequential circuit are flip-flops. These flip-flops store binary data and their states are changed by the logic input signals in accordance with the current information stored in them. The sequential logic operations are generally sequenced by a periodic logic signal known as a clock. The clock is an oscillator that generates rectangular pulses at a fixed frequency.

As with combinational logic circuits, there can exist almost an infinite variety of sequential logic circuits. However, in practice only a few special types seem to reoccur regularly. The two most commonly used sequential circuits are counters and shift registers. Because these two types of sequential circuits are the most widely used, we will emphasize their operation and application in this unit. In addition, clock circuits and special sequential circuit components such as one shot multivibrators and the delay lines will also be considered. Like combinational logic circuits, most sequential circuits are implemented with integrated circuits. In fact, most of the commonly used sequential circuits are available as a single ready to use MSI logic package. We will concentrate on these popular devices in this unit.

COUNTERS

A binary counter is a sequential logic circuit made up of flip-flops that is used to count the number of binary pulses applied to it. The pulses or logic level transitions to be counted are applied to the counter input. These pulses cause the flip-flops in the counter to change state in such a way that the binary number stored in the flip-flops is representative of the number of input pulses that have occurred. By observing the flip-flop outputs you can determine how many pulses were applied to the input.

There are several different types of counters used in digital circuits. The most commonly used is the binary counter. This type of counter counts in the standard pure binary code. BCD counters which count in the standard 8421 BCD code are also widely used. In addition, counters can be developed to count in any of the special binary or BCD codes in common use. Both up and down counters are available.

BINARY COUNTERS

A binary counter is a sequential logic circuit that uses the standard pure binary code. Such a counter is made up by cascading JK flip-flops as shown in Figure 7-1. The normal output of one flip-flop is connected to the toggle (T) input of the next flip-flop. The JK inputs on each flip-flop are open or high. The input pulses to be counted are applied to the toggle input of the A flip-flop.

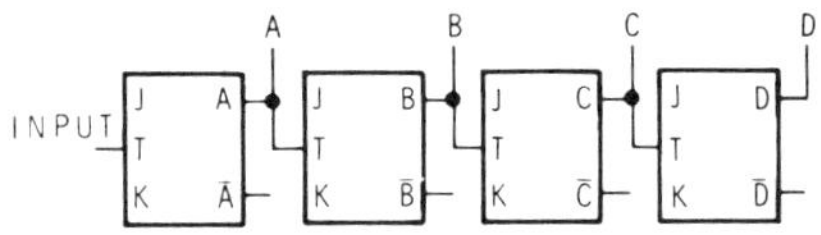

Figure 7-1
Four bit binary counter.

To see how this binary counter operates, remember that a JK flip-flop toggles or changes state each time a trailing edge transition occurs on its T input. The flip-flops will change state when the normal output of the previous flip-flop switches from binary 1 to binary 0. If we assume that the counter is initially reset, the normal outputs of all the flip-flops will be binary 0. When the first input pulse occurs, the A flip-flop will become set. The binary number stored in the flip-flops indicates the number of input pulses that have occurred. To read the number stored in the counter you simply observe the normal outputs of the flip-flops. The A flip-flop is the least significant bit of the word. Therefore, the four bit number stored in the counter is designated DCBA. After the first input pulse, the counter state is 0001. This indicates that one input pulse has occurred.

When the second input pulse occurs the A flip-flop toggles and this time becomes reset. As it resets its normal output switches from binary 1 to binary 0. This causes the B flip-flop to become set. Observing the new output state, you see that it is 0010 or the binary equivalent of the decimal number 2. Two input pulses have occurred.

When the third input pulse occurs the A flip-flop will again set. The normal output switches from binary 0 to binary 1. This transition is ignored by the T input of the B flip-flop. The number stored in the counter at this time then is 0011 or the number 3 indicating that three input pulses have occurred.

When the fourth input pulse occurs, the A flip-flop is reset. Its normal output switches from binary 1 to binary 0 thereby toggling the B flip-flop. This causes the B flip-flop to reset. As it does, its normal output switches from binary 1 to binary 0 causing the C flip-flop to become set. The number now in the counter is 0100 or a decimal 4. This process continues as the input pulses occur. The count sequence is the standard 4 bit binary code as indicated in Figure 7-2.

D	C	B	A
0	0	0	0
0	0	0	1
0	0	1	0
0	0	1	1
0	1	0	0
0	1	0	1
0	1	1	0
0	1	1	1
1	0	0	0
1	0	0	1
1	0	1	0
1	0	1	1
1	1	0	0
1	1	0	1
1	1	1	0
1	1	1	1

RECYCLE

Figure 7-2
Count sequence of four bit binary counter.

An important point to consider is the action of the circuit when the number stored in the counter is 1111. This is the maximum value of a four bit number and the maximum count capacity of the circuit. When the next input pulse is applied, all flip-flops will change state. As the A flip-flop resets, the B flip-flop resets. As the B flip-flop resets it, in turn, resets the C flip-flop. As the C flip-flop resets, it toggles the D flip-flop which is also reset to zero. The result is that the contents of the counter becomes 0000. As you can see from Figure 7-2, when the maximum content of the counter is reached it simply recycles and starts its count again.

The complete operation of the four bit binary counter is illustrated by the input and output waveforms in Figure 7-3. The upper waveform is a

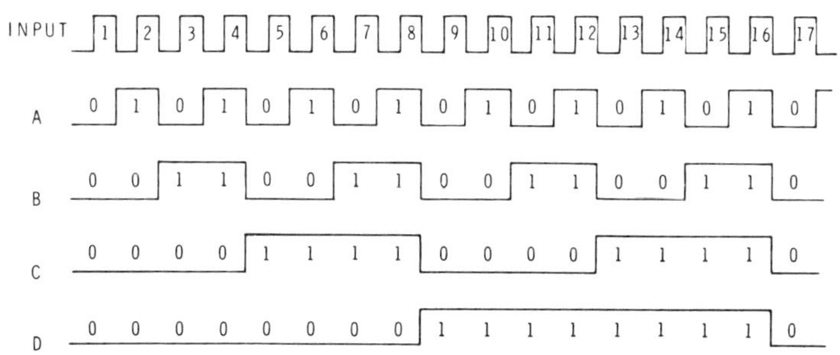

Figure 7-3 Input and output waveforms of a 4 bit binary counter.

series of input pulses to be counted. Here they are shown as a periodic binary waveform but of course it is not necessary for the input signal to be of a constant frequency or have equally spaced input pulses. The waveforms also show the normal output of each flip-flop.

In observing the waveforms in Figure 7-3, you should note several important things. First, all the flip-flops toggle (change state) on the trailing edge or the binary 1 to binary 0 transition of the previous flip-flop. With this in mind you can readily trace the output of the first (A) flip-flop by simply observing when the trailing edges of the input occur.

The output of the B flip-flop is a function of its input which is the output of the A flip-flop. Note that its state change occurs on the trailing edge of the A output. The same is true of the C and D flip-flops. The binary code after each input pulse is indicated on the waveforms. Of course, this corresponds to the binary count sequence in Figure 7-2.

Frequency Divider. Another important fact that is clear from the waveforms in Figure 7-3 is that the binary counter is also a frequency divider. The output of each flip-flop is one half the frequency of its input. If the input is a 100 kHz square wave, the outputs of the flip-flops are:

A — 50 kHz
B — 25 kHz
C — 12.5 kHz
D — 6.25 kHz

The output of a pure binary counter is always some sub-multiple of two. The four bit counter divides the input by 16, (100 kHz ÷ 16 = 6.25 kHz).

Maximum Count. The maximum count capability of a binary counter is a function of the number of flip-flops in the counter. The maximum number that can be contained in a binary counter before it recycles is determined in the same way that we can determine the maximum binary number that can be represented by a word with a specific number of bits. The formula below expresses the relationship between the number of flip-flops in a counter and its maximum count capability.

$$N = 2^n - 1$$

Here N is the maximum number that occurs prior to the counter recycling. The number of flip-flops is designated by n. For example, the maximum number that can be contained in a counter using four flip-flops is

$$N = 2^4 - 1 = 16 - 1 = 15 \text{ (binary 1111)}$$

Another example is a binary counter with nine flip-flops. A maximum count capability here then is

$$N = 2^9 - 1 = 512 - 1 = 511 \text{ (binary 111111111)}$$

To determine the number of flip-flops required to implement a counter with a known or required count capability, use the formula given below.

$$n = 3.32 \log_{10} N$$

For example, if you wish to implement a binary counter capable of counting to 100, you could determine the number of flip-flops as follows:

$$n = 3.32 \log_{10} 100 = 3.32\ (2) = 6.64$$
$$n = 7$$

Since there is no such thing as a fractional part of a flip-flop, the next higher whole number value is used. A counter with 7 flip-flops has a maximum count capability of

$$2^7 - 1 = 127.$$

When the binary counter is used as a frequency divider, the factor by which the counter divides is a function of the number of flip-flops used. To determine the divide ratio, the expression below is used.

$$N = 2^n$$

For example, if the counter contained six flip-flops it would divide an input signal by

$$N = 2^6 = 64$$

What this means is that the output of the sixth flip-flop will be $1/64$ the frequency of the input signal applied to the first flip-flop. With a binary counter, the frequency division ratio is always some power of 2. As you will see later, it is possible to implement frequency dividers that can divide the frequency by any integer value.

Down Counters. The binary counter just described is referred to as an up-counter. Each time that an input pulse occurs, the binary number in the counter is increased by one. We say that the input pulses increment the counter. It is also possible to produce a down counter where the input pulses cause the binary number in the counter to decrease by one. The input pulses are said to decrement the counter.

A four bit binary down counter is shown in Figure 7-4. It is practically identical to the up counter described earlier. The only difference is that

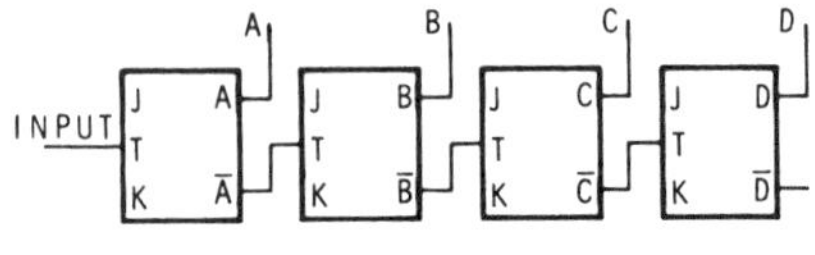

Figure 7-4
Four bit binary down counter.

the complement output rather than the normal output of each flip-flop is connected to the toggle input of the next flip-flop in sequence. This causes the count sequence to be the exact reverse of the up counter. The count sequence is illustrated in Figure 7-5. The waveforms associated with this counter are shown in Figure 7-6.

D	C	B	A
1	1	1	1
1	1	1	0
1	1	0	1
1	1	0	0
1	0	1	1
1	0	1	0
1	0	0	1
1	0	0	0
0	1	1	1
0	1	1	0
0	1	0	1
0	1	0	0
0	0	1	1
0	0	1	0
0	0	0	1
0	0	0	0

RECYCLE

Figure 7-5 Count sequence for four bit down counter.

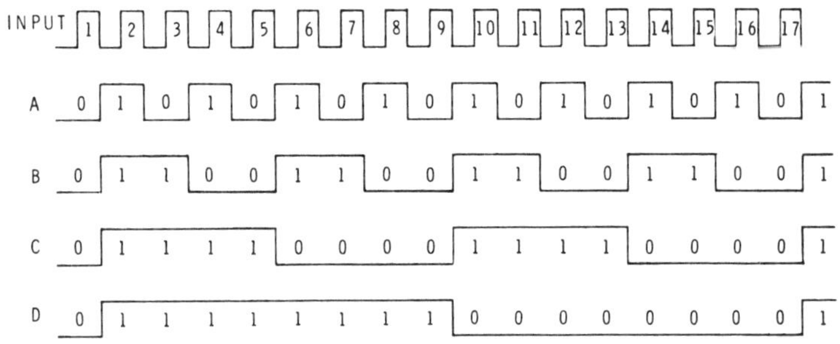

Figure 7-6
Input and output waveforms of a four bit binary down counter.

In analyzing the operation of this counter, keep in mind that we still determine the contents of the counter by observing the normal outputs of the flip-flops as we did with the up-counter. Assuming the counter is initially reset and its contents is 0000, the application of an input pulse will cause all flip-flops to become set. With the A flip-flop reset, its complement output is high. When the first input pulse is applied, the A flip-flop will set. As it does its complement output will switch from binary 1 to binary 0 thereby toggling the B flip-flop. The B flip-flop becomes set and its complement output also switches from binary 1 to binary 0. This causes the C flip-flop to set. In the same way the complement output of the C flip-flop switches from high to low thereby setting the D flip-flop. The counter recycles from 0000 to 1111.

When the next input pulse arrives, the A flip-flop will again be complemented. It will reset. As it resets the complement output will switch from binary 0 to binary 1. The B flip-flop ignores this transition. No further state changes take place. The content of the counter then is 1110. As you can see this input pulse causes the counter to be decremented from 15 to 14.

Applying another input pulse again complements the A flip-flop. It now sets. As it sets, its complement output switches from binary 1 to binary 0. This causes the B flip-flop to reset. As it resets, the complement output switches from binary 0 to binary 1. The C flip-flop ignores this transition. The new counter contents then is 1101 or 13. Again the input pulse caused the counter to be decremented by one. By using the table in Figure 7-5 and the waveforms in Figure 7-6, you can trace the complete operation of the 4 bit binary down counter.

Up-Down Counter. The up counting and down counting capabilities can be combined within a single counter as illustrated in Figure 7-7. AND and OR gates are used to couple the flip-flops. The normal output of each flip-flop is applied to gate 1. The complement output of each flip-flop is connected to gate 2. These gates determine whether the normal or complement signals toggle the next flip-flop in sequence. The count control line determines whether the counter counts up or down.

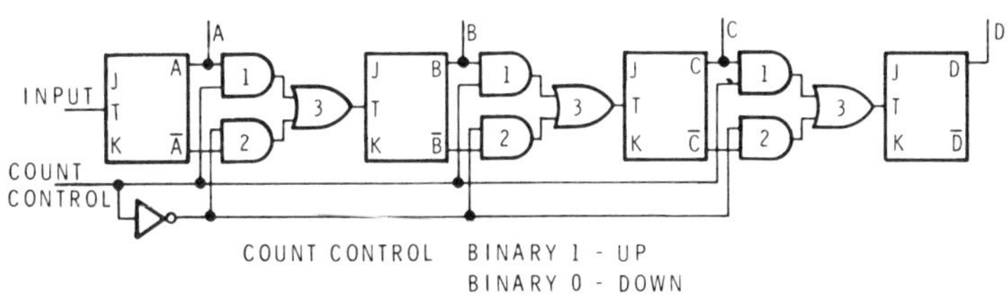

Figure 7-7
Binary Up/Down Counter.

If the count control input is binary 1, all gate 1s are enabled. The normal output of each flip-flop then is coupled through gates 1 and 3 to the T input of the next flip-flop. The counter therefore counts up. During this time, all gate 2s are inhibited.

By making the count control line binary 0, all gate 2s are enabled. The complement output of each flip-flop is coupled through gates 2 and 3 to the next flip-flop in sequence. With this arrangement, the counter counts down.

Synchronous Counters. The counters that we have discussed so far are known as ripple counters or asynchronous counters. The term ripple is derived from the fact that the flip-flops in the counter are cascaded with the output of one driving the input of the next. As the count pulses are applied to the first or input flip-flop, the count, in effect, ripples through the flip-flops. The term asynchronous comes as a result of the flip-flops not being controlled by a single common clock pulse. Synchronous digital circuits are ones in which all elements are synchronized to a master timing signal known as a clock.

The primary advantage of a ripple counter is its simplicity. Its primary limitation is its counting speed. The counting speed of a binary counter is limited by the propagation delay of the flip-flops in the counter. In a ripple counter, the propagation delay of the flip-flops is additive. Since each flip-flop in the counter is triggered by the preceding circuit, it can take a significant amount of time for an impulse to ripple through all of the flip-flops and change the state of the last flip-flop in the chain. This worse case condition occurs when all flip-flops in the counter change state simultaneously. Referring back to the waveforms for the four bit binary counter in Figure 7-3 you can see that this worse case condition occurs in two places. It occurs when the count changes from 0111 to 1000 and when the count changes from 1111 to 0000. If each flip-flop in the four bit circuit has a propagation delay of 50 nanoseconds, it can take as long as $4 \times 50 = 200$ nanoseconds for the D flip-flop to change state upon the application of an input pulse. Should the input pulses occur at a rate faster than 200 nanoseconds, the binary number stored in the counter will not truly represent the number of input pulses that have occurred. The counter state will lag the input signal. The upper frequency limit (f) of the ripple counter is approximately equal to

$$f = \frac{1}{n\,t} \times 10^9$$

where n is the number of flip-flops in the counter and t is the propagation delay time of a single flip-flop in nanoseconds.

For a flip-flop with a 35 nanosecond propagation delay in a four bit counter, the maximum counting speed is

$$f = \frac{1}{4(35)} \times 10^9 = 7.1428 \text{ MHz}$$

This means that the counter can count at speeds up to about 7 MHz without counting errors. At higher frequencies, the states of the flip-flops cannot keep pace with the rapid occurrence of the input pulses. The counter may actually lag several pulses depending upon the input frequency and the exact propagation delay. Counting errors can occur if the counting is periodically stopped and continued.

The direct solution to this problem, of course, is to use flip-flops with a lower propagation delay. Flip-flops are available to count at high frequencies up to approximately 500 MHz. The propagation delay is very small. Such flip-flops generally employ a non-saturating logic circuit such as ECL to achieve this fast counting rate. Such circuits are expensive and have high power consumption and are undesirable for many applications.

It is possible to reduce the propagation delay effects and increase the counting speed of a binary counter by using a special circuit arrangement. Counters employing this technique are known as synchronous counters.

A synchronous counter is one where all of the flip-flops are triggered simultaneously by a clock pulse or the signal to be counted. Since all flip-flops change state at the same time, the total propagation delay for the circuit is essentially equal to that of a single flip-flop. The propagation delays are not additive and therefore much higher counting speeds can be achieved.

A typical synchronous binary counter is shown in Figure 7-8. Notice that all of the flip-flop T inputs are connected together to a common count input line. This connection is what makes the counter synchronous. All of the flip-flops are synchronized to the input signal to be counted.

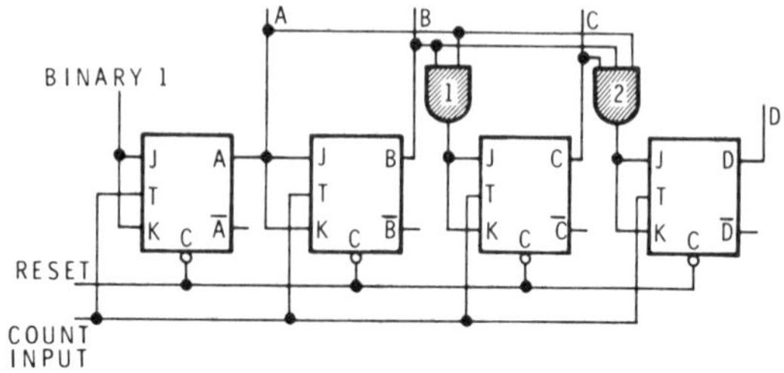

Figure 7-8
A synchronous binary counter.

The operation of the flip-flops is controlled by the states of the JK inputs. As you recall, the J and K inputs can be used as controls for the flip-flop. Up to this point in our discussion of counters we have assumed that the JK inputs were open or connected to a binary 1 level. This essentially enables the flip-flop and permits it to be toggled or complemented each time the trailing edge of an input signal appears on the T input line. If the JK inputs are brought to binary 0, the signal applied to the T input will be ignored. The flip-flop will simply remain in the state to which it was set prior to making the JK input lines low. By using the JK input lines we can then enable or inhibit the toggling of the flip-flops.

In Figure 7-8 the JK inputs on the A flip-flop are connected to binary 1 to enable the flip-flop permanently. Each time a count input signal appears the flip-flop will toggle or change state just as the A flip-flop on the ripple counters discussed earlier operated.

The JK inputs to the B flip-flop are controlled by the normal output of the A flip-flop. This means that the only time that the B flip-flop can change state is when the output of the A flip-flop is binary 1. The JK inputs to the C flip-flop are controlled by the normal outputs of both the A and the B flip-flops. The A and B signals are ANDed together in gate 1 and its output used to control the JK inputs. In order for the C flip-flop to change state, both A and B must be set. The C flip-flop will change state at the first count pulse occurring after A and B are high.

The D flip-flop is controlled by the A, B and C flip-flops. The A, B and C signals are ANDed in gate 2 and the output of gate 2 used to control the JK inputs.

With the circuit arrangement shown in Figure 7-8, the counting sequence is identical to that given in the Table of Figure 7-2. The waveforms of Figure 7-3 are also applicable. In other words, this circuit is still a binary counter but its mode of operation is somewhat different.

The benefit of this binary counter can be best seen by analyzing the state changes in the flip-flop. Assume that the counter contains the number 0111. This means that the A, B and C flip-flops are set. The JK inputs to the B, C and D flip-flops are enabled. This means that upon the occurrence of the next count input pulse, all flip-flops will toggle. When this pulse occurs, flip-flops A, B and C will reset. The D flip-flop will be set. The new number in the counter will be 1000. The important point to note here is that all flip-flops change state simultaneously. The maximum delay between the occurrence of the count input pulse and the change of state of the outputs is only as long as the longest propagation time of the flip-flops in the circuit. All flip-flops of the same type will have approximately the same propagation delay.

Considering this same state change in the binary ripple counter, we can illustrate the effect of the accumulative propagation delay. With the number 0111 stored in the ripple counter of Figure 7-1, the output states will change as follows when an input count pulse is applied. The A flip-flop will change state first. As it does it will toggle the B flip-flop. The B flip-flop must then change state and it in turn will then toggle the C flip-flop. The C flip-flop will change state a short time later and it in turn will set the D flip-flop. Because of the finite propagation delay of each flip-flop the effect of an input count pulse does indeed ripple through the counter and it takes a specific amount of time for the correct binary number to appear in the counter.

In summary then we can say that the advantages of the synchronous counter over the ripple or asynchronous counter are as follows:

1. The synchronous counter is much faster. For the same type of flip-flop, the counting speed of the synchronous counter is significantly higher than that of the ripple counter.
2. All flip-flops in the synchronous counter change states at the same time. As the counter changes from one state to the next, there are no ambiguous states that occur because of the accumulative propagation delays.

Like the ripple counter, the synchronous counter can also be expanded to as many bits as required by the application. Each flip-flop in the counter must be controlled by all previous flip-flops in the counter through an AND gate as indicated. The higher order flip-flops will require AND gates with as many inputs as there are previous flip-flops. Keep in mind that the propagation delays of these gates while small will also have a minor effect on the counting speed of the circuit. Generally, the propagation delay of a gate is significantly lower than that of a flip-flop. This technique can also be extended to down counters as well.

Counter Control Functions. There are several common control functions that are often associated with the use of binary counters. These are reset and preset. Resetting a counter is a process of putting all of the flip-flops in the binary 0 state. In many counter applications it is necessary to clear, reset or zero the counter prior to the start of a counting operation. This process ensures that the counter starts its count sequence with no prior counts stored in the flip-flops. It ensures an accurate count of the input.

Resetting a counter is easily accomplished when JK flip-flops are used. The asynchronous clear input on the flip-flops as you recall are normally used to put the flip-flop into its binary 0 state. By bringing the clear input low, the flip-flop is reset. By connecting all of the asynchronous clear inputs together, all of the flip-flops will be reset to binary 0 simultaneously when a reset pulse is applied. Figure 7-9 shows a binary up-counter with all of the asynchronous clear inputs connected together. In order for the counter to perform normally, the reset line will rest in a high or binary 1 state. To reset the counter we momentarily bring this line to a binary 0. For typical TTL JK flip-flops, a pulse whose duration is 100 nanoseconds or more can be used to reset the counter.

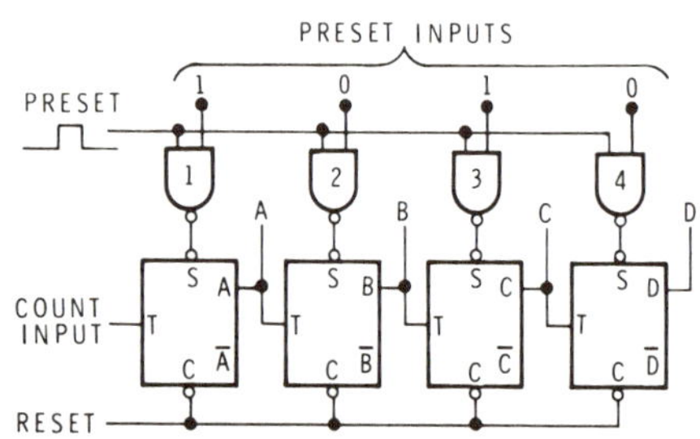

Figure 7-9 A binary counter with reset and preset inputs.

Presetting a counter is a process of loading some binary number into the counter prior to the count sequence. It is sometimes desirable to program a counter to start counting at a particular point. The point at which the counter is to start is determined and then loaded into the counter prior to the start of the count operations.

A counter can be preset by using the asynchronous set input when JK flip-flops are used. When this input on a JK flip-flop is made binary 0, the flip-flop is set. By first clearing the counter and then setting the desired flip-flops, any binary number can be preset into the counter. The circuit shown in Figure 7-9 can be used for this purpose. In order to preset the counter to a given number, the counter is first reset by applying a binary 0 to the reset input line. Next, the desired binary number is applied to the preset inputs. There is one input for each of the flip-flops in the counter. A parallel binary number from any source can be used. When the preset input line is made high, the outputs of the gates to which the parallel input number are applied will cause the asynchronous set lines on the JK flip-flops to assume the correct states to preset the number into the counter. For example, assume that we wish to preset the number 5 into the counter. To do this we would apply the binary number 0101 to the counter as indicated. The counter is then reset, and the preset line is brought high momentarily. The parallel inputs that are binary 0 are applied to gates 2 and 4. These binary 0 inputs hold the outputs of gates 2 and 4 high regardless of the state of the preset input. This keeps the set inputs to the B and D flip-flops high. With these inputs high the flip-flops are not affected. The binary 1 states of the desired input number are applied to gates 1 and 3. When the preset input goes high, the outputs of gates 1 and 3 will go low. This will cause flip-flops A and C to become set. The number 0101 is then stored in the counter.

The method of presetting a counter shown in Figure 7-9 is somewhat awkward in that it requires two operations. First the counter must be reset and then the desired preset number is loaded. It is desirable to have the preset operation take place with a single operation. This can be accomplished by the circuits shown in Figure 7-10. Only one JK flip-flop is shown to simplify the discussion. One of these circuit arrangements would be used on each flip-flop in a counter if presetting were desired. In Figure 7-10A, gates 1 and 2 are connected to the asynchronous set and clear inputs of the JK flip-flop. The desired parallel input (IN) is applied to gate 1. A preset line is tied to both gates 1 and 2. To the input line is applied a binary 0 or binary 1 state which will specify the state of the flip-flop after the preset input is enabled. When the preset input goes high, the JK flip-flop is preset to the desired state. For example, assume that the input is binary 1. When the preset line goes high, the output of gate 1 will go low. This will cause the set(s) input of the JK flip-flop to go low thereby setting the flip-flop. The low on the output of gate 1 will cause the output of gate 2 to remain high. This has no effect on the C input. A low input to gate 1 will reset the flip-flop. With a low input the output of gate 1 will be high. This high output does not affect the S input to the JK flip-flop. It does however enable gate 2. When the preset line goes high, the output of gate 2 will go low. This causes the JK flip-flop to be put in a binary 0 state. The preset operation takes place with only the single operation of applying the preset input pulse. Note that since the asynchronous inputs are used, the presetting operation will over-ride all other flip flop operations.

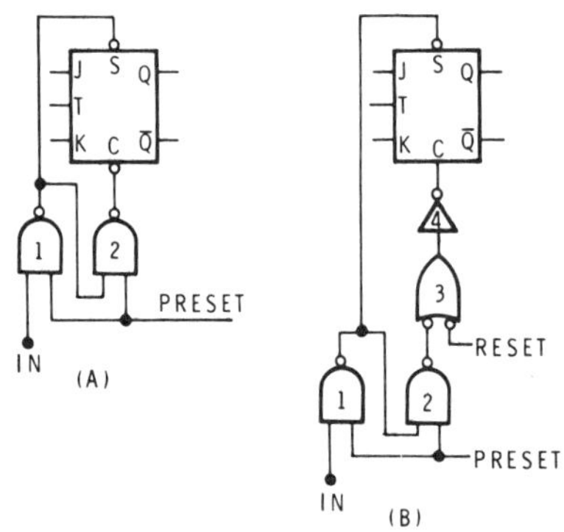

Figure 7-10
Methods of presetting a counter
(A) preset only (B) preset with reset.

Sometimes it is desirable to combine both the reset and preset functions in a counter. This can be accomplished with the circuit shown in Figure 7-10B. Here gates 1 and 2 perform the same basic operation as they do in the circuit of Figure 7-10A. They are used to preset the flip-flops to the desired state. Gate 3 and inverter 4 provide an ORing function to permit the reset function to be incorporated. If the reset input line is brought low, the output of gate 3 will go high and the output of inverter 4 will go low. This will reset the flip-flop. This occurs regardless of the conditions of the preset inputs.

To preset the flip-flop, the desired binary state is applied to the input line on gate 1. When the preset line is brought high, the flip-flop will be put into the states specified by the input. The operation is identical to the circuit in Figure 7-10A. The important point to note about this circuit is that the reset and preset operations are independent of one another. It is not necessary to reset the flip-flop prior to presetting it as was the case in the circuit of Figure 7-9.

An important point to remember is that these circuits for resetting and presetting flip-flops can be applied to any type of counter: synchronous, asynchronous, up-counter, down-counter, binary or BCD counter.

Typical Integrated Circuit Counters

While it is still sometimes necessary and desirable to implement counters with individual IC JK flip-flops, most counter applications can be met with a variety of available MSI integrated circuit counters. All of the most often used binary counters have been implemented in integrated circuit form thereby eliminating the necessity for designing such counters for each application. A variety of counter types and specifications are available. In designing digital equipment, it is desirable to first investigate the types of MSI IC counters available. In most cases you will find one suitable for your application. Only in rare cases where an unusual or peculiar type of counter for unique applications is required, will it be necessary for you to design a special counter. However, for such applications, versatile integrated circuit JK flip-flops are available.

In this section we are going to consider one of the most popular and widely used integrated circuit binary counters available.

Figure 7-11 shows the logic diagram of the 74193 TTL MSI IC counter. This is a four bit synchronous counter that can be used for either up or down counting. It also has separate clear or reset inputs as well as the ability to be preset from some external four bit parallel source. In other words, this particular device incorporates all of the features we have considered in a binary counter up to this point.

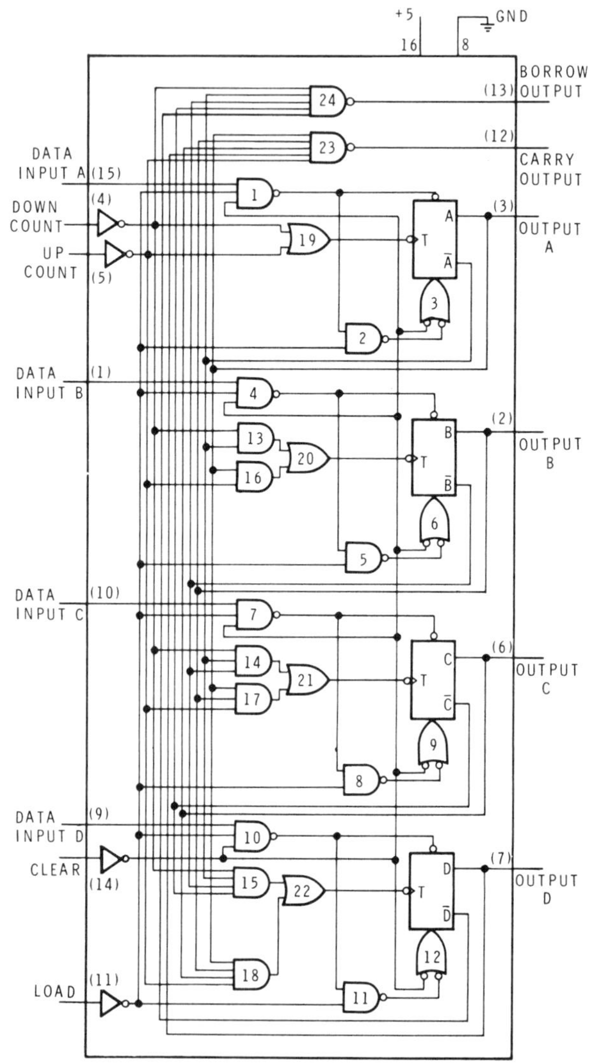

Figure 7-11
Type 74193 TTL MSI binary counter.

As shown in the diagram in Figure 7-11, the counter consists of four JK flip-flops. Gates 1 through 12 make up the logic circuitry used in the reset and preset operations. This circuitry is similar to the reset and preset circuit operations discussed previously. To reset this counter, you apply a high or binary 1 level to the clear input line. This forces all four flip-flops into the binary 0 state. The clear operation is asynchronous and overrides all other counter functions.

The counter can be preset by applying a parallel 4-bit binary number to the data inputs. Data input A is the LSB. When the load input is brought low, the 4-bit input number will be loaded into the flip-flops. This preset function is also asynchronous and will override any synchronous counting functions that occur.

The input pulses to be counted are applied to either the up-count input or the down-count inputs. Instead of having a single count input and an up/down control line as described in the previous discussion of up/down counters, this IC counter uses two separate inputs. To increment the counter, pulses are applied to the up-count input. To decrement the counter, pulses are applied to the down-count input. The counter changes state on the leading edge of the applied input pulse. In other words, it is the binary 0-to-binary 1 transition at the count input that causes the counter to change state. In order for the counter to operate properly, the unused count input must be in the high or binary 1 state while count pulses are applied to the other input. The up and down count sequences are identical to those considered previously.

Synchronous operation is used in this counter by having the input count pulses clock all flip-flops simultaneously so that all outputs change coincidentally with one another. Instead of controlling each flip-flop by use of the JK inputs as in the previously discussed synchronous counter, gates are used ahead of the T inputs to the flip-flops for this purpose. The outputs of the flip-flops control the states of the gates ahead of the T inputs to permit the count pulse to be applied at the appropriate time. Gates 16, 17, and 18 are used to control the application of the up count pulses to the T inputs. The up count input is applied to these gates simultaneously. Note that the outputs of the previous flip-flops are connected to the inputs of these gates in order to control when the count pulse is allowed to toggle the flip-flop. Gates 13, 14 and 15 perform the same function for the down-count operation. Gates 20, 21, and 22 are simply OR gates that permit either the up or down count pulses to appear at the flip-flop T inputs.

This counter has both carry and borrow output gates which are used for cascading these counters. When it is necessary to use a counter with more than 16 states, several of these ICs can be cascaded to provide counters whose lengths are some multiple of 4.

The carry output is developed by gate 23. This NAND gate monitors the normal outputs of the flip-flops. When all of the normal outputs are binary 1 and the up-count pulse occurs, the carry output line will go low. The duration of the carry output pulse is equal to the duration of the count input pulse. This pulse indicates that the counter is in its maximum count state (1111) and that the next count input pulse will cause it to recycle to 0000. This carry output pulse is connected to the count up input of the next counter in sequence when counters are cascaded.

The borrow output is produced by gate 24. This gate monitors the complement outputs of the four flip-flops. The down count input is also applied to gate 24. When the counter has been decremented to the 0000 state and with the down count input high, the output of gate 24 will go low. This indicates the counter is in its lowest count state (0000) and that upon application of the next count input pulse it will recycle to 1111. To cascade 74193 counters for down counting applications, the borrow output is connected to the down count input of the next counter in series.

As you can see, this device is a very flexible counting unit. It can perform nearly any of the required basic counting functions often encountered in digital work. Figure 7-12 illustrates the operation of this counter. The waveform for the clear, preset, count up and count down operations are shown.

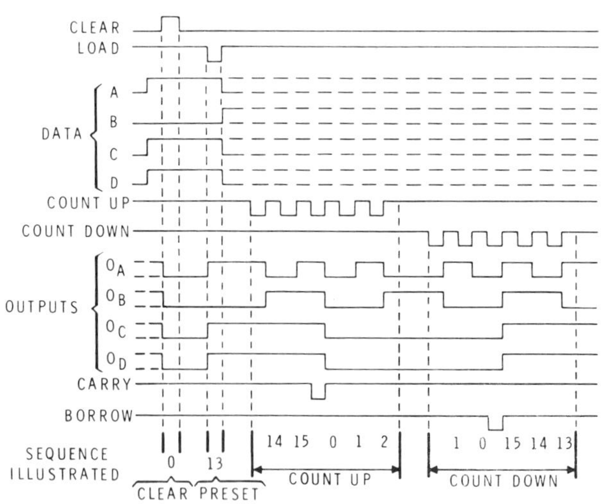

Figure 7-12
Typical clear, load, and count sequences for the 74193 counter.

BCD COUNTERS

A BCD counter is a sequential circuit that counts by tens. The BCD counter has ten discrete states which represent the decimal numbers 0 through 9. Because of its ten state nature, a BCD counter is also sometimes referred to as a decade counter.

The most commonly used BCD counter counts in the standard 8421 binary code. The table in Figure 7-13 shows the count sequence. Note that a four bit number is required to represent the ten states 0 through 9. These ten four bit codes are the first ten of the standard pure binary code. As count pulses are applied to the binary counter, the counter will be incremented as indicated in the table. Upon the application of a tenth input pulse, the counter will recycle from the 1001 (9) state to the 0000 state.

D	C	B	A
0	0	0	0
0	0	0	1
0	0	1	0
0	0	1	1
0	1	0	0
0	1	0	1
0	1	1	0
0	1	1	1
1	0	0	0
1	0	0	1

RECYCLE

Figure 7-13
Count sequence of 8421 BCD counter.

An asynchronous 8421 BCD counter constructed with JK flip-flops is shown in Figure 7-14. This counter will generate the BCD code given in the table of Figure 7-13. Note that the counter consists of four flip-flops like the four bit pure binary counter discussed earlier. The output of one flip-flop drives the T input to the next in sequence thereby, making this BCD counter a ripple or asynchronous type. Unlike the binary counter discussed earlier, however, this circuit has several modifications which permit it to count in the standard 8421 BCD sequence. The differences consist of a feedback path from the complement output of the D flip-flop back to the J input of the B flip-flop. Also a two input AND gate monitors the output states of flip-flops B and C and generates a control signal that is used to operate the J input to the D flip-flop. These circuit modifications in effect *trick* the standard four bit counter and cause it to recycle every ten input pulses.

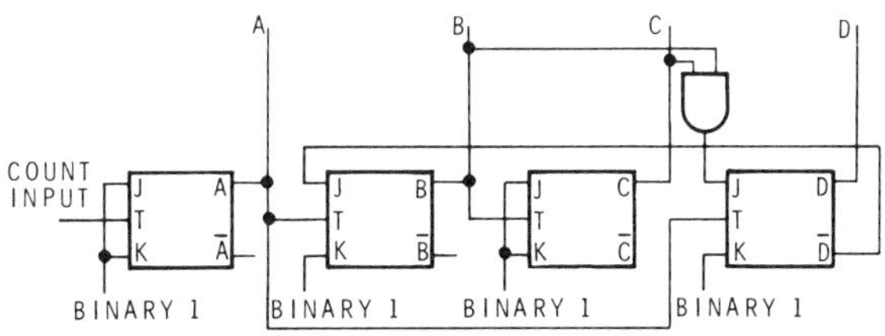

Figure 7-14
An asynchronous 8421 BCD counter.

The waveforms shown in Figure 7-15 illustrate the operation of the 8421 BCD counter. The count sequence is identical to that of the standard four bit pure binary counter discussed earlier for the first 8 input pulses. The operations that occur during the 9th and 10th pulses are unique to the BCD counter.

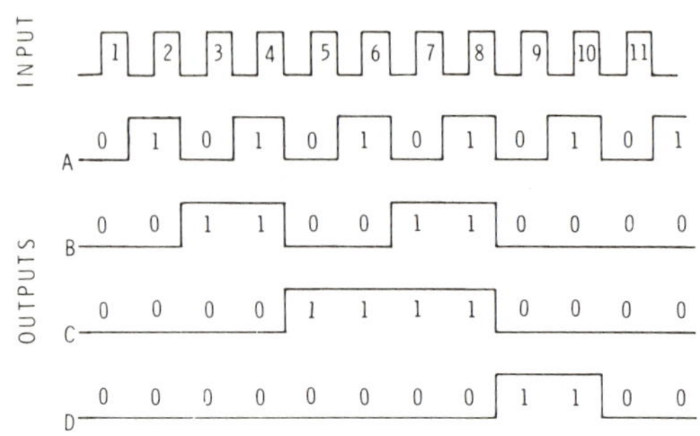

Figure 7-15
Waveforms of the 8421 BCD counter.

Assume that the counter in Figure 7-14 is initially reset. The outputs of flip-flops B and C will be binary 0 at this time. This makes the output of the AND gate low and causes the J input of the D flip-flop to be held low. The D flip-flop cannot be set by the toggle input from the A flip-flop until the J input goes high. Note also that the complement output of the D flip-flop which is binary 1 during the reset state is applied to the J input of the B flip-flop. This enables the B flip-flop permitting it to toggle when the A flip-flop changes state.

If count pulses are now applied, the states of the flip-flops will change as indicated in Figure 7-15. The count ripples through the first three flip-flops in sequence as in the standard 4 bit binary counter considered earlier. However, consider the action of the counter upon the application of the 8th input pulse. With flip-flops A, B, and C set and D reset, the B and C outputs are high thereby enabling the AND gate and the J input to the D flip-flop. This means that upon the application of the next count input that all flip-flops will change state. The A, B and C flip-flops will be reset while the D flip-flop is set. The counter state changes from 0111 to 1000 when the trailing edge of the 8th input pulse occurs.

In this new state, the B and C outputs are low therefore causing the J input to the D flip-flop to again be binary 0. With the J input 0 and the K input binary 1 and the D flip-flop set, the conditions are right for this flip-flop to be reset when the T input switches from binary 1 to binary 0. In addition, the complement output of the D flip-flop is low at this time thereby keeping the J input to the B flip-flop low. The B flip-flop is reset at this time and therefore the occurrence of a clock pulse at the T input will not affect the B flip-flop.

When the 9th input pulse occurs, the A flip-flop sets. No other state changes occur at this time. The binary number in the counter is now 1001. The transition of the A flip-flop switching from binary 0 to binary 1 is ignored by the T input of the D flip-flop.

When the 10th input pulse occurs, the A flip-flop will toggle and reset. The B flip-flop will not be affected at this time since its J input is low. No state change occurs in the C flip-flop since the B flip-flop remains reset. The changing of the state of the A flip-flop however, does cause the D flip-flop to reset. With its J input binary 0 and K input binary 1, this flip-flop will reset when the A flip-flop changes state. This 10th input pulse therefore causes all flip-flops to become reset. As you can see by the waveforms in Figure 7-15, the counter recycles from the 1001 (9) state to the 0000 state on the 10th input pulse.

Numerous variations of the basic BCD counter in Figure 7-14 are possible. Using the same basic count modifying techniques, synchronous BCD counter can be constructed. All of the flip-flops are toggled simultaneously by the common count input. As in the binary counter, the counting speed of the BCD counter can be significantly increased by this synchronous technique. In addition, it is also possible to construct a BCD down counter. Each time an input pulse is applied, the BCD counter is decremented. The count sequence is from 9 through 0.

Cascading BCD Counters. A single BCD counter has a maximum of ten discrete states and therefore can only represent the numbers 0 through 9. When the counter must count more than ten pulses, several BCD counters must be cascaded. Each BCD counter in the counting chain will represent one decimal digit. The number of BCD counters used determines the maximum count capabilities.

Figure 7-16 shows a counter chain with four BCD counters. Each BCD counter is represented by a single block to simplify the drawing. The count input line and the four flip-flop output lines are designated for each counter. In each case, the A output is the least significant bit and the D output is the most significant bit of that counter. The input BCD counter contains the least significant digit of the count contained in the counter. The most significant digit is represented by the counter on the far right. Since this counter contains four BCD counters, the maximum count capability is 9999.

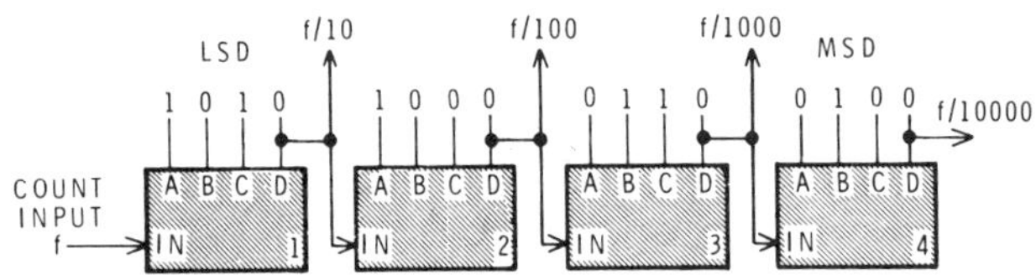

Figure 7-16
Cascading BCD counters to increase count capability.

As count input pulses are applied to BCD counter number 1, it will be incremented as indicated previously. The output states will change in accordance with the 8421 BCD code. Note that the most significant bit output (D) of this first counter is connected to the count input of the second BCD counter. Each time the input counter counts ten pulses and recycles it will trigger the next counter in sequence. By referring back to the waveforms in Figure 7-15, you can see that the trailing edge of the D output occurs on the trailing edge of the 10th input pulse. As this 10th input pulse occurs, the input counter recycles to 0 and the trailing edge increments the next counter in sequence to 1. The remaining counters in the chain are connected in the same way. As you can see then the counter does perform a decimal counting function with each BCD counter representing one of the decimal digits. The decimal contents of the counter can be determined by observing the flip-flop outputs. In Figure 7-16, the counter contains the decimal number 2615. This means that 2615 pulses have occurred at the input assuming the counter was initially reset.

The BCD Counter as a Frequency Divider. Like any counter, the BCD counter can also be used as a frequency divider. Since the BCD counter has ten discrete states, it will divide the input frequency by ten. The output of the most significant bit flip-flop in the BCD counter will be one tenth of the input frequency. From Figure 7-14, you can see that only a single output pulse occurs at the D output for every ten input pulses. While the D output does not have a 50 percent duty cycle, the frequency of the signal is nevertheless one tenth of the input frequency.

By cascading BCD counters, the input frequency can be reduced by any desired factor of ten. For example, in the counter of Figure 7-16, the output of the 4th BCD counter will be 1/10000th of the input frequency. The output of the third counter will be 1/1000th of the input frequency. The output of the second, of course, will be 1/100th of the input frequency. If an input signal of 2 MHz is applied to the counter in Figure 7-16, the D output of the MSD counter will be 200 Hz. When used as a frequency divider, the BCD counter is often referred to as a decade scaler.

Typical Integrated Circuit BCD Counter. The most widely used integrated circuit BCD counter is the type 7490A. This TTL MSI counter is an asynchronous or ripple counter that counts in the standard 8421 BCD code. The logic diagram of this counter is shown in Figure 7-17. Logically, it is identical to the BCD counter discussed earlier. The counter is made up of four JK flip-flops and the associated gating to permit the 8421 BCD sequence.

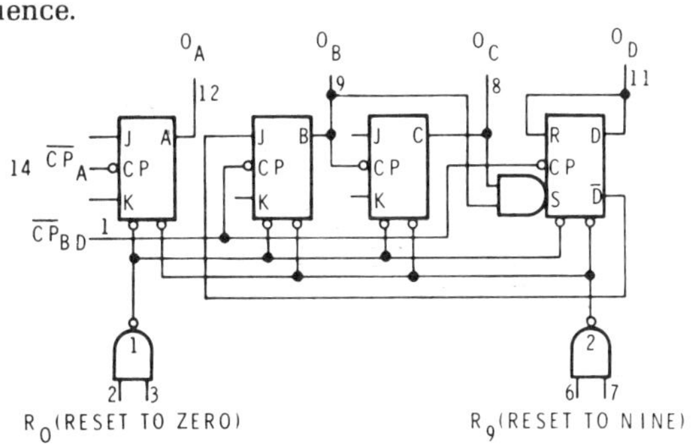

Figure 7-17

Logic diagram of 7490A BCD counter.

A look at the logic diagram of the 7490A counter in Figure 7-17 shows that the A flip-flop is not internally connected to the other three flip-flops. In order to produce an 8421 BCD count, the A output must be connected to the B input. This must be done externally. The input pulses to be counted then are applied to input A.

Gate 1 in Figure 7-17 is used to reset the flip-flop. When both inputs to gate 1 are high, all four flip-flops in the counter will be put in the binary 0 state. This permits two or more inputs to control the resetting of the flip-flop. Normally, only one input will be necessary and therefore both of the reset inputs can be simply tied together.

Gate 2 in the 7490A counter is used to preset the counter to the binary state 1001 or 9. When both inputs to gate 2 are high, a 9 will be preset into the counter. This particular function is useful in applications requiring arithmetic operations to be performed with BCD counters.

Despite the fact that the 7490A counter is an asynchronous counter, its maximum count frequency is approximately 32 MHz. This TTL MSI counter comes in a 14 pin DIP and is widely used in scaling and counting applications.

SPECIAL COUNTERS

Binary and BCD counters are by far the most commonly used counters in digital systems. Most counting applications can be implemented with MSI binary and BCD counters. However, there are some applications where a special counter may be required. It may be necessary to count in a peculiar sequence or to have the counter sequence through the states of some special code. In other applications it may be desirable to divide or scale an input frequency by some value other than even powers of two or by ten. Special counters can be constructed to perform all of these applications. Some typical examples are a counter that counts in the Gray code or a frequency divider that divides the input by seven.

Because of the flexibility of the JK flip-flop, such special counters are relatively easy to implement. The basic approach is to construct a standard binary counter and then by using feedback and input gating controls on the JK inputs a counter can be developed to count in any sequence with as many individual states as desired.

We refer to the number of discrete states that a counter can assume as the modulus of that counter. A modulo N counter is one that has N states. A BCD or decade counter is a modulo 10 counter since it can assume one of ten discrete binary states. The binary counters that we discussed earlier are modulo N counters where N is some power of two. A counter containing four flip-flops is a modulo 16 counter. As you have seen it is easy to construct a binary counter whose modulo is some power of two. BCD counters with a modulo of 10 are also easily assembled. However, there are other applications that require counters with modulos of other integer values.

Modulo 3 Counter. A modulo three counter is shown in Figure 7-18. Since the T inputs to both JK flip-flops are both connected to the count input, the circuit is synchronous. The feedback line from the B output back to the J input of the A flip-flop causes the circuit to count by three.

To determine the operation of the counter assume that both flip-flops are initially in the reset state. The A and B outputs are binary 0. The low output of flip-flop A holds the J input to flip-flop B low. When the trailing edge of the first input pulse occurs, flip-flop A will set. The $\overline{B}$ output is holding the J input to A flip-flop high thereby enabling it to be set when the proper T input pulse occurs. When the first input pulse occurs, A is set. The B flip-flop is not affected. With the A output high, the J input to the B flip-flop is now high thereby permitting that flip-flop to toggle upon application of the next input pulse. The $\overline{B}$ output of the B flip-flop is still high thereby continuing to enable the J input of the A flip-flop.

When the trailing edge of the second input pulse occurs, the A flip-flop will again toggle. This time it will reset. At the same time the B flip-flop will set. The J input to the B flip-flop is now low while the J input to the A flip-flop is also low. When the next clock pulse occurs, the B flip-flop will reset. The A flip-flop having been previously reset simply remains reset. On the application of the trailing edge of the third input pulse, the counter state cycles back to its original reset condition. If further input pulses are applied, the counter will simply repeat the cycle just described. The count sequence for the modulo three counter is shown in Figure 7-18. The waveforms showing the operation of the modulo three counter are illustrated in Figure 7-19. Trace through the operation of the circuit using these waveforms as a guide to be sure you fully understand how the circuit operates. This circuit can be used for simple counting applications requiring a three state counter. The circuit can also be used as a divide by 3 scaler. The output of either the A or B flip-flops has a frequency that is one third of the count input frequency. A single output pulse occurs for every three input pulses as indicated by the waveforms in Figure 7-19.

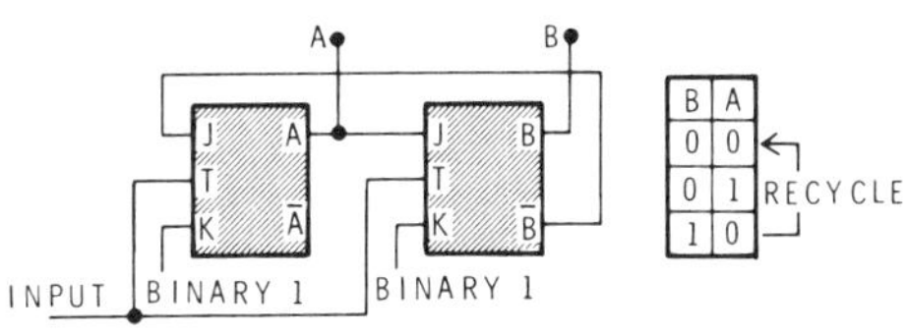

Figure 7-18
A modulo 3 counter and its count sequence.

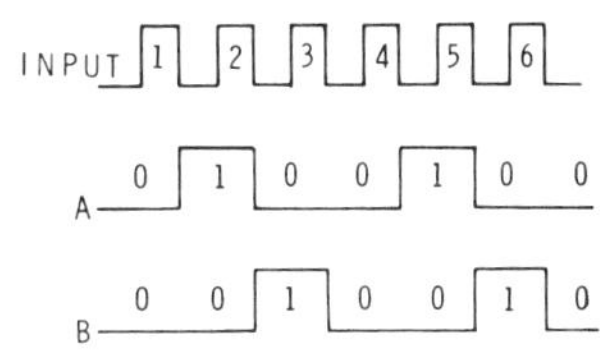

Figure 7-19
Waveforms for the modulo 3 counter.

By cascading a modulo three counter with additional JK flip-flops, modulo six and modulo twelve counters are easily formed. This is shown in Figure 7-20. By connecting a JK flip-flop to the output of the modulo three counter, a modulo six counter is formed. The modulo three will cycle through its three states twice, once with the JK flip-flop C reset and again with the JK flip-flop C set. This produces a total of six discrete states as indicated in the table shown in Figure 7-21A. If you will look closely at the sequence of the states you will find that this does not correspond to the standard binary code. Since the code produced is not the pure binary code or the BCD code, it is referred to as an unweighted code.

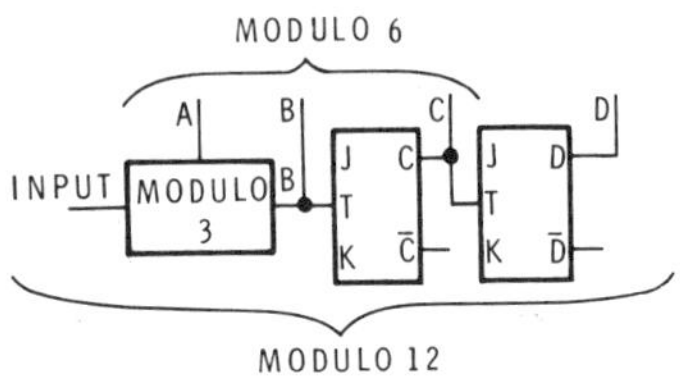

Figure 7-20

Forming modulo 6 and modulo 12 counters using a modulo 3 circuit as a base.

(A)

C	B	A
0	0	0
0	0	1
0	1	0
1	0	0
1	0	1
1	1	0

RECYCLE

COUNT SEQUENCE MODULO 6 COUNTER

(B)

D	C	B	A
0	0	0	0
0	0	0	1
0	0	1	0
0	1	0	0
0	1	0	1
0	1	1	0
1	0	0	0
1	0	0	1
1	0	1	0
1	1	0	0
1	1	0	1
1	1	1	0

RECYCLE

COUNT SEQUENCE FOR MODULO 12 COUNTER

Figure 7-21

Count sequence for a (A) modulo 6 counter and a (B) modulo 12 counter.

By adding another JK flip-flop (flip-flop D) as shown in Figure 7-20, a modulo 12 counter is formed. Here the modulo 6 counter cycles through its six states two times, once while the D flip-flop is reset and again while it is set. This produces the 12 discrete states shown in Figure 7-21B. The counter shown in Figure 7-20 can be used to perform frequency division by three, six or twelve.

Counters with a modulo 6 and a modulo 12 can also be formed by putting the JK flip-flops ahead of the modulo three counter instead of after it. The counter thus formed still has six or twelve states respectively. However, the binary code produced by this arrangement is different from that obtained by the connection shown in Figure 7-20. When using this arrangement in frequency divider applications, it generally doesn't matter which connection is used as the output frequency will always be either $^{1}/_{6}$th or $^{1}/_{12}$th of the input frequency. Where the specific code sequence is critical however, these two arrangements should be carefully considered. It should be pointed out that the counter shown in Figure 7-20 produces a 50 percent duty cycle when the C and D outputs are used. By putting the modulo 3 counter after the cascaded JK flip-flops, a duty cycle other than 50 percent will be produced.

Modulo 5 Counter. A modulo five counter is shown in Figure 7-22. The counter produces five distinct three bit states. Synchronous operation is obtained by applying the count input to the T input of all three JK flip-flops. A combination of feedback and external logic gates are used to cause the counter to count by five.

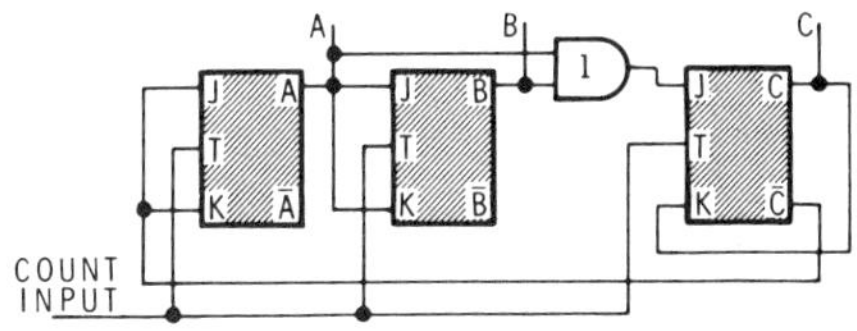

C	B	A
0	0	0
0	0	1
0	1	0
0	1	1
1	0	0

RECYCLE

COUNT SEQUENCE

Figure 7-22

Modulo 5 counter and its count sequence.

The count sequence for this circuit is shown in Figure 7-22. Note that the count sequence simply recycles for the application of each five input pulses.

The input and output waveforms of the modulo five counter are illustrated in Figure 7-23. Compare the output states of these waveforms for each of the flip-flops to the count sequence table in Figure 7-22.

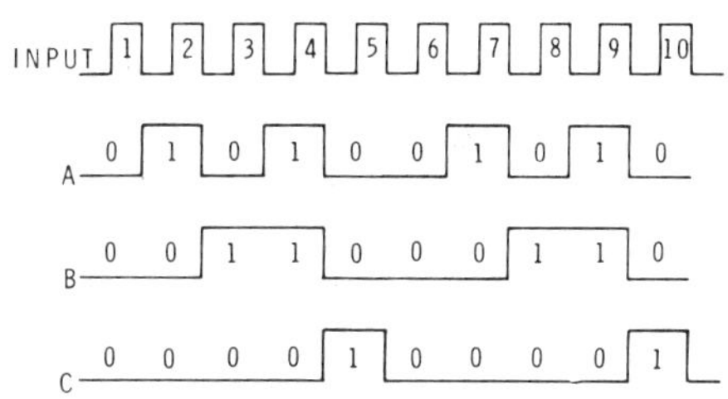

Figure 7-23

Waveforms for the modulo 5 counter.

The detailed operation of this counter is not included here. It will be excellent practice for you to reason out the count sequence of this circuit yourself. Remembering the operation of the JK flip-flop and using the count sequence table and the waveforms in Figure 7-23, trace the operation of the counter for all five states until it recycles. As a starting point, assume that all three flip-flops are initially reset.

If you will refer back to Figure 7-17 showing the diagram of the type 7490A decade counter you will see that flip-flops B, C and D in this circuit form a modulo 5 counter by themselves. As indicated earlier a separate input is used for this circuit. Normally, it is tied to the output of the A flip-flop to produce BCD counting. However, the modulo 5 section of this counter can be used independently by simply applying the count input to the B terminal. The A flip-flop is not used.

Modulo N Counters with MSI. While JK flip-flops can be interconnected by the use of feedback and external logic gates to form a counter with any desired modulo, the availability of MSI integrated circuit counters greatly simplifies the construction of modulo N counters. The type 74193 TTL MSI synchronous up/down counter discussed earlier is an excellent choice for implementing modulo N counters.

Figure 7-24 shows the type 74193 TTL MSI counter connected as a modulo N counter. The counter is connected to count down. For frequency divider applications, it does not matter whether the counter

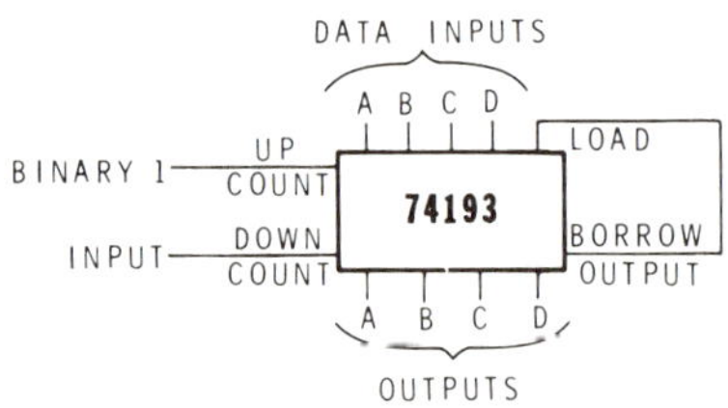

Figure 7-24
A type 74193 TTL MSI
Counter used as a modulo N counter.

counts up or down. The up count input is held to a binary 1 while the input pulses are applied to the down count input. The borrow output line which essentially detects the 0000 state is connected back to the load input line of the counter. To the four parallel data input lines is connected a binary word that will determine the modulus of the counter. The modulus of the counter is equal to the binary equivalent of the decimal number applied to the data inputs. This is indicated by the table in Figure 7-25. For example to obtain a modulo 7 counter, the binary number for the decimal number seven (0111) is connected to the data inputs.

MODULO	DATA INPUTS			
	D	C	B	A
1	0	0	0	1
2	0	0	1	0
3	0	0	1	1
4	0	1	0	0
5	0	1	0	1
6	0	1	1	0
7	0	1	1	1
8	1	0	0	0
9	1	0	0	1
10	1	0	1	0
11	1	0	1	1
12	1	1	0	0
13	1	1	0	1
14	1	1	1	0
15	1	1	1	1

Figure 7-25
Parallel data input code and related decimal modulos.

In operation, the counter is preset to the binary number applied to the data inputs. The counter is then decremented by the input pulses. It down counts in binary until the zero state is reached. At this time the borrow output line goes low and again presets the counter to the number applied to the data inputs. This sequence repeats as long as pulses are applied to the count input.

As soon as the borrow output line goes low, the binary number applied to the data inputs will be immediately (asynchronously) loaded into the counter. Of course as soon as the new number is loaded into the counter, the borrow output will disappear since the counter state will no longer be 0000. What this means is that the duration of the borrow output pulse must be long enough in order to ensure that the data input is loaded before the zero output state disappears. In order to ensure this condition happens, the input duration of the clock pulse must be greater than the total propagation delay of the gates in the counter associated with presetting the number on the data inputs. Recall from the previous discussion of the operation of the 74193 counter that the borrow output is also derived from the down count input signal.

Even though the borrow output pulse disappears as soon as the counter is preset to the data input states, the internal propagation delays of the circuit are such that all flip-flops become preset before the borrow output disappears. The load input signal must propagate through both the flip-flops and the gates in the circuit. Since the propagation delay through the various circuits in the counters vary from one device to the next, it is possible that erratic operation can occur if the propagation delays are too short. The reliability of the counter can therefore be improved by adding some propagation delay between the borrow output and the load input pins. This can be accomplished by cascading a number of inverters between these two pins on the circuit. Be sure to use an even number of inverters so the proper polarity binary signal will be applied.

SHIFT REGISTERS

Another widely used type of sequential logic circuit is the shift register. Like a counter, a shift register is made up of binary storage elements. While flip-flops are the most commonly used storage element in shift registers, other types of circuits are also used. The storage elements in a shift register are cascaded in such a way that the bits stored there can be moved or shifted from one element to another adjacent element. All of the storage registers are actuated simultaneously by a single input clock or shift pulse. When a shift pulse is applied, the data stored in the shift register is moved one position in one of two directions. The shift register is basically a storage medium where one or more binary words may be stored. However, because of the ability to move the data one bit at a time from one storage element to another makes the shift register valuable in performing a wide variety of logic operations.

Shift Register Operation

The illustration in Figure 7-26 shows how a shift register operates. Here the shift register consists of four binary storage elements such as flip-flops. The binary number 1011 is currently stored in the shift register. Another binary word, 0110, is generated externally and is available to the shift register serially. As shift pulses are applied, the number stored in the register will be shifted out and lost while the external number will be shifted into the register and retained.

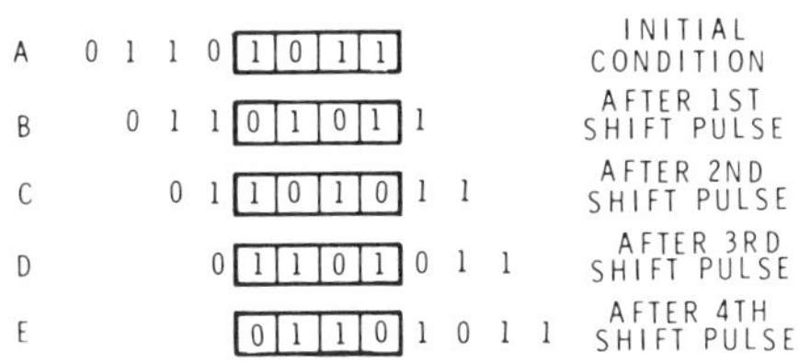

Figure 7-26
Operations of a shift register.

The initial conditions for this shift register are illustrated in Figure 7-26A. After one clock pulse, the number stored in the register initially is shifted one bit position to the right. The right most bit is shifted out and lost. At the same time, the first bit of the externally generated serial number is shifted in to the left most position of the shift register. This is illustrated in Figure 7-26B. The remaining three illustrations in C, D, and E show the results after the application of additional shift pulses. After four shift pulses have occurred, the number originally stored in the register has been completely shifted out and lost. The serial number appearing at the input on the left has been shifted into the register and now resides there.

This figure illustrates several important points about a shift register. First, it indicates that the basic shift register operations are serial in nature. That is data is moved serially, a bit at a time, into and out of the register. Most shift register operations are serial operations but many circuits are provided with both parallel inputs and parallel outputs. Such shift registers permit data to be preset in parallel and data to be read out in parallel. The ability to combine both serial and parallel operations makes the shift register an ideal circuit for performing serial to parallel and parallel to serial data conversions.

Another important point to note is that the data is shifted one bit position for each input clock or shift pulse. Clock pulses have full control over the shift register operation. In this shift register, the data was shifted to the right. However, in other shift registers it is also possible to shift data to the left. The direction of the shift is determined by the application. Most shift registers are of the shift right type.

The shift register is one of the most versatile of all sequential logic circuits. It is basically a storage element used for storing binary data. A single shift register made up of many storage elements can be used as a memory for storing many words of binary data. Such memories are referred to as serial memories since the data stored in them is entered and removed in serial form.

Shift registers can also be used to perform arithmetic operations. Shifting the data stored in a shift register to the right or to the left a number of bit positions is equivalent to multiplying or dividing that number by a specific factor. As indicated earlier, the shift register is also widely used for serial to parallel and parallel to serial data conversions. Shift registers can also be used for generating a sequence of control pulses for a logic circuit. And in some applications shift registers can be used to perform counting and frequency dividing.

In this section you are going to study the basic operation of a shift register. One of the most commonly used types of shift registers is the bipolar shift register which is made up of flip-flops. These can be constructed with individual JK flip-flops or are available in a variety of configurations in MSI form. Another type of shift register is the MOS shift register. These registers made with MOSFETs are available in two basic types, static and dynamic. Static shift registers are made up of MOSFET flip-flops. Dynamic shift registers are made up of storage elements that take advantage of the unique characteristics of MOSFETs, namely their high impedance and capacitive nature. Because of the small size of the MOSFET structure, many storage elements can be made on a single chip of silicon. Therefore, long shift registers capable of storing many words can be made very small and economical. Both types of shift registers are widely used in digital systems.

Bipolar Logic Shift Registers

Shift registers constructed from bipolar logic circuits such as TTL and ECL circuits are usually implemented with JK flip-flops. Type D flip-flops can also be used, but shift registers implemented with JK flip-flops are far more versatile. A typical shift register constructed with JK flip-flops is shown in Figure 7-27. The serial input data and its complement are applied to the JK inputs of the input (A) flip-flop. From there the other flip-flops are cascaded with the outputs of one connected to the JK inputs of the next. Note that the clock (T) input lines to all flip-flops are connected together. The clock or shift pulses are applied to this line. Of course, since all flip-flops are toggled simultaneously, the shift register is definitely a synchronous circuit. Note that the asynchronous clear inputs on each flip-flop have been connected together to form a reset line. Application of a low or binary 0 level to this line causes the shift register to be reset. This shift register can also be preset by using any of the techniques described earlier for presetting binary counters. Data applied to the input will be shifted to the right through the flip-flops. Each clock or shift pulse will cause the data at the input and that stored in the flip-flops to be shifted one bit position to the right.

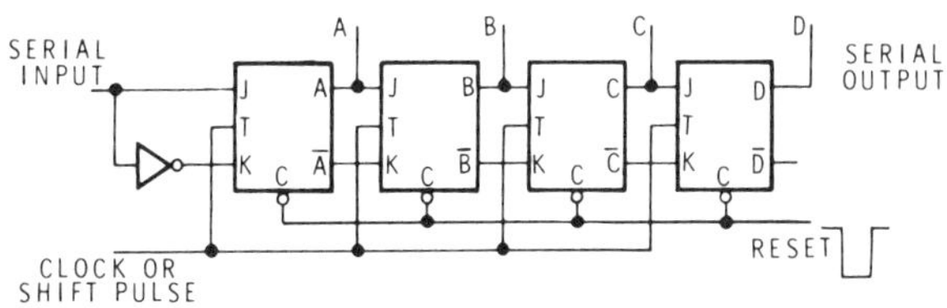

Figure 7-27
Four bit shift register made with JK flip-flops.

The waveforms in Figure 7-28 illustrate how a serial data word is loaded into the shift register of Figure 7-27. As the waveforms show, the binary number 0101 in serial form occurs in synchronization with the input clock or shift pulses. In observing the waveforms in Figure 7-31, keep in mind that time moves from left to right. This means that the clock pulses on the right occur after those on the left. In the same way, the state of the serial input shown on the left occurs prior to the states to the right. With this in mind, let's see how the circuit operates.

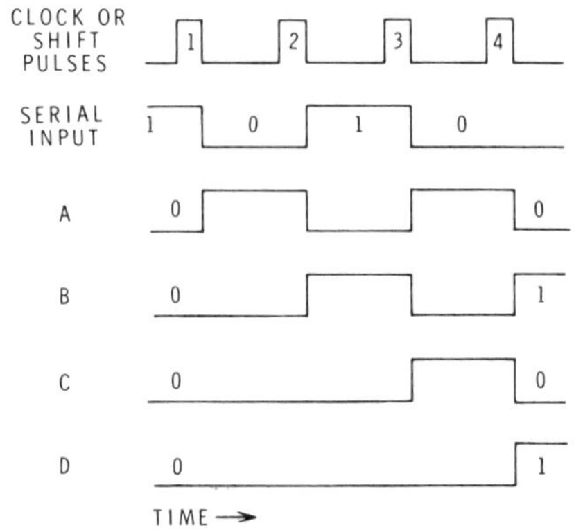

Figure 7-28

Waveforms illustrating how the serial binary number 0101 is loaded into a shift register.

Note that the shift register is originally reset. The A, B, C, and D outputs of the flip-flops therefore are binary 0 as indicated in the waveforms. Prior to the application of the number 1 shift pulse, the serial input state is binary 1. This represents the first bit of the binary word to be entered. On the trailing edge of the first clock pulse, the binary one will be loaded into the A flip-flop. The JK inputs of the A flip-flop are such that when the clock pulse occurs the flip-flop will become set. This first shift pulse is also applied to all other flip-flops. The state stored in the A flip-flop will be transferred to the B flip-flop. The states stored in the B and C flip-flops will be transferred to the C and D flip-flops respectively. Since all flip-flop states are initially zero, naturally no state changes in the B, C or D flip-flops will take place when the first clock pulse occurs.

After the first clock pulse the A flip-flop is set while the B, C, and D flip-flops are still reset. The first clock pulse also causes the serial input word to change. The clock or shift pulses are generally common to all other circuits in the system and therefore any serial data available in the system will generally be synchronized to the clock.

The input to the A flip-flop is now binary 0. When the trailing edge of the second clock pulse occurs, this binary 0 will be written into the A flip-flop. The A flip-flop which was set by the first clock pulse causes the JK inputs to the B flip-flop to be such that it will become set when the second clock pulse occurs. As you can see by the waveforms, when the second clock pulse occurs, the A flip-flop will reset while the B flip-flop will set. The 0 state previously stored in the B flip-flop will be transferred to the C flip-flop, and the C flip-flop state will be shifted to the D flip-flop. At this point the first two bits of the serial data word have been loaded into the shift register.

The serial input is now binary 1 representing the third bit of the serial input word. When the third clock pulse occurs the A flip-flop will set. The zero previously stored in the A flip-flop will be transferred to the B flip-flop. The binary 1 stored in the B flip-flop will now be shifted into the C flip-flop. The D flip-flop remains reset.

The serial input to the A flip-flop is now binary 0. When the trailing edge of the fourth clock pulse occurs, the A flip-flop will reset. The binary 1 stored there previously will be transferred to the B flip-flop. The 0 stored in the B flip-flop will be shifted into the C flip-flop. The binary 1 in the C flip-flop now moves to the D flip-flop. As you can see, after four clock pulses have occurred, the complete four bit binary word 0101 is now shifted into the register as indicated by the states shown in the waveforms. A glance at the flip-flop output waveforms will show the initial binary 1 bit moving to the right with the occurrence of each shift pulse.

While we have illustrated the operation of the shift register with only four bits, naturally as many flip-flops as needed can be cascaded to form longer shift registers. Most shift registers are made up to store a single binary word. In most modern digital systems, shift registers have a number of bits that is some multiple of four.

While shift registers are readily implemented with JK or D type flip-flops, in most applications, MSI shift registers are used. MSI shift registers are available in four and eight bit sizes. Here we are going to discuss a typical four bit MSI TTL shift register. You will see how it can perform the basic shift right operation described previously and how it can be connected to shift left or be parallel loaded.

Figure 7-29 shows the logic diagram of a type 7495 TTL shift register. It is made up of four flip-flops with the appropriate gating on the JK inputs. A mode control input line controls this input gating. The mode control also operates the clock input selection circuitry. Two clock signals can be used depending upon the state of the mode control.

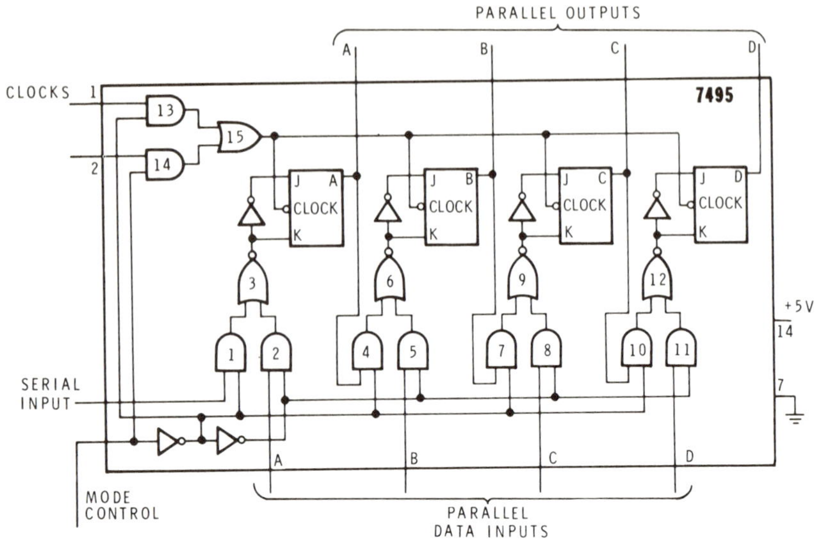

Figure 7-29

Logic diagram of 7495 TTL MSI shift register.

When the mode control input is a binary 0, gates 1, 4, 7, and 10 are enabled. This causes the shift register to be set up to perform the basic shift right operation. Serial input is applied to gate 1 and passes through gate 3 to the JK inputs of the flip-flop. The output of the A flip-flop is connected to the inputs of the B flip-flop through gates 4 and 6. In the same way, the outputs of the B and C flip-flops are connected to the inputs of the C and D flip-flops respectively. Also notice that a binary 0 on the mode control input also enables gate 13. This permits clock pulse 1 to pass through gates 13 and 15 to control the flip-flops. In this mode, the shift register performs the standard shift right operation.

When the mode control is placed in the binary 1 state, gates 2, 5, 8, and 11 are enabled. With these gates enabled and gates 1, 4, 7, and 10 inhibited, the parallel data inputs are recognized. Also note that a binary 1 on the mode control input also enables gate 14 so that clock pulse 2 can actuate the flip-flops. When a clock pulse occurs, an external 4-bit parallel word will be loaded into the flip-flops. In this mode, the shift register can be parallel loaded or preset to some desired value.

Shift left operations can be performed with the mode control in the binary 1 position if the parallel data input lines are connected to the appropriate flip-flop outputs. To perform a left shift, the D flip-flop output is connected to the C data input, the C output is connected to the B data input, and the B flip-flop output is connected to the A data input. The D data input is used as the serial input line for external data. When clock pulses are applied to gate 14, data will be shifted left from D to C, C to B, and B to A. External serial data is shifted into D. With this connection, the mode control input acts as a shift right, shift left control line.

SHIFT REGISTER APPLICATIONS

The shift register is basically a storage element for a binary word. Data can be conveniently shifted into and out of the register in serial form. Despite its simplicity, the shift register has many applications. In this section we are going to look at some of the more popular uses for shift registers.

Serial to Parallel Conversion. One of the most common applications of a shift register is in serial to parallel or parallel to serial data conversions. There are many occasions in digital systems where it is necessary to convert an existing parallel word into a serial pulse train. The shift register can readily perform both of these operations.

Figure 7-30 shows how a shift register is used in serial to parallel and parallel to serial data conversion. In Figure 7-30A the shift register is shown being loaded by a parallel input. The number 1101 is preset into the shift register. Then four clock pulses are applied so that the data is shifted out serially. In Figure 7-30B, the shift register is used for serial to parallel conversion. Here the serial input number 1001 is shifted into the shift register by four clock pulses. Once the data is in the register, the outputs of the individual flip-flops may be monitored simultaneously to obtain the parallel ouput data.

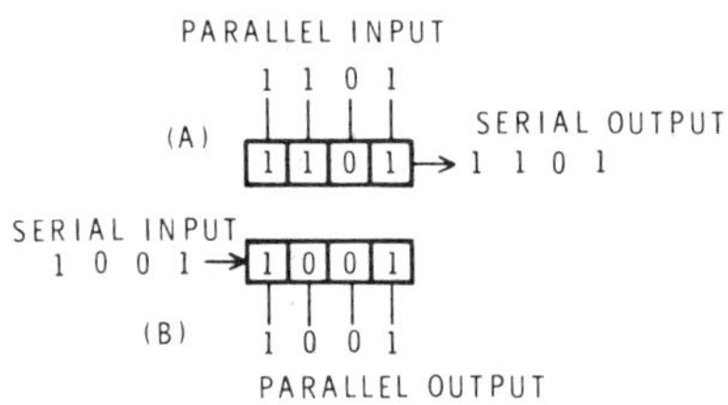

Figure 7-30
Parallel to serial and serial to parallel data conversions with a shift register.

Scaling Operations. A shift register can be used to perform arithmetic operations such as multiplication and division. Shifting the binary number stored in a shift register to the left has the effect of multiplying that number by some power of 2. Shifting the data to the right has the effect of dividing the number in the register by some power of 2. Shifting operations are a simple and inexpensive way of performing multiplication and division with binary numbers.

Figure 7-31A shows a shift register containing a binary number. Assuming the binary point is located to the far right, we can then convert the binary number in to its decimal value. In the initial condition state this number is 3. Now if we perform a shift left operation and move the binary word one position to the left you can see immediately from Figure 7-31B that a new binary number has been formed. As we shift the data to the left one bit position binary 0s are entered on the right. The new number stored in the register is 6. As you can see, shifting the data one position to the left has the effect of multiplying the original number by 2.

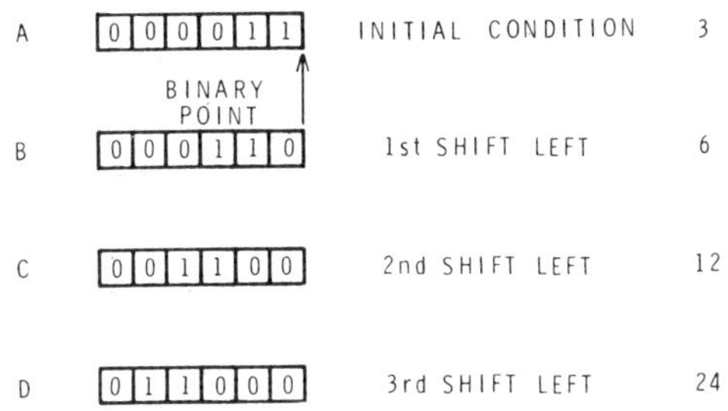

Figure 7-31
Multiplication by factors of 2 by shifting left.

If we perform another shift left operation the number in the register becomes as shown in Figure 7-31C. Converting it to decimal we see that it is 12. The additional shift left operation has again multiplied the number in the register by 2. Two shift left operations have caused the initial number to be multiplied by four.

A third shift left operation will further verify this effect. The number shown in Figure 7-31D is now 24. Shifting the word one additional position to the left has multiplied the number previously by 2. With three shift left operations the initial number in the register has been multiplied by eight. As you can see then, the factor by which the number in the register is multiplied is some power of 2. The multiplying factor is 2^N where N is the number of shift left operations that take place. With three shift left operations as in Figure 7-31, the original number is multiplied by $2^3 = 8$. The important thing to remember about this operation is the shift register must be large enough to accommodate the largest number expected by multiplication. In addition, note the binary 0s are shifted into the right most position as the data is moved to the left.

Division by some power of 2 is accomplished by shifting right. This is illustrated in Figure 7-32. Here the number initially stored in the register is 20.0. Note in this 6 bit register the binary point has been specified as being between the right most bit position and the next most significant bit.

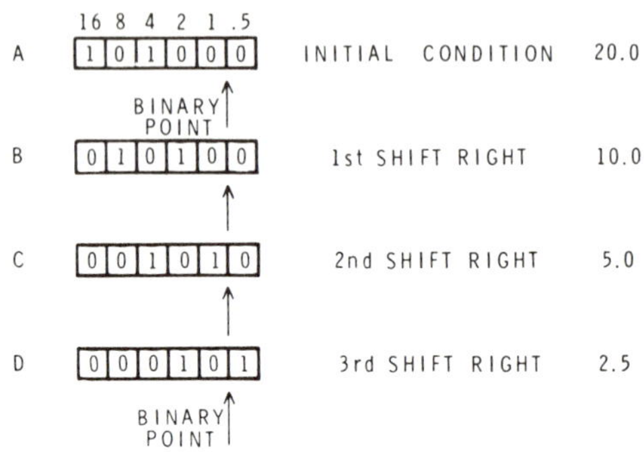

Figure 7-32

Division by powers of 2 by shifting right.

Applying one clock pulse causes the data in the register to be shifted one position to the right. Evaluating the new decimal value of this number we find that it is 10.0. The number initially stored in the register has been divided by 2. Applying another shift pulse causes the data to be moved one more position to the right. Evaluating this number we find that it is 5.0. Again the number has been divided by 2 while the overall division accomplished by two shift pulses is 4. Performing a third shift right operation moves the data again one bit to the right. Evaluating the new number we find it to be 2.5. Again the data has been divided by 2.

The value by which the number in the register is divided is some power of 2. The divide ratio is 2^N where N is the number of shift right operations. Here the original number 20 was divided by $2^N = 2^3 = 8$ or $20 \div 8 = 2.5$. Again an important consideration is that the register be large enough to accommodate the numbers resulting from the scaling operations by shifting. If the register is not large enough, data will be shifted out of the register and lost thereby making the arithmetic operation incorrect.

Shift Register Memory. Shift registers, because they store binary data, are often used for temporary memories in digital equipment. Such a shift register memory is usually capable of storing at least one binary word. Many such memories are made long enough to store many binary words. In such an application, there are two operations that the memory must perform. First, it must be able to accept data and then store it. In other words, we must be able to write new data into the memory. Second, we must be able to retrieve that data or read it out upon command. One of the requirements of the memory is that when we do read the data out that it will not be lost. A shift register can accomplish both of these operations by providing external logic circuitry as shown in Figure 7-33. Here an eight bit shift register is used to store a single binary word. The external control gates are used to select a read or write operation. To store or write data into the memory, the write/recirculate line is set to the binary 1 state. This causes gate 1 to be enabled. Serial data applied to the other input of gate 1 is passed on into the shift register. Once data is stored in the memory, the write/recirculate control line is set to the binary 0 condition. This inhibits gate 1 and prevents other data appearing at the input from being recognized by the shift register. Instead gate 2 is now enabled. Note that the shift register output is connected to gate 2. As shift pulses are applied, the data in the register is shifted out serially and may be used by some external circuit. As the data is being shifted out it is also being shifted back into the input of the shift register through gates 2 and 3. In other words, we are recirculating the data in the register. The read out operation is accomplished and at the same time the data is restored. When we wish to write a new word into the shift register, the write/recirculate line is again made binary 1 and 8 shift pulses are applied.

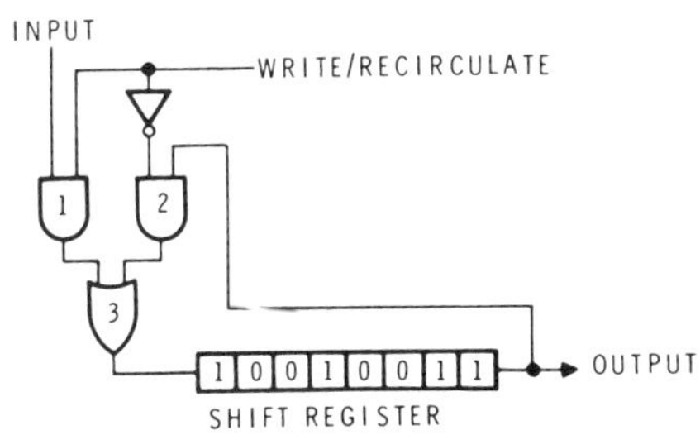

Figure 7-33
Shift register memory.

Sequencer/Ring Counter. Another popular application of the shift register is a sequencer or ring counter. Many logic circuits require a sequence of equally spaced timing pulses for initiating a series of operations. A properly connected shift register can be used to serve this purpose.

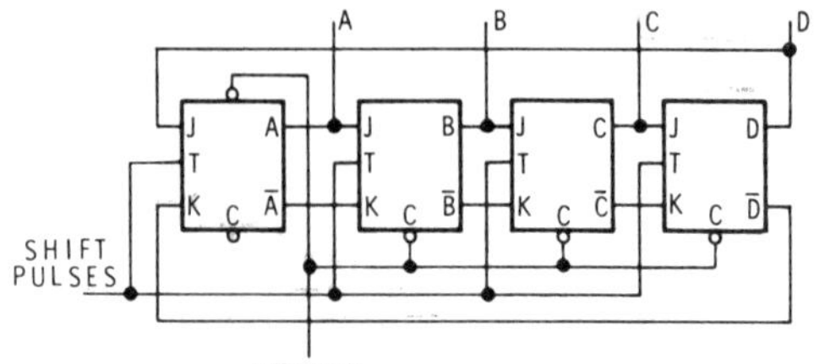

Figure 7-34
A shift register connected as a sequencer or ring counter.

A shift register connected as a ring counter is shown in Figure 7-34. This is the standard shift register circuit we discussed earlier. However, note that the outputs of the shift register from the D flip-flop are connected back to the JK inputs of the A flip-flop. This provides feedback that causes the shift register to continue to rotate or sequence the data in the register. The asynchronous set and clear inputs of the flip-flop are used to preset a single bit in the shift register. A binary 0 level applied to the preset line causes the A flip-flop to become set and the other three flip-flops to become reset. As shift pulses are applied, the binary 1 in the A flip-flop is shifted to the B, C and D flip-flops, and then back around to the A flip-flop. This sequence repeats as long as the shift pulses are applied. Since the one bit initially programmed into the shift register is continually recirculated, the name ring counter definitely applies. The waveforms and the state table in Figure 7-35 show the operation of the four bit shift register as a ring counter.

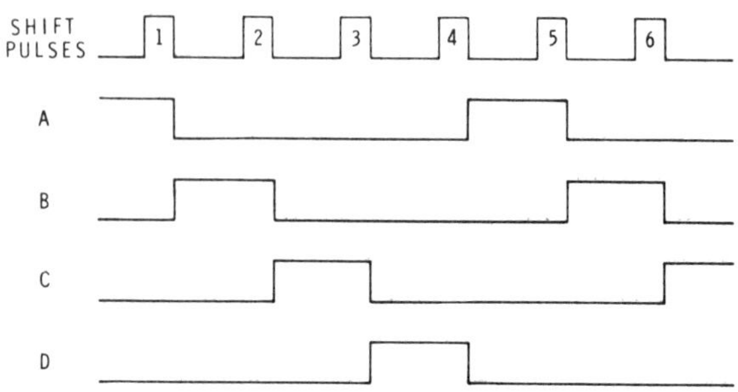

STATE	A	B	C	D
0	1	0	0	0
1	0	1	0	0
2	0	0	1	0
3	0	0	0	1

RECYCLE

Figure 7-35
Waveforms for a 4 bit shift register ring counter. State table for ring counter.

A close look at the waveforms for the four bit ring counter in Figure 7-35 shows that the output at any flip-flop has a frequency one fourth that of the shift pulse frequency. In other words, the shift register connected as a ring counter produces frequency division by four since four shift input pulses occur for each flip-flop output pulse. As you can see then the shift register when connected as a ring counter can be used as a frequency divider. By presetting one of the flip-flops in the shift register, frequency division by any integer value can be accomplished by simply using as many flip-flops as needed. For example, to divide by seven, seven flip-flops would be required in the ring counter.

One of the disadvantages of the ring counter circuit shown in Figure 7-34 is that it must be initially preset in order for it to function properly. When power is initially applied to this circuit, the flip-flops in the register can come up in any state. If any random state is allowed in the shift register, the operation previously described will not occur. The preset operation must take place to initially load one of the flip-flops with a binary 1 and the others with binary 0. This disadvantage can be overcome by using a self correcting circuit as shown in Figure 7-36. Here the NAND gate monitors the outputs of flip-flops A, B, and C. The output of the NAND gate and its complement are connected to the JK inputs of the A flip-flop. With this circuit arrangement, any of the sixteen possible states can occur in the shift register and the shift register will automatically correct itself to where only one of the flip-flops is set and the others are reset. Regardless of the initial states of the flip-flops, after a maximum of two shift pulses, the contents of the shift register will automatically be corrected so that only a single flip-flop is set. From that point on the shift register will simply recirculate the single one bit stored there.

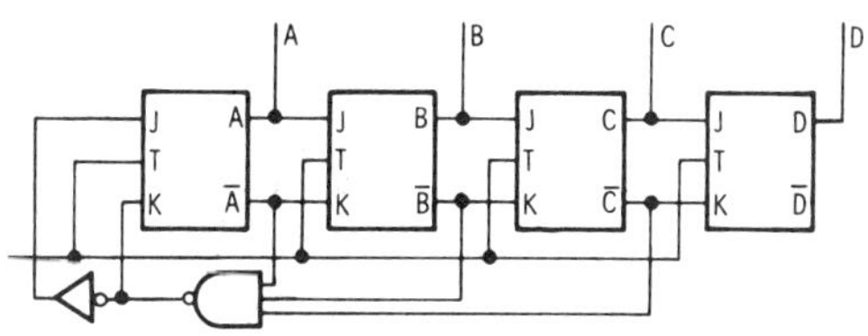

Figure 7-36

Self correcting shift register ring counter.

To use the ring counter described here as a sequencer, the output pulses from the flip-flops are simply connected to the logic circuits whose sequence is to be controlled. Since the shift pulses are normally derived from a fixed frequency clock, the timing interval of the shift registers is precise and thereby permitting very exacting control of the external logic circuits. In addition, it is not necessary to use all of the pulses derived by the shift register. Only those required by the circuit need be used. Keep in mind that as more circuits must be controlled or sequenced, additional flip-flops can be added to the shift register to produce them.

Counters. When connected as a ring counter, shift registers can also be used as a counter. In some special applications they may replace binary counters for certain operations. The four bit ring counter circuit described previously has four distinct states and these states repeat or recycle as the clock pulses are applied.

A popular type of shift register counter is the Johnson counter shown in Figure 7-37. While any number of flip-flops may be cascaded to form a Johnson counter, a five bit circuit is often used. Note that like in the ring counter, the output of the last flip-flop is connected back to the inputs of the first flip-flop in order to recirculate the data. However, note that in this case the normal output of the E flip-flop is connected to the K input and the complement output is connected to the J input. Because of this connection the Johnson counter is often referred to as a twisted ring counter or switch tail counter. With this arrangement, the counter or shift register will have 2N different states where N is the number of flip-flops in the shift register. The five bit Johnson counter shown in Figure 7-37 therefore, will have ten discrete states.

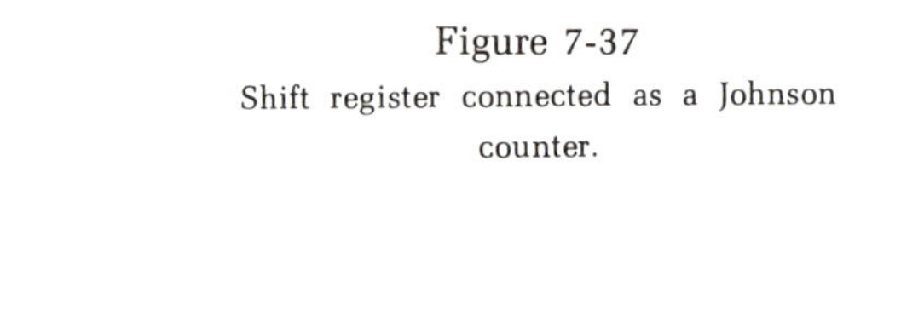

Figure 7-37

Shift register connected as a Johnson counter.

Like the ring counter shift register discussed previously, it is necessary to initialize the counter after power is applied in order to have the counter operate properly. A self correcting circuit similar to that shown in Figure 7-36 can be used to initialize the circuit. Otherwise initialization can be accomplished by simply resetting all of the flip-flops to zero. When shift pulses are applied, the binary state sequences shown in the table of Figure 7-38 are generated. Note the ten individual states. The counter recycles every tenth input pulse. Since the Johnson counter has ten individual states, it is often used as a divide by 10 frequency divider. Other counting operations can often be implemented with this type of counter. However, because of the nonweighted codes generated by the Johnson counter and the ring counter shift register, many counting applications are difficult and inconvenient to implement.

STATE	A	B	C	D	E
0	0	0	0	0	0
1	1	0	0	0	0
2	1	1	0	0	0
3	1	1	1	0	0
4	1	1	1	1	0
5	1	1	1	1	1
6	0	1	1	1	1
7	0	0	1	1	1
8	0	0	0	1	1
9	0	0	0	0	1

RECYCLE

Figure 7-38
State table for Johnson counter.

MOS SHIFT REGISTERS

In applications requiring shift registers with a limited bit capacity, bipolar integrated circuits such as TTL or ECL are used. CMOS shift registers are also available. Such registers are generally used for storing a single binary word. This word may be as small as four bits but could be as long as 32 bits in some applications. Where more data storage is needed, additional flip-flops or MSI shift registers can be cascaded to form memories which can be used for storing many binary words. When implemented with standard bipolar and CMOS integrated circuits, such shift register memories become very large and expensive. In these applications MOS shift registers can be of value.

MOS integrated circuit shift registers using P or N channel enhancement mode MOSFETs contain many storage elements. Because of the very high component density and very low power dissipation of the MOS structure, very large shift registers can be made on a very tiny silicon chip. MOS shift registers with thousands of storage elements are available for memory applications. Such shift registers are commonly used to store many binary words in a serial format. For example, to store 128 eight bit binary words we would need an $8 \times 128 = 1024$ bit shift register.

MOS shift registers this large are very practical and are commonly used for temporary data storage and for delay operations. Any application requiring the temporary storage of a large volume of binary data can use MOS shift registers. In addition, most MOS LSI shift registers are of the serial in/serial out type. The parallel loading and readout of data is not generally performed with MOS shift registers. These MSI and LSI circuits provide a very economical memory source.

There are two basic types of MOS shift registers: static and dynamic. A static shift register is one in which the clock may be stopped without loss of data. This is the type of shift register that we have discussed in the previous sections. Clock signals are applied to shift data into or out of the register. When the clock pulses are stopped, the data in the shift register is retained in the storage elements. Data is not lost if the clock is stopped.

In another type of MOS shift register the data will be lost if the clock is stopped. This type of shift register is known as the dynamic type. Because of the characteristics of the storage element used in a dynamic register, the clock pulses must run continuously if data is to be retained. Data must be continuously recirculated or refreshed in order to prevent its loss. Naturally, the static shift register is generally more desirable from a standpoint of operation and convenience. However, static MOS shift registers are generally more complex and consume much more power.

Dynamic shift registers can be made smaller and more simply, can operate at higher speeds and have far lower power dissipations. Such trade offs must be considered when designing with MOS shift registers. Most MOS shift registers are fully compatible with TTL and CMOS circuits. No special interfacing is required.

Dynamic MOS Shift Register. The basic storage element in a MOS shift register, whether it is dynamic or static, is the capacitance that exists between the gate and the channel of the MOSFET transistors used. While this capacitance is very small (on the order of several tenths of a picofarad), the high impedance nature of the MOSFET permits a charge voltage to be placed on this capacitance and retained for a relatively long period of time. The impedance between the gate and the source of an enhancement mode MOSFET is on the order of 10^{15} ohms or greater. Such a high impedance is virtually an open circuit and has a minimum effect on the gate capacitance. If we apply a voltage between the gate and source of a MOSFET, the gate capacitance will charge and remain there until it leaks off through the very high impedance between the source and gate. In high quality MOSFETs, this discharge time can be as long as one millisecond.

The storage element circuit used in a MOS shift register is a MOSFET inverter. The input capacitance of the inverter transistor stores the data. Figure 7-39 shows how two MOSFET inverters (I1 and I2) are combined with MOSFET transmission gates (Q_1 and Q_2) to form a one bit storage element. The input data is applied to inverter I1 through transmission

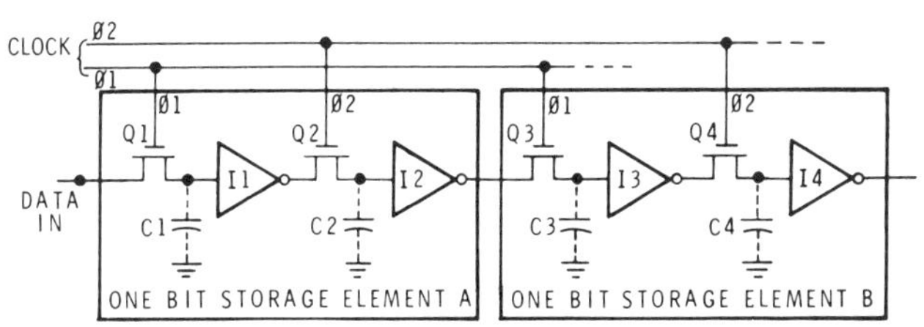

Figure 7-39
Dynamic MOS Shift Register.

gate Q1. MOSFET Q1 is simply used as an on/off switch to connect the input to capacitor C1 and disconnect it. The output of inverter I1 is connected to the input of inverter I2 through transmission gate Q2 which again is used as a simple on/off switch. The switching of the transmission gate transistors is controlled by two clock signals designated phase 1 ($\phi 1$) and phase 2 ($\phi 2$). These two phase clock signals are illustrated in Figure 7-40. Note that when $\phi 1$ is on $\phi 2$ is off and vise-versa.

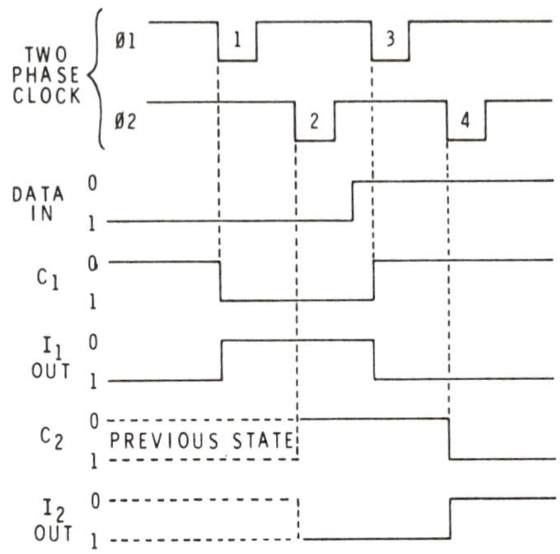

Figure 7-40
Clock and circuit waveforms for the dynamic MOS shift register.

For our discussion here let's assume the use of P channel MOS circuits where a binary 0 is equal to zero volts or ground and a binary 1 is some negative voltage level.

The data to be stored (written) into storage element A is applied to the data input line. Assume that we apply a binary 1 input which is some negative voltage level. When clock pulse $\phi 1$ occurs, transmission gate Q1 will conduct. This will cause capacitor C1 to charge to the input voltage. Applying a binary 1 input voltage to inverter I1 causes a binary 0 level to appear at its output. After the occurrence of the $\phi 1$ clock pulse, capacitor C1 retains the charge and acts as the input voltage source for inverter I1.

The $\phi 2$ clock pulse occurs next. This causes transmission gate Q2 to conduct. The state of the output of I1 is, therefore, transferred to capacitor C2. This is a binary 0, so capacitor C2 has zero charge. The input to inverter I2 is a binary 0. The output of I2, therefore, is a binary 1. After one $\phi 1$ clock pulse and one $\phi 2$ clock pulse, the binary 1 that was at the input to storage element A appears at the output of storage element A. On the next cycle of $\phi 1$ and $\phi 2$ clock pulses, this binary 1 value will be transferred to the next storage element (B) of the shift register. Any new data appearing at the input of the first storage element will be shifted in at this time. The waveforms in Figure 7-40 illustrate the storage of a binary 1 in element A and its transfer to element B as a binary 0 is entered.

The inverters in Figure 7-39 can be any one of several different types of MOS logic inverters. Figure 7-41 shows the two types most commonly used. In Figure 7-41A, a static inverter is used. Here Q2 is the inverter element while Q1 is a MOSFET biased into conduction to act a load resistance. This type of device dissipates power because Q1 is continuously conducting. In long MOS shift registers this power dissipation is additive and can produce a significant amount of heat.

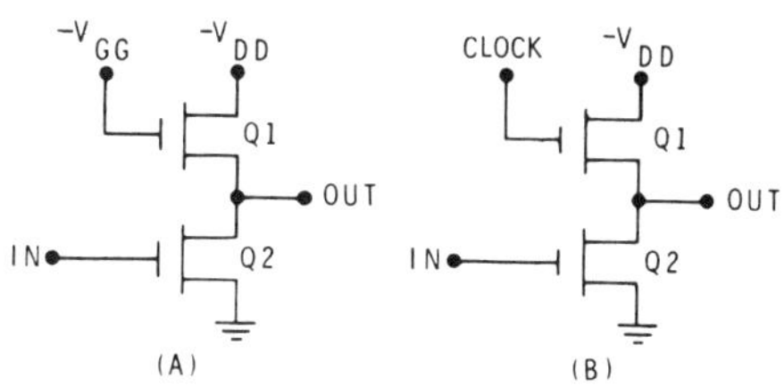

Figure 7-41
MOS inverters (A) static and (B) clocked.

Another type of inverter shown in Figure 41B uses a clock load device. Here, Q2 is the inverter element while Q1 is the load element. Q1, however, does not conduct, except during the time a clock pulse is applied. When such an inverter is used in the dynamic shift register circuit at Figure 7-39, the load element is clocked during $\phi 1$ or $\phi 2$ along with the associated output transmission gate. For example, the load element in I1 would be clocked at $\phi 2$ time while the load element in I2 would be clocked at $\phi 1$ time. This arrangement greatly reduces the power dissipation of the device.

In order to keep the data from being lost during the shifting process, the clock must run continuously and the data must be continually recirculated from output to input. Write/recirculate logic at the input of the shift register is used to select the mode of operation. Should the clock pulses stop, the data stored as charges on the capacitances in the circuits will leak off and be lost. The loss of data can occur in only several hundred microseconds depending upon the circuitry used. For that reason the minimum clock rate is approximately 5 kHz for most typical dynamic MOS shift registers. Dynamic shift registers with minimum clock rates in the 100 Hz range are available.

Static MOS Shift Registers. There are some applications where it is desirable to stop the clock in a digital system. For such applications, static MOS shift registers can be used. Such shift registers employ storage elements that retain the data even after the clock has stopped. In addition, it is not necessary to continually recirculate the data in order to retain it.

A typical static storage element for a static MOS shift register is shown in Figure 7-42. It also uses the gate capacity of a MOSFET inverter for temporary data storage. The storage element uses two such inverters. One inverter is transistor Q3 with its load device Q2. The other inverter is Q7 with its load device Q6. These two inverters are cross coupled through transmission gates Q4 and Q5. When Q4 and Q5 conduct, a latch type flip-flop is formed. Naturally, a latch will retain data even when the clocking signals are removed as long as Q4 and Q5 conduct. Transistor Q1 is used as a transmission gate to load data into the storage element. To explain the operation of the circuit, assume that P channel devices and negative logic are used.

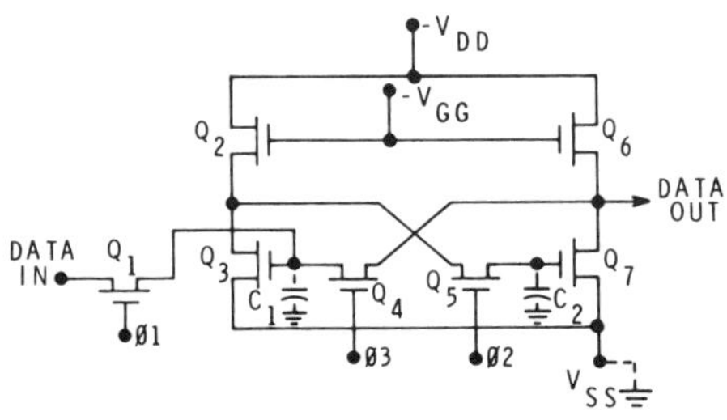

Figure 7-42
One bit MOS storage element for a static MOS shift register.

The proper operation of this shift register requires a three phase clock signal. These clock signals are shown in Figure 7-42. The $\phi 3$ clock is a delayed replica of the $\phi 2$ clock signal. In some static MOS shift registers, the $\phi 3$ and sometimes the $\phi 2$ and $\phi 3$ clock pulses are generated on the chip. Therefore, the circuit requires only a single or double phase external clock for proper operation.

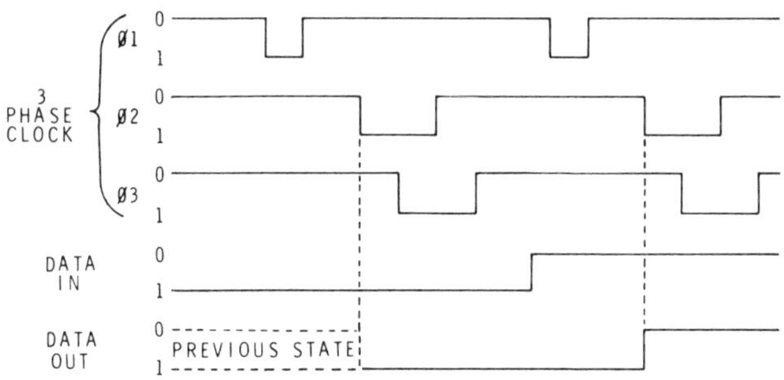

Figure 7-43
Waveforms for static MOS shift register.

To load data into the circuit, the desired bit is applied to the data input line. When the $\phi 1$ clock occurs, transmission gate Q1 conducts. This causes the gate capacitance of Q3 (C1) to charge to the proper state. If a binary 1 is applied to the data input, C1 will assume a negative charge. This negative charge is applied to inverter Q3. Q3 conducts and its drain goes low representing a binary 0.

The $\phi 2$ clock occurs next and Q5 conducts, thereby, transferring the state of Q3 to capacitor C2. In our example, this is a binary 0. The output of inverter Q7 then is as a negative level binary 1. The $\phi 3$ clock signal occurs and Q4 conducts. This applies the negative signal back to the gate of Q3 to keep it on. At this time, the data is latched. The shift register will remain in this state until the state of the input has changed and the next clocking cycle has been completed.

CLOCKS AND ONE SHOTS

Most sequential logic circuits are driven by a clock. The clock is a periodic signal that causes logic circuits to be stepped from one state to the next. The clock steps the sequential circuits through their normal operating states so that they perform the function for which they were designed.

The clock signal is generated by a circuit known as a clock oscillator. Such an oscillator generates rectangular output pulses with a specific frequency, duty cycle and amplitude. The most commonly used clock oscillator is some form of astable multivibrator. Such circuits can be constructed with discrete components or with logic gates.

Another circuit widely used to implement sequential logic operations is the one shot. The one shot or monostable multivibrator produces a fixed duration output pulse each time it receives an input trigger pulse. The duration of the pulse is usually controlled by external components. By cascading one shot circuits, a wide variety of sequential circuits can be implemented.

Clock Oscillator Circuits

Practically all digital clock oscillator circuits use some form of astable multivibrator circuit for generating a periodic pulse waveform. Such a circuit has two unstable states, and the circuit switches repeatedly between these two states. Both discrete component and integrated circuit clocks are used in digital equipment.

Discrete Component Circuits. The most commonly used clock oscillator is the astable multivibrator circuit shown in Figure 7-44. It consists of two transistor inverters Q1 and Q2 with the output of one connected to the input of the other. Resistors R2 and R3 are used to bias the transistors into saturation. Capacitor C1 and C2 couple the output of one inverter to the input of the other. In normal operation, one transistor is conducting while the other is cut off. The frequency of oscillation is determined by the values of R2, R3, C1 and C2.

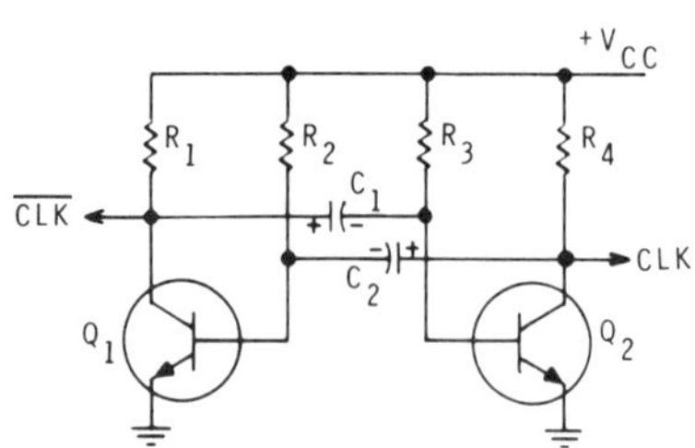

Figure 7-44

Astable multivibrator clock oscillator.

Assume that Q2 is conducting and Q1 is cut off. Capacitor C1 then charges through the emitter-base junction of Q2 and R1 to the supply voltage $+V_{CC}$. Capacitor C2 which was previously charged to the supply voltage with the polarity shown keeps Q1 cut off as it discharges through resistor R2. As soon as it discharges to zero it begins to charge in the opposite direction. When the charge on C2 reaches about 0.7 volt, Q1 conducts. As soon as Q1 conducts, it effectively connects the positive side of C1 to ground. This puts a negative voltage between the base and the emitter of Q2 causing it to switch off quickly. C1 then discharges through R3. At this time C2 is recharged through the emitter-base junction of Q1 and R4. As soon as the charge on C1 has reached zero, it will begin to charge to the supply voltage. However, as soon as the voltage is high enough, Q2 again conducts and the state of the circuit reverses. This cycle continues at a rate determined by the discharge time of C1 and C2 which in turn depends upon the values of R2 and R3. R2 and R3 are generally selected in order to ensure saturation of Q1 and Q2. Capacitors C1 and C2 are then chosen to produce the desired operating frequency with the given base resistors. The frequency of oscillation (f) is approximately equal to:

$$f = \frac{1}{1.4\ RC}$$

This formula assumes that R = R2 = R3 and C = C1 = C2. With this arrangement the circuit will produce a 50 percent duty cycle output square wave. Unequal values of capacitors can be used to produce a duty cycle more or less than 50 percent.

Figure 7-45 shows the circuit outputs. The outputs are taken from the collectors of the two transistors and they are complementary. The outputs switch between the supply voltage $+V_{CC}$ and the V_{CE} (sat) of each transistor. This clock oscillator circuit can drive most standard logic families such as TTL and CMOS directly. For other types of logic, interface circuitry may be required between the clock oscillator and the logic circuitry.

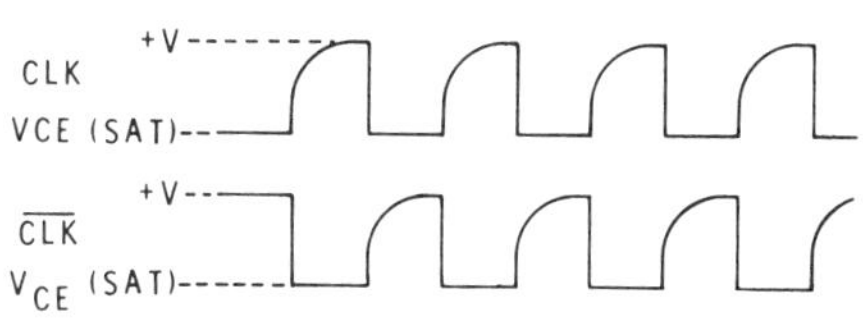

Figure 7-45

Waveforms for the astable multivibrator.

Another form of clock circuit is the relaxation oscillator shown in Figure 7-46A. This circuit uses a programmable unijunction transistor (PUT). The PUT is a four layer semiconductor device that is used as a threshold sensitive switching device. It has three terminals designated the cathode (K), the anode (A), and the gate (G).

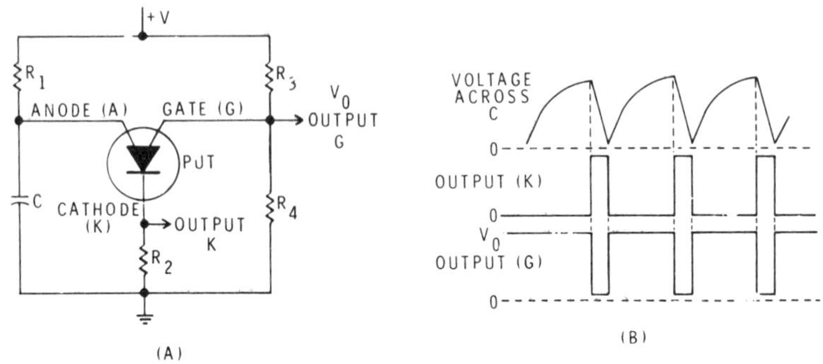

Figure 7-46
(A) Clock oscillator using a PUT. (B) Waveforms for the PUT clock oscillator.

The PUT will conduct current between its cathode and gate when the anode is properly biased. Normally, the device will not conduct if the anode potential is equal to or less than the voltage at the gate. The voltage at the gate is set by external resistors R3 and R4. In most circuits R3 is made equal to R4 making the voltage at the gate approximately one half the supply voltage +V. In operation, capacitor C charges through R1 toward the supply voltage. As soon as the charge on capacitor C is equal to 0.7 volt more positive than the gate voltage, the PUT conducts and current flows between the cathode and the gate. The PUT becomes a very low resistance and capacitor C discharges through it and R2. R2 is a very low value of resistance therefore, the capacitor discharges quickly. As the capacitor discharges it produces a voltage across R2 which is used as the output. The waveforms for the PUT relaxation oscillator are shown in Figure 7-46B. The upper waveform shows the charging voltage across the capacitor C. The remaining waveform shows the output signals from the cathode and from the gate. The duration of the output pulse produced by this circuit is very narrow because of the very rapid discharge time of C when the PUT conducts.

The frequency of oscillation in this circuit is a function of the RC time constant (R_1 C) and the voltage at the gate of the PUT.

The PUT can be simulated by using complementary bipolar transistors as shown in Figure 7-47. Transistors Q1 and Q2 are connected to form a switch that is controlled by the charging capacitor C and the voltage divider R2/R3. The operation of the circuit is identical to the PUT circuit discussed earlier. The emitter of the PNP transistor Q1 is equivalent to the anode of the PUT. The base of the PNP transistor Q1 tied to the collector of Q2 is equivalent to the gate of the PUT. The emitter of the NPN transistor Q2 is equivalent to the cathode of the PUT. The voltage divider made up of R2 and R3 sets the bias voltage on the base of transistor Q1. As in the PUT circuit, the voltage divider resistors are generally made equal so that the voltage at the base of Q1 is approximately one half the supply voltage. Capacitor C then charges through R1 to the supply voltage. When the charge on C exceeds the emitter-base voltage threshold of Q1 (about 0.7 volts), Q1 will conduct. It causes base current to flow in Q2. As a result, Q2 saturates. Capacitor C discharges quickly through Q2 and Q1. As with the PUT circuit, the duration of the output pulse is very short due to the very short discharge time of the capacitor. The output can be taken from the collector of Q2. Alternately, a low value resistance can be connected in the emitter lead of Q2 to develop a positive going output pulse if required. The waveforms for this circuit are similar to those for the PUT circuit discussed earlier.

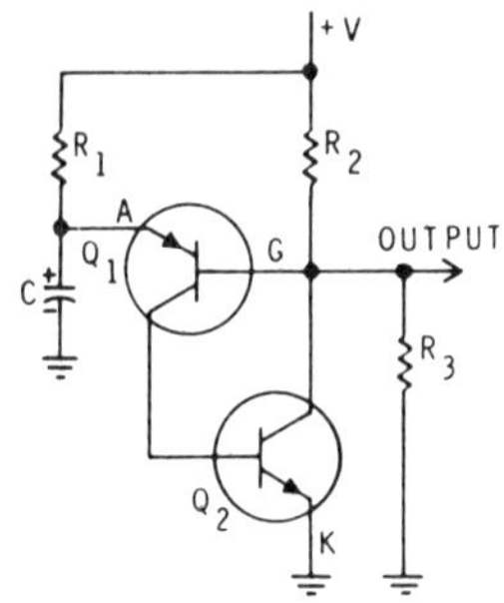

Figure 7-47
NPN-PNP simulation of a PUT in an astable clock oscillator.

IC Clock Circuits. The astable multivibrator of Figure 7-44 can also be implemented by using integrated circuit gates or inverters. Figure 7-48A shows the astable circuit implemented with TTL inverter circuits. The operation of this circuit is practically identical to the circuit in Figure 7-44. The frequency of oscillation is a function of the values of resistance and capacitance in the circuit. The resistors provide a charge path for the capacitors and are also used to provide bias to the inverter circuits. The frequency of oscillation of this circuit is approximately equal to:

$$f = \frac{1}{2\ RC}$$

Where f is in kHz, R is in k ohms and C is in microfarads. The output of the circuit will be a 50 percent duty cycle square wave. Inverter 3 is used to buffer the circuit output and to isolate the load from the frequency determining components.

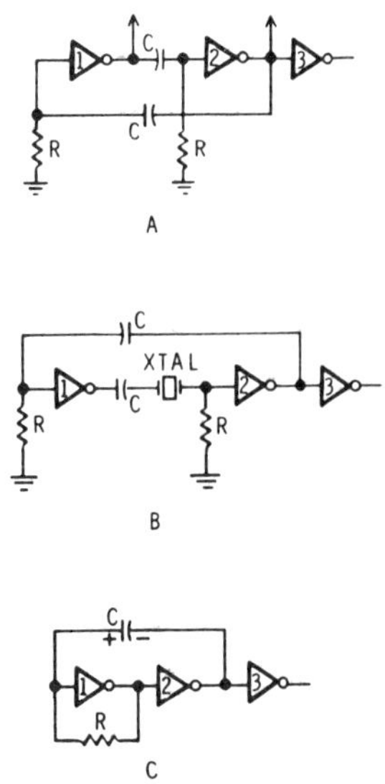

Figure 7-48
Astable multivibrators made with TTL inverters.
(A) Conventional astable,
(B) crystal controlled astable,
(C) simplified astable.

The basic IC astable multivibrator can be modified as shown in Figure 7-48B to include a frequency determining crystal. The values of R and C are selected to perform oscillation near the desired frequency. The circuit will oscillate at the crystal frequency. This circuit is desirable when the clock frequency must be very accurate and remain stable. Again inverter 3 is used to buffer the output and isolate the load from the frequency determining components.

Figure 7-48-C shows another version of the basic astable multivibrator This circuit uses only a single RC network. Resistor R is used to bias inverter 1 close to the linear region. In this circuit, the value of R is very critical and should be somewhere in the 150 to 220 ohm range when standard TTL inverters or OR gates are used.

The operation of this astable multivibrator is somewhat different from the discrete component circuit and its integrated circuit counterparts discussed earlier. This is how it operates.

Refer to Figure 7-48C. Assume the output of inverter 2 goes low. This low will be coupled through capacitor C to the input of inverter 1, therefore, the output of inverter 1 will go high. The high input to inverter 2 ensures that its output remains low. At this time capacitor C charges through R to the output voltage of inverter 1. When the voltage to the input of inverter 1 reaches approximately 1.5 volts, the output of inverter 1 will go low forcing the output of invertor 2 high. The high output from inverter 2 plus the charge on capacitor C ensures a high level to the input of inverter 1 keeping its output low. Capacitor C now begins to discharge through R. As soon as the charge on C becomes low enough, the output of inverter 1 will switch high causing the output of inverter 2 to go low. The cycle will then repeat itself. The period (p) of oscillation of this circuit is approximately 3 RC. As in the other circuits, inverter number 3 is used to isolate the load from the frequency determining components and to ensure a clean square wave output.

$$p = 3RC$$

$$f = \frac{1}{3\ RC}$$

Two-Phase Clocks. The circuits we have discussed generate a single-phase clock. For most MOS integrated circuits, a two-phase clock is required. Bipolar logic circuits and CMOS usually operate from a single-phase clock. There are several different methods of generating two-phase clock signals, but one of the most commonly used methods is shown in Figure 7-49. Two TTL JK flip-flops are connected to form a

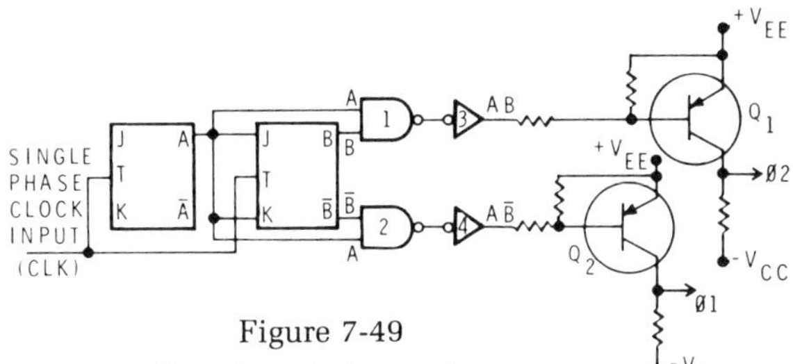

Figure 7-49
Two-phase clock generator.

synchronous 2-bit binary counter. The count sequence is 00, 01, 10, and 11. Gates 1 and 2 are used to detect the AB and $A\overline{B}$ states of the counter. These gates are used to drive transistors Q1 and Q2 which form the interface circuitry for developing the proper logic levels for use with PMOS integrated circuits. The phase 1 ($\phi 1$) and phase 2 ($\phi 2$) clock signals switch between $+V_{EE}$ and $-V_{CC}$. These levels are typically +5 and −5 volts. The waveforms for the two-phase clock circuitry are shown in Figure 7-50.

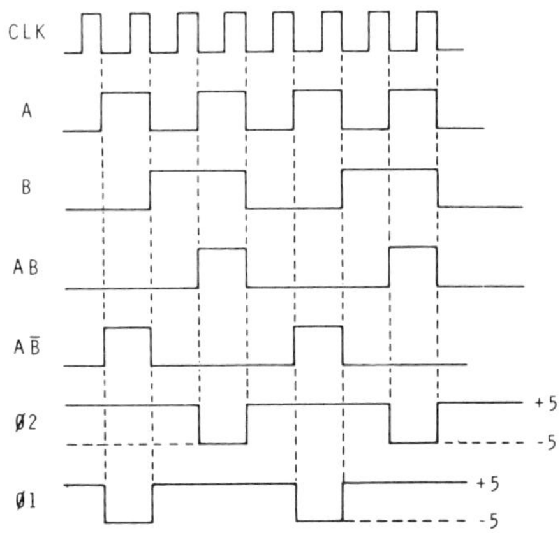

Figure 7-50
Waveforms for two-phase clock generators.

When both flip-flops are set, the output of inverter 3 goes high. Q1 cuts off and the $\phi 2$ output becomes $-V_{CC}$. When the output of inverter 3 goes low, Q1 conducts and the output becomes $+V_{EE}$. The operation of Q2 is similar. When flip-flop A is set and B is reset, the output of inverter 4 goes high. Q2 cuts off and the $\phi 1$ output goes to $-V_{CC}$. When the output of inverter 4 is low, Q2 conducts and the $\phi 1$ output is $+V_{EE}$. The waveforms show the sequence.

One Shot Multivibrators

The one shot or monostable multivibrator is a circuit that generates a rectangular output pulse of a specific time duration each time it receives an input trigger pulse. The output pulse duration is usually adjustable by varying the value of external circuit components. By cascading these circuits, a variety of sequential logic operations can be implemented.

Discrete Component One Shot Circuit. Figure 7-51 shows the schematic diagram of a one shot multivibrator. This circuit has two states: a stable state where Q2 conducts and Q1 is cut off and an unstable state where Q1 conducts and Q2 is cut off. The circuit normally rests in its stable state when it is not being triggered. The unstable state is initiated when the circuit receives an input trigger pulse. The one shot then goes into its unstable state and for a period of time depending upon the values of C and R2 it generates an output pulse. The circuit then returns to its stable state.

In the stable state, resistor R2 forward biases the emitter-base junction of Q2. Q2 saturates and its output is near zero volts or ground. Therefore, the voltage applied to R4 is insufficient to cause Q1 to conduct. Q1 therefore, is cut off and its collector is at $+V_{CC}$. Capacitor C_1 charges through the emitter-base junction of Q2 and resistor R1 to a voltage approximately equal to supply voltage $+V_{CC}$.

The circuit will remain in this stable state until it receives an input trigger pulse. To trigger the circuit, an input pulse is applied. The network consisting of C2 and R5 differentiates the input pulse. The sharp positive and negative pulses occurring at the leading and trailing edges of the input waveform are then applied to diode D1. D1 permits only the negative going pulse to be coupled to the base of Q2. The negative going pulse reverse biases the emitter-base junction of Q2. Q2 switches off and its output voltage rises to $+V_{CC}$. This causes Q1 to become forward biased. It receives base current through R3 and R4 from $+V_{CC}$. With Q1 saturated its collector is near zero volts. C1 begins to discharge through

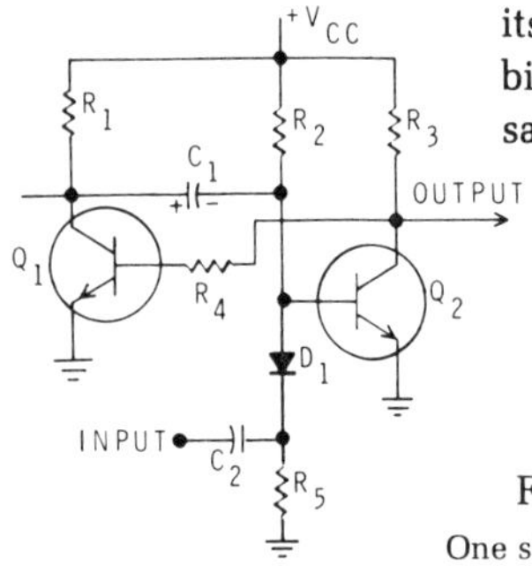

Figure 7-51
One shot multivibrator.

resistor R2. The negative voltage from this capacitor at the base of Q2 keeps Q2 cut off. As C1 discharges through R2, however, its voltage drop becomes smaller. Soon C1 will be completely discharged and will begin to charge to the opposite polarity. When the voltage across it is high enough, it forward biases Q2. As soon as Q2 switches on, the output pulse is terminated. C1 then recharges through the emitter-base junction of Q2 and R1. Figure 7-52 shows the input and output waveforms generated by the one shot circuit.

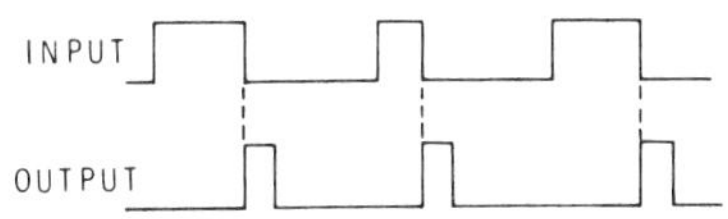

Figure 7-52
Waveforms for the one shot multivibrator.

There are several important facts about the one shot that should be considered in its application. First, the output pulse duration is a function of the values of C1 and R2. The value of R2 is rather critical since it must be low enough in value to ensure the complete saturation of Q2 during normal operation. The value of C1 can be almost any value. The time duration of the output pulse (t) is approximately $t = 0.69R2\ C1$. In most practical monostable multivibrators, the output pulse can be adjusted from nanoseconds to seconds.

The duty cycle is generally limited to a maximum of approximately 90 percent. A duty cycle greater than approximately 90 percent will generally cause the circuit to operate unreliably. The reason for this is that sufficient time must be provided for the circuit to recover between input trigger pulses. This is the time required for capacitor C1 to completely recharge through the emitter-base junction of Q2 and R1 after a pulse has been generated. This finite charge time for C1 can be reduced by making R1 smaller. However, there is a limitation because of practical circuit considerations. As for minimum duty cycle, there is no practical lower limit. Duty cycles of only a few percent can be achieved with such a multivibrator.

DUTY CYCLE

Duty cycle is the ratio of the output pulse duration to the total period of the trigger pulse input expressed as a percentage.

$$\text{Duty cycle} = \frac{t}{p} \times 100 \text{ percent}$$

Here t is the pulse duration and p is the period.

As an example, assume the pulse duration is 5 milliseconds and the input frequency is 50 Hz. See Figure 7-53. The input period is $1/50 = .02$ seconds or 20 milliseconds. The duty cycle then is:

$$\text{Duty cycle} = \frac{5}{20} \times 100 = 25 \text{ percent}$$

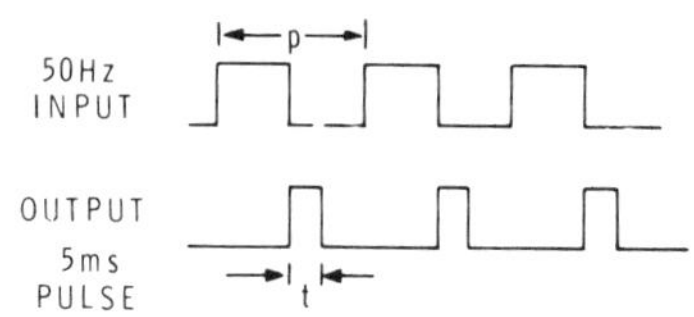

Figure 7-53

Integrated One Shots. Most one shot circuits in use today are in integrated circuit form. Their operation is virtually identical to the discrete component circuit just discussed.

Figure 7-54 shows the logic symbol use to represent these one shots. This circuit has three inputs by which the one shot may be triggered. Inputs A1 and A2 can trigger the one shot if the B input is held high. Inputs A1 or A2 will trigger the one shot on the trailing edge. When A1 or A2 switches from high to low, the one shot will generate an output pulse. Complementary output pulses appear at the Q and $\overline{Q}$ outputs. The duration of the output pulse is a function of the external components C and R. The manufacturer provides guidelines for selecting these values and charts for computing the pulse switch for given values of C and R. Generally, the external value of R is limited to approximately 50 k ohms while practically any value of capacitance from 10pf to 100μf can be used. Duty cycles as high as 90 percent are possible.

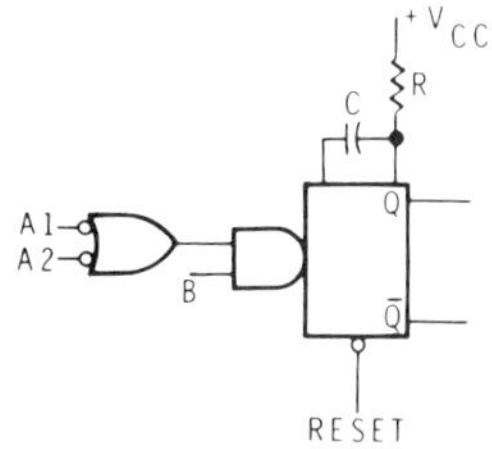

Figure 7-54
Integrated circuit one shot.

Input B can also be used to trigger the one shot if inputs A1 and A2 are not used (held low). The one shot will be triggered when input B switches from low to high. In other words, input B triggers the one shot on the leading edge of the input. This input is used primarily for inhibiting or enabling of inputs A1 and A2. Note also that the one shot has a reset input. This is similar to the asynchronous direct clear input of a JK flip-flop. Bringing this input low automatically terminates the output pulse during a timing period. When the one shot is not triggered, the Q output is binary 0 while $\overline{Q}$ is binary 1. When a trigger pulse is received, the one shot goes into its unstable state where Q is binary 1 and $\overline{Q}$ is binary 0. A reset pulse applied during the timing period will cause the normal output to switch to binary 0, immediately terminating the timing sequence. The waveforms of Figure 7-55 illustrate these operations of the IC one shot. Input pulses 1 and 2 trigger the circuit into operation on the trailing edge. The output is a pulse whose duration is defined by the values of R and C. Note that the timing interval terminates prior to the application of each new input pulse. On the third input pulse, the one shot is triggered but the timing interval is cut short because of the occurrence of a reset pulse.

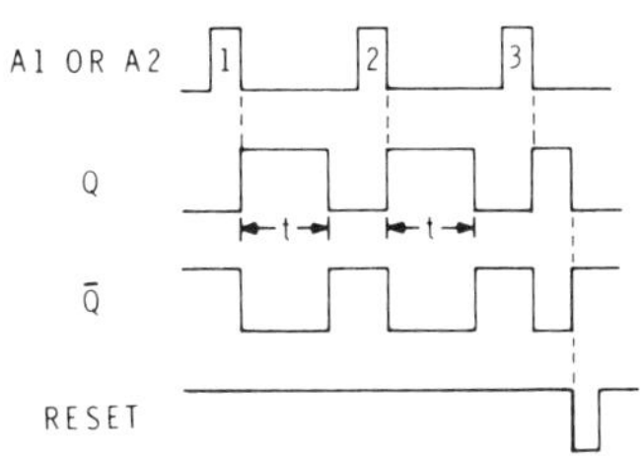

Figure 7-55
Waveforms of an IC one shot.

Another type of IC one shot circuit available to the digital designer is the retriggerable monostable or negative recovery monostable. Most one shots require a finite period of time in order to recover from a trigger pulse. Once a one shot has been triggered and times out, it will take a short period of time for the capacitor to become recharged through the circuitry resistances. It is this recovery time that limits the upper duty cycle limit of most one shots to approximately 90 percent. The retriggerable monostable eliminates this problem. Its recovery time is practically instantaneous thereby making 100 percent duty cycle outputs a possibility. A 100 percent duty cycle represents a constant binary 1 output at Q.

One of the benefits of the retriggerable monostable is its ability to generate very long duration output pulses. By adjusting the external resistor and capacitor values of the one shot to provide an output pulse duration that is longer than the interval between the input trigger pulses, the retriggerable one shot will remain in the triggered state for a substantial period of time. The waveforms in Figure 7-56 illustrate this effect. Initially, the one shot is in its normal stable state. When the trailing edge of input pulse 1 occurs, the monostable is triggered. However, before it can complete its output pulse, whose duration is a function of the external component values, input pulse 2 occurs. When its trailing edge occurs the first timing interval is automatically terminated and a new timing interval initiated. This happens quickly so that the output remains high. Note that another input pulse does not occur after input pulse 2 and therefore the one shot is then allowed to time out and generate its normal output pulse width.

In addition to generating very long output pulses, the retriggerable one shot can also be used as a missing pulse detector. By making the pulse width of the multivibrator longer than the period of input trigger pulses, the one shot will remain triggered during the sequence of input pulses. If one of the input pulses should disappear or be lost due to a malfunction or noise interference, the one shot will time out. Its output will go low and will therefore indicate the missing pulse.

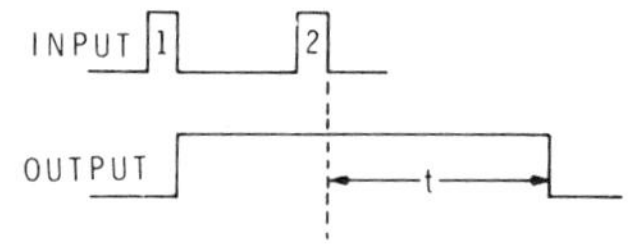

Figure 7-56

Input and output waveforms of a retriggerable monostable.

One Shot Applications

Because of the flexibility of an integrated circuit one shot, many sequential operations can be quickly and easily implemented. The ability to adjust the output pulse width with external components to a desired value plus the retriggerable and reset features makes the one shot a versatile component. As a result, digital designers find many applications for it. Because of the nature of the one shot, these applications involve pulse generation, timing and sequencing. To generate a pulse of specific width, all that the designer needs to do is to add a one shot with the appropriate size external resistor and capacitor.

Figure 7-57 shows how one shots can be used for generating a sequence of timing pulses. Here one shots labeled A, B, and C trigger one another. Assume that the one shots trigger on the trailing edge of the input. The waveforms for this circuit are shown in Figure 7-58. When an input pulse occurs, it triggers one shot A. This one shot generates a pulse width t_1, that is a function of its external component values. At the termination of its output pulse, it triggers one shot B. One shot B generates another output pulse of a specific duration t_2. Upon its termination this pulse triggers one shot C which produces another output pulse, t_3. Such a chain of one shots provides a simple method of sequencing and timing digital operations.

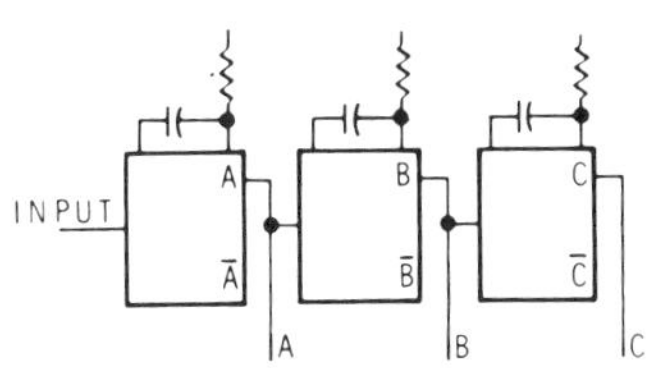

Figure 7-57
One shot pulse sequence generator.

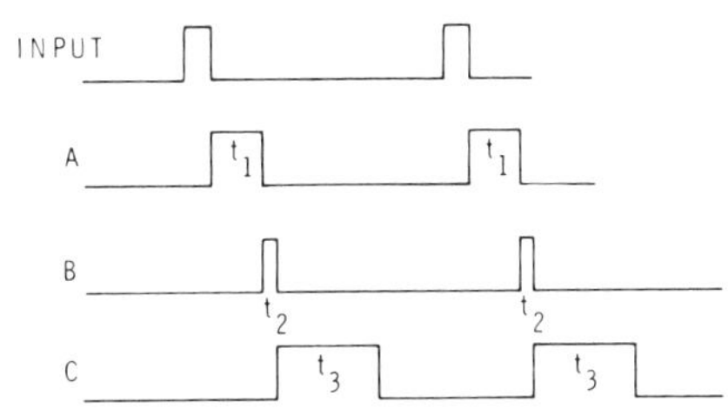

Figure 7-58
Waveforms of the one shot pulse sequencer.

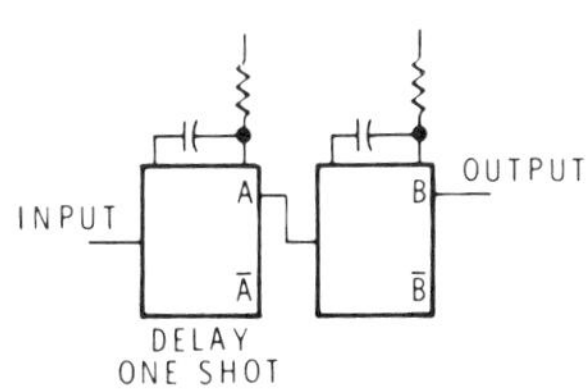

Figure 7-59
Pulse delay using one shots.

Another common application for the one shot is in implementing a delay. In some circuits it is necessary to delay the operation of a particular portion of a circuit. This is essentially a timing operation. A one shot can provide this delay. The input signal to be delayed is applied to the A one shot as shown in Figure 7-59. The A one shot generates the desired delay time. At the end of its delay interval it triggers one shot B which then produces the output pulse that initiates the desired operation. The waveforms in Figure 7-60 illustrate this delay function. The pulse width of one shot B can be adjusted to equal that of the input pulse if desired.

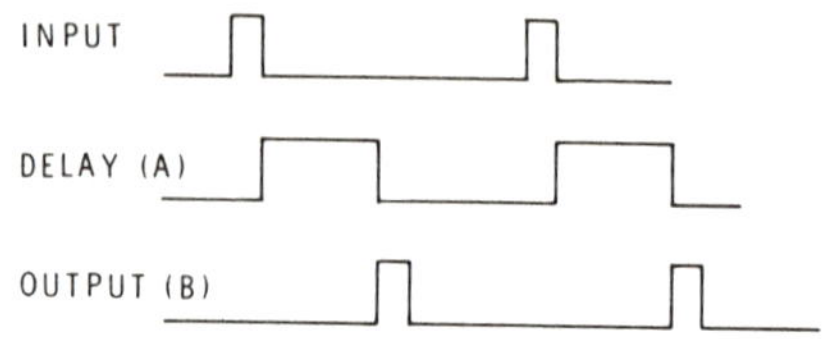

Figure 7-60
Using a one shot to delay the occurrence of a pulse.

While the one shot appears to be a very flexible and versatile component, it has a poor reputation among digital designers. Before the availability of the high quality integrated circuit one shots, the monostable functions were implemented with discrete component circuits like the one discussed earlier. Such circuits were generally unstable and unreliable. In order to provide a very stable fixed output pulse width, high quality timing resistors and capacitors had to be used. In addition, one shots of this type were very susceptible to false triggering by noise on the power supply line, at the normal trigger input or on the circuit ground. Any stray noise or "glitch" can effectively trigger the one shot and cause timing operations to occur at times when it is not wanted. Because of these problems most digital designers attempt to design without one shots. In most cases, timing functions can be implemented with other types of logic circuits such as counters and shift registers combined with logic gates. In synchronous logic circuits under the control of a master timing clock signal, sequencing pulses with the proper time intervals and durations can be readily generated without the use of one shots. Normally this method is preferred.

The modern integrated circuit one shot has overcome most of the problems associated with the early unreliable circuits. However, the timing pulse stability is still largely a function of the quality of the external resistor and capacitor used to set the output pulse duration. The noise problems have essentially been taken care of by providing high threshold noise immunity at the input. By the use of proper grounding and power supply decoupling networks, false triggering can be kept to a minimum. A good general rule of thumb is to design sequential logic circuits using counters, registers, and gates and developing the timing pulses based on synchronous clock signals. However, you will find some applications where one shots are necessary and desirable.

Unit 8

COMBINATIONAL LOGIC CIRCUITS

INTRODUCTION

Combinational logic circuits are digital circuits that are made up of gates and inverters. The output of a combinational logic circuit is a function of the states of its inputs, the types of gates used, and how they are interconnected. As you saw in a previous unit on Boolean algebra, there are many different ways to interconnect logic gates to form combinational circuits. Any unique binary function can be implemented.

An analysis of a wide variety of different types of digital equipment reveals that there are certain combinational logic circuits that regularly reoccur. Despite the large possible number of combinational circuits, most digital equipment can be implemented with just a few basic types. These circuits are called functional logic circuits. The most common functional logic circuits are decoders, encoders, multiplexers, comparators, and code converters.

In this unit, you are going to study the most common types of functional combinational logic circuits. You will learn how they operate and how they are used. You will see that even though you can construct these common functional circuits from gates and inverters, in most cases these functional logic circuits are already available as a completely wired and ready to use MSI integrated circuit. The availability of these functional circuits in MSI form, usually eliminates the need to design them. In designing digital equipment, you will find that the job is largely one of identifying the functional circuits, selecting appropriate MSI devices and interconnecting them properly.

DECODERS

One of the most frequently used combinational logic circuits is the decoder. A decoder is a logic circuit that will detect the presence of a specific binary number or word. The input to the decoder is a parallel binary number and the output is a binary signal that indicates the presence or absence of that specific number.

The basic decoding circuit is an AND gate. The output of an AND gate is a binary 1 only if all inputs are a binary 1. By properly connecting the inputs on an AND gate to the source of the data, the presence of any binary number will be detected when the output is binary 1.

Figure 8-1A shows a two input AND gate used to detect the presence of the two bit binary number 01. The number to be detected consists of two bits, A and B, with B the least significant bit (LSB). When A is 0 and B is 1, both inputs to the AND gate will be high and the output C will be a binary 1 indicating the presence of the desired number. The inverter on the A input causes the upper input to the AND gate to be binary 1 when the A input is binary 0. For any other combination of input bits the decoder output will be binary 0.

The truth table accompanying the circuit in Figure 8-1A illustrates the performance of the circuit. Note that when the input number is 01 the output C is binary 1. For all other two input combinations the output is binary 0. The circuit does indeed detect the presence of the number 01.

Figure 8-1B shows the AND gate decoder for detecting the number 01 where the binary number input source is a flip-flop register. Since the complement outputs of the flip-flops are available, the inverter is not needed. When the A flip-flop is reset and the B flip-flop is set, the number stored in the register is 01. At this time the $\overline{A}$ and B outputs are high. The decode gate output will be high at this time.

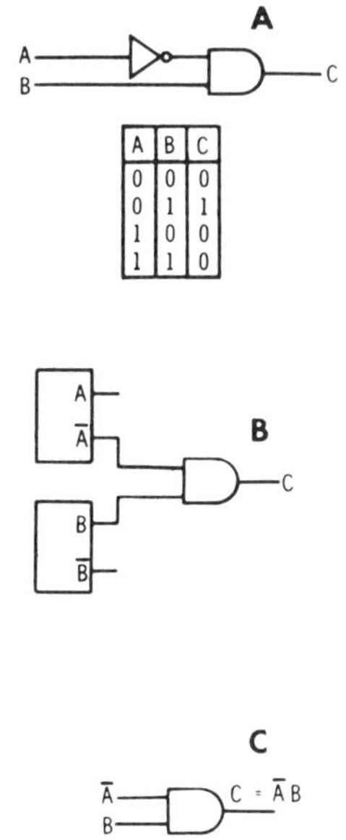

A	B	C
0	0	0
0	1	1
1	0	0
1	1	0

Figure 8-1. Two input AND gate decoders used for detecting the number 01.

To simplify the drawing of a decoder circuit, the AND gate input source is often omitted. See Figure 8-1C. Only the input states are shown at the gate inputs. Note the output equation which can be written from the circuit or the truth table.

An AND gate can be used to detect the presence of any binary number regardless of size. The number of inputs to the gate will be equal to the number of bits in the binary word. Figure 8-2 shows how a four input gate can be used to detect the binary number ABCD = 0101. Note that the decode gate receives its inputs from a 4 bit register. When the number 0101 is present in the register the output of the decode gate will be a binary 1. For any other 4 bit number in the register, the decoder output will be a binary 0.

While there are some situations where the presence of a single binary word must be recognized, most applications require the detection of all possible states that can be represented by the input word. For example, with a two bit input word there are a total of $2^2 = 4$ different input combinations that exist. A practical decoder will recognize the existence of each of these states.

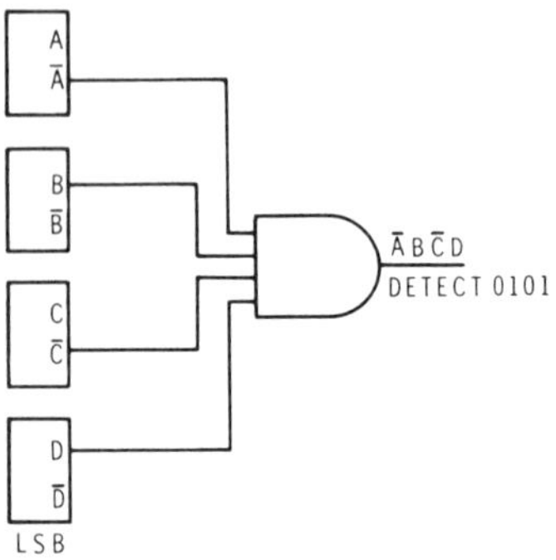

Figure 8-2. Four input AND gates used to decode 0101.

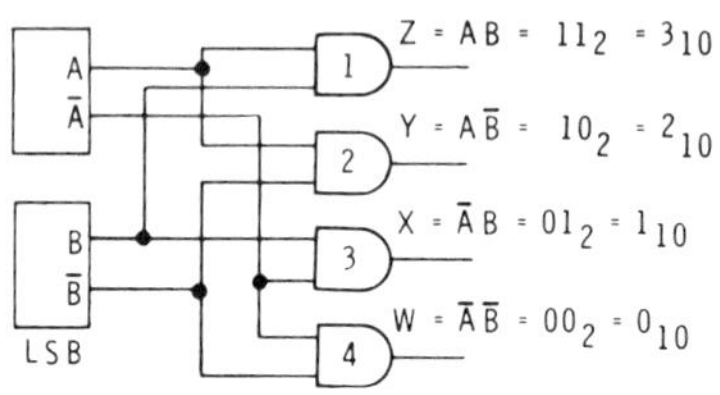

INPUTS		OUTPUTS			
A	B	W	X	Y	Z
0	0	1	0	0	0
0	1	0	1	0	0
1	0	0	0	1	0
1	1	0	0	0	1

Figure 8-3. 1 of 4 decoder.

Figure 8-3 shows such a decoder. A two bit binary word with bits A and B (B is the LSB) is stored in a flip-flop register. Four AND gates are used to decode the four possible combinations. For example, gate 4 detects the 00 input state. If the binary number stored in the flip-flops is 00, the $\overline{A}$ and $\overline{B}$ outputs will be high. Gate 4 will produce a binary 1 output. Gates 1, 2, and 3 will have a binary 0 on at least one of their inputs thereby keeping their outputs low. The truth table in Figure 8-3 shows the four possible input states and the outputs of each of the decoder gates. Such a decoder circuit is called a one of four decoder since only one of the four possible outputs will be a binary 1 at any given time.

Another way of looking at the decoder circuit in Figure 8-3 is as a binary to decimal converter. It converts a binary number into an output signal representing one of the four decimal numbers 0, 1, 2, or 3. If flip-flops A and B are both set, the register is storing the binary number 11_2. Gate 1 will be enabled at this time and its output will indicate the presence of that particular number in the register. The output of this gate could then be used to turn on an indicator light marked with the decimal number 3.

BCD to Decimal Decoder

One of the most common applications for decoder circuits is binary to decimal conversion. A widely used type of decoder is the BCD to decimal decoder. The input to the decoder is a parallel four bit number representing the BCD digits 0000 through 1001. Ten AND gates are used to look at or observe the inputs and decode the ten possible output states 0 through 9. When a BCD number is applied to this decoder, one of the ten output lines will go high indicating the presence of that particular BCD number. The output of such a decoder is generally used to operate a lighted decimal number read-out or display.

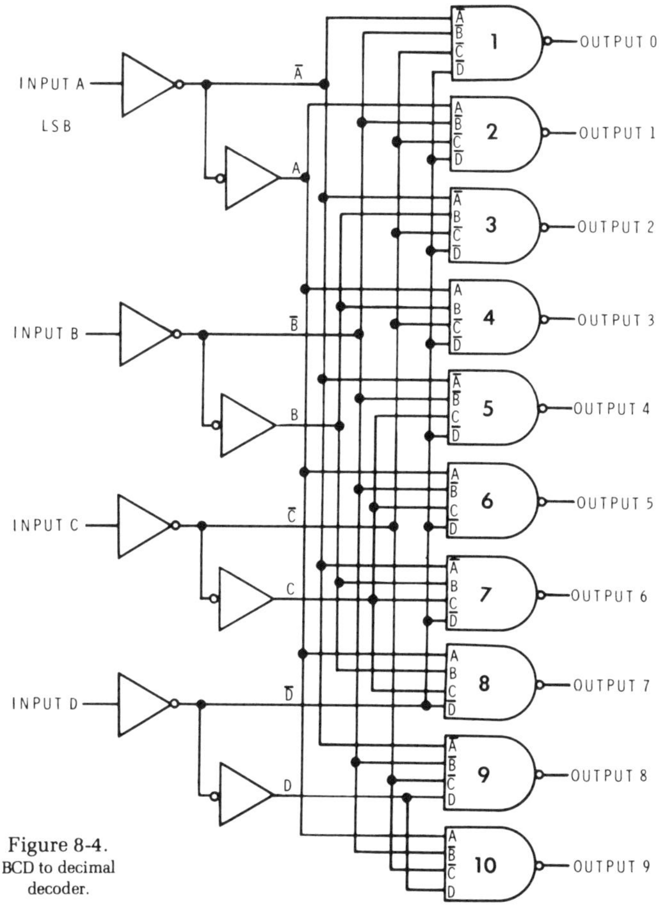

Figure 8-4.
BCD to decimal decoder.

A typical BCD to decimal decoder is shown in Figure 8-4. The four bit BCD number with bits designated A, B, C, and D are applied to the inverters which generate the normal and complement versions of the inputs to be applied to the decode gates. Bit A is the LSB. Note also that NAND gates are used instead of AND gates for the decoding process. When all of the inputs to a NAND gate are binary 1, its output will be binary 0. For all other input combinations the output will be binary 1. For that reason, all of the outputs from the gates in this decoder are high except the one decoding a specific input state.

Figure 8-5 shows the truth table for the BCD to decimal decoder. When one of the ten 8421 BCD codes is applied to the input, the appropriate output will go low. For example, when the input 0110 is applied to the decoder, all inputs to gate 7 will be binary 1. The output of gate 7 will go low indicating the presence of the four bit BCD number representing a decimal 6 is at the input. All other gate outputs will be high at this time. Notice in the truth table that if any one of the six invalid four bit BCD code numbers is applied to the input of the decoder, all outputs will remain high. This BCD decoder simply does not recognize the four bit words that are not included in the standard 8421 BCD code.

NO.	BCD INPUT				DECIMAL OUTPUT									
	D	C	B	A	0	1	2	3	4	5	6	7	8	9
0	L	L	L	L	L	H	H	H	H	H	H	H	H	H
1	L	L	L	H	H	L	H	H	H	H	H	H	H	H
2	L	L	H	L	H	H	L	H	H	H	H	H	H	H
3	L	L	H	H	H	H	H	L	H	H	H	H	H	H
4	L	H	L	L	H	H	H	H	L	H	H	H	H	H
5	L	H	L	H	H	H	H	H	H	L	H	H	H	H
6	L	H	H	L	H	H	H	H	H	H	L	H	H	H
7	L	H	H	H	H	H	H	H	H	H	H	L	H	H
8	H	L	L	L	H	H	H	H	H	H	H	H	L	H
9	H	L	L	H	H	H	H	H	H	H	H	H	H	L
INVALID	H	L	H	L	H	H	H	H	H	H	H	H	H	H
	H	L	H	H	H	H	H	H	H	H	H	H	H	H
	H	H	L	L	H	H	H	H	H	H	H	H	H	H
	H	H	L	H	H	H	H	H	H	H	H	H	H	H
	H	H	H	L	H	H	H	H	H	H	H	H	H	H
	H	H	H	H	H	H	H	H	H	H	H	H	H	H

H = BINARY 1
L = BINARY 0

Figure 8-5. Truth table for BCD to decimal decoder.

The decoder circuit in Figure 8-4 can be readily constructed with individual logic gates. A typical SSI logic gate provides two four input NAND gates in a single dual in line package. To decode the 10 output states, five of these integrated circuits would be required. The inverters could be implemented with a hex inverter IC. A typical unit contains six inverter circuits in a single DIP. Since 8 inverters are required, two of these hex inverter ICs would be required. This makes a total of seven integrated circuits required to implement the BCD to decimal decoder. Some form of printed circuit board or other interconnecting medium would be required to wire the circuit as indicated.

Fortunately, modern integrated circuit technology has eliminated the necessity for constructing such a circuit with SSI logic circuits. The entire BCD to decimal decoder circuit shown in Figure 8-4 is available in a single 16 pin dual in line package. Because of the complexity of this circuit it is considered to be a medium scale integrated circuit. This is a classical example of a functional MSI logic circuit.

Octal and Hex Decoders

Two other widely used decoder circuits are the one of eight (octal) decoder and the one of sixteen (hexadecimal) decoder. The one of eight decoder accepts a parallel three bit input word and decodes all eight output states representing the numbers 0 through 7. The BCD to decimal decoder circuit in Figure 8-4 can be used as a one of eight or octal decoder by simply using the A, B and C inputs only. The D input is simply wired to a binary 0 condition and the 8 and 9 outputs from gates 9 and 10 are ignored. A hex decoder is a one of sixteen decoder. All 16 states represented by four input bits are recognized by this circuit.

Instead of drawing the complex logic circuitry for a decoder, a simplified block diagram like the one shown in Figure 8-6 is often used. The decimal weights of the inputs and the decimal equivalent of the decoded outputs are indicated within the block to identify its function.

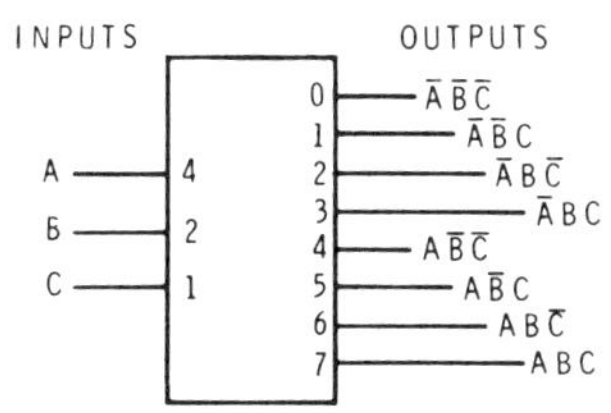

Figure 8-6.
1 of 8 or octal decoder.

BCD to 7 Segment Decoder

A special form of decoder circuit is the popular BCD to 7 segment decoder-driver. This is a combinational logic circuit that accepts the standard 8421 BCD input code and generates a special 7 bit output code that is used to operate the widely used 7 segment decimal readout display. This functional circuit is available as a single package MSI device.

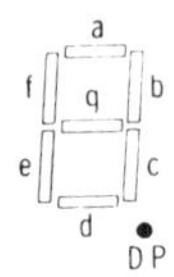

Figure 8-7.
7 segment display format.

A 7 segment readout is an electronic component used to display the decimal numbers 0 through 9 and occasionally special letter combinations by illuminating two or more segments in a specially arranged 7 segment pattern. The standard 7 segment display configuration is shown in Figure 8-7. The segments themselves can be constructed with a variety of light emitting elements such as an incandescent filament, a light emitting diode, a gas discharge glow diode or a liquid crystal segment. By illuminating the appropriate segments, the numbers 0 through 9 and many letters can be displayed. Figure 8-8 shows the typical 7 segment presentation of these numbers.

Figure 8-8. 7 segment format for numbers 0 through 9.

Seven segment displays are widely used in electronic equipment such as test instruments, electronic calculators and digital clocks. Several examples are shown in Figure 8-9. The digital counter shown in Figure 8-9A uses light emitting diode displays. A light emitting diode is a special semiconductor junction diode that emits light when it is forward biased. Most LED displays emit a brilliant red light. Yellow and green LED displays are also available.

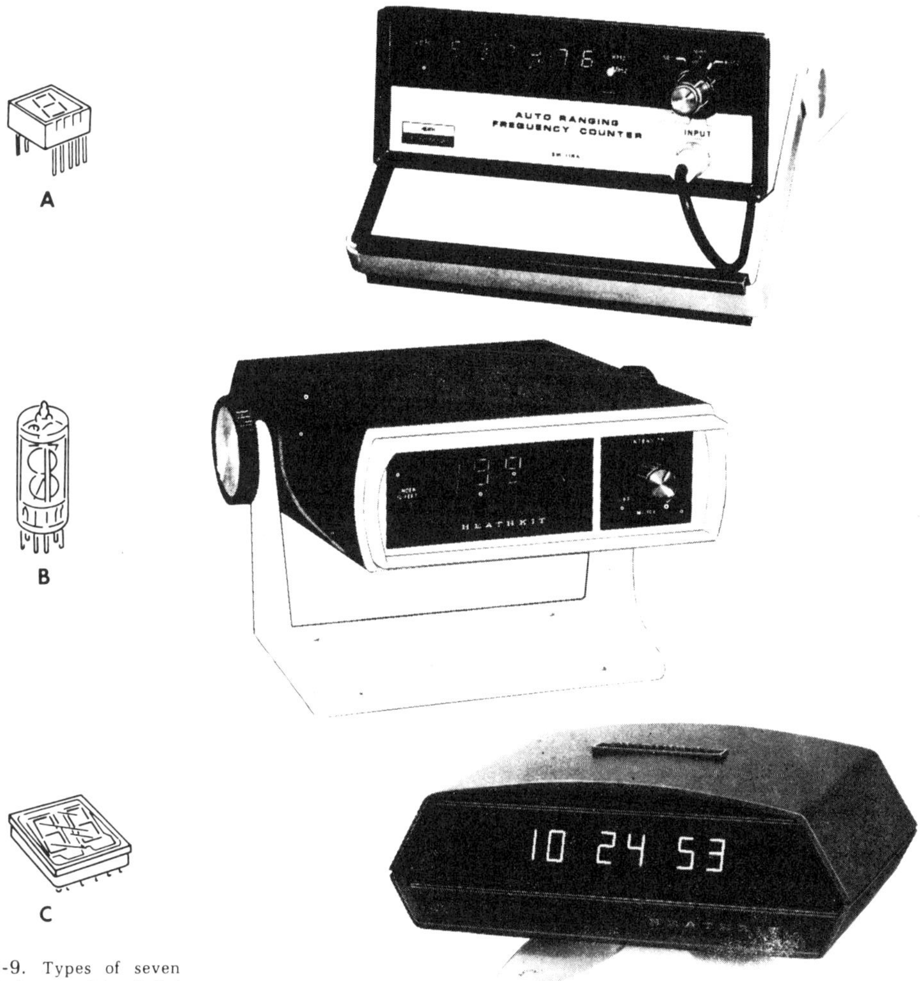

Figure 8-9. Types of seven segment displays used in digital equipment (A) LED, (B) incandescent, (C) gas discharge.

The digital depth sounder in Figure 8-9B uses an incandescent-filament, 7-segment display tube. Each segment is a thin tungsten wire that emits a brilliant white light when current is passed through it. The seven segments are mounted in a single plane within a glass tube. A filter or window in front of the display can make the light any color desired.

The digital clock in Figure 8-9C uses a 7-segment, gas-glow discharge display. Each element of the 7-segment readout is a cathode of a special gas-discharge diode. When a high voltage is applied to the element, the gas surrounding the element is ionized and emits a red-orange glow.

These are only a few of the many different types of 7-segment display devices available. A BCD to 7-segment decoder-driver circuit is used to operate these display devices. This is an MSI logic circuit that decodes the decimal states 0 through 9 and develops the seven output signals that operate the display device.

Figure 8-10 shows the truth table for this decoder circuit. The BCD inputs (ABCD) are in the standard 8421 code form. The outputs are designated a, b, c, d, e, f, g, and correspond to the elements shown in Figure 8-7. A binary 0 in the segment output columns indicates that the corresponding segment is illuminated. You can check this code against the segment letters illustrated in Figure 8-7.

INPUTS					SEGMENT OUTPUTS						
DECIMAL	A	B	C	D	a	b	c	d	e	f	g
0	0	0	0	0	0	0	0	0	0	0	1
1	0	0	0	1	1	0	0	1	1	1	1
2	0	0	1	0	0	0	1	0	0	1	0
3	0	0	1	1	0	0	0	0	1	1	0
4	0	1	0	0	1	0	0	1	1	0	0
5	0	1	0	1	0	1	0	0	1	0	0
6	0	1	1	0	1	1	0	0	0	0	0
7	0	1	1	1	0	0	0	1	1	1	1
8	1	0	0	0	0	0	0	0	0	0	0
9	1	0	0	1	0	0	0	1	1	0	0

Figure 8-10. Truth table for BCD to seven segment decoder driver.

A logic diagram for one particular type of BCD to 7 segment decoder-driver is shown in Figure 8-11. In addition to the four BCD inputs, this circuit also has a blanking input, a ripple blanking input and a lamp test input. When the lamp test input is binary 0 all seven segments of the display are turned on in order to see that none have failed. When the blanking input is low, all segments are turned off. This feature is used where a number of displays are grouped to readout a multi-digit number. This feature blanks or suppresses all leading zeros automatically. For example, in an eight digit display without leading zero suppression, the number 1259 would be displayed as 00001259. With leading zero suppression, only the desired digits 1259 will show. The other four displays will be automatically blanked. By applying a variable duty cycle pulse signal to the ripple blanking input, the intensity of the display can be varied without resorting to supply voltage or current controls.

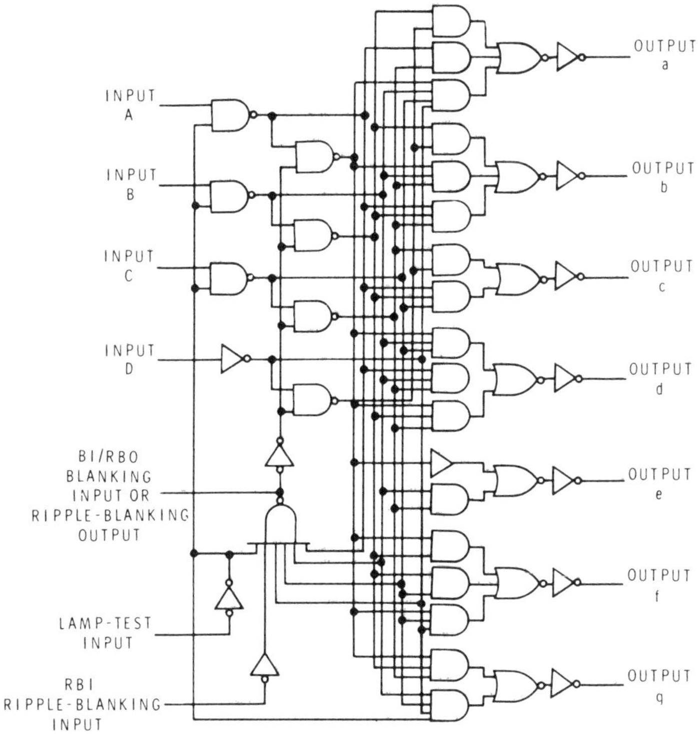

Figure 8-11. BCD-to-seven segment decoder-driver IC.

Figure 8-12 shows the typical output circuit for one segment of the decoder-driver. Besides decoding the BCD input states this circuit also drives or operates the light producing segments. The output is usually a saturated transistor switch with an open collector. Figure 8-12 shows one LED segment connected to the output. When the transistor conducts, the collector output goes low and current flows through the LED turning it on. A series dropping resistor sets the LED current and hence the intensity of the light it produces.

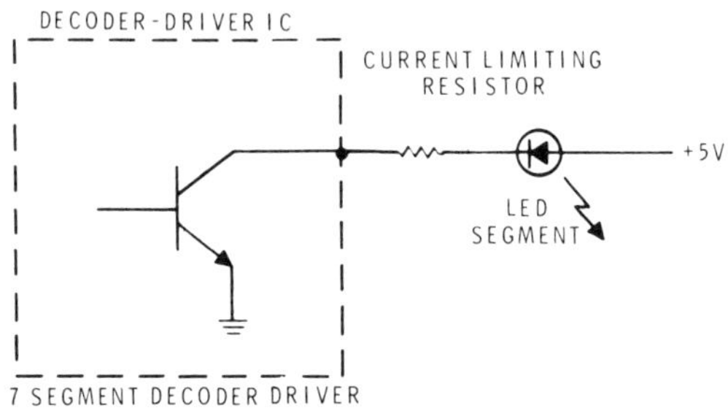

Figure 8-12. Typical 7 segment decoder-driver output circuit and external connections.

ENCODERS

An encoder is a combinational logic circuit that accepts one or more inputs and generates a multi-bit binary output code. In a sense, encoders are exactly the opposite of decoders. Decoders detect or identify specific codes while encoders generate specific codes.

Figure 8-13 shows a simple encoder circuit. The inputs are three pushbuttons labeled 1, 2 and 3. The encoder circuit consists of two positive NAND/negative NOR gates. The outputs AB form a two bit binary code. When pushbutton 1 is depresed, the output of gate 2 goes high. At this time both inputs to gate 1 are high therefore its output is low. By depressing button 1, the output code 01 is generated.

Depressing the number 2 pushbutton forces the output A of gate 1 high. The B output of gate 2 is low therefore the output code is 10. Depressing button 3 forces the outputs of both gates high generating the code 11. As you can see, the binary code corresponding to the decimal number given to each input switch is generated when that switch is closed. The truth table in Figure 8-19 summarizes the operation of the circuit. When all of the switches are open (not depressed) the output code is 00.

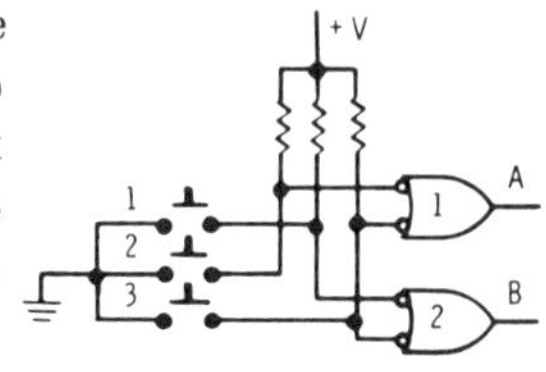

INPUTS	OUTPUT A	OUTPUT B
1	0	1
2	1	0
3	1	1

Figure 8-13
Simple encoder circuit.

A typical application for an encoder circuit is in translating a decimal keyboard input signal into a binary or BCD output code. Figure 8-14 shows a decimal to BCD encoder circuit. When any one of the input lines is brought low, the corresponding 4 bit BCD output code is generated. For example, bringing the number 5 input line low with a pushbutton forces the outputs of gates 1 and 3 high. Gates 2 and 4 have low outputs at this time. The output code on lines DCBA then is 0101 or the binary equivalent of the decimal number 5. This circuit, like all encoders, generates a unique output code for each individual input.

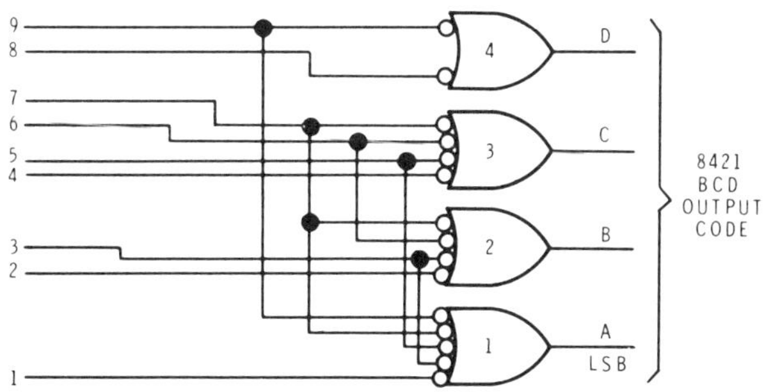

Figure 8-14
Decimal to BCD encoder.

A typical example of a modern integrated circuit binary encoder circuit is shown in Figure 8-15. This is a TTL MSI 8-input priority encoder. The encoder accepts data from eight input lines and generates the binary code corresponding to the number assigned to the input. The input must be brought low in order to generate the corresponding output code. We say that the inputs are active low. Unlike the two previously discussed encoder circuits, the outputs of the circuit in Figure 8-15 (labeled $\overline{A0}$, $\overline{A1}$, and $\overline{A2}$) are also active low. For this circuit what this means is that a low output represents binary 1. This circuit generates a negative logic output code.

A unique feature of this particular circuit is that a priority is assigned to each input so that when two or more inputs are low simultaneously, the input with the highest priority is represented at the output. In this case the inputs with the higher numerical value have the highest priority. This means that if the 3 and 6 inputs are low simultaneously, the binary code representing 6 will be generated at the output.

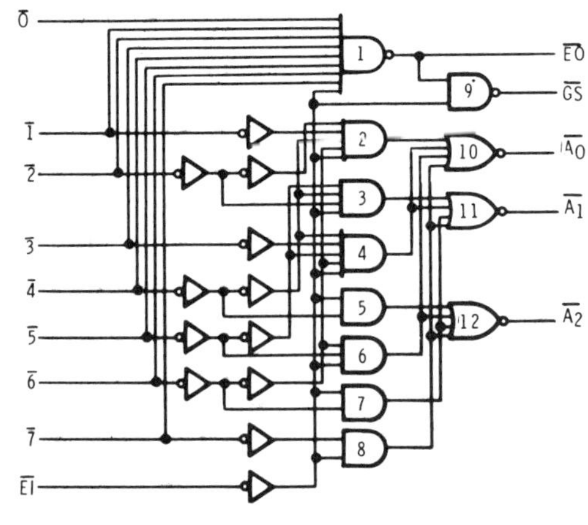

Figure 8-15
Eight input priority encoder circuit.

The $\overline{EI}$ input on this circuit is an enabling input. When this input is high, the output of the inverter connected to it is low. This inhibits gates 1 through 8 and forces the three binary output lines high. When the $\overline{EI}$ input line goes low, the output of the inverter goes high thereby enabling all of the circuitry.

The 8 input NAND gate number 1 monitors all 8 input lines. If any one of them should go low, the $\overline{EO}$ output goes high indicating that one or more of the input lines has been activated. The $\overline{GS}$ output also goes low. If all inputs are high or open (not activated), the $\overline{EO}$ output line is low indicating this state. By using the $\overline{EI}$ input and $\overline{EO}$ and $\overline{GS}$ outputs, several of these devices may be combined to encode N different input states. A truth table for this circuit is shown in Figure 8-16.

TRUTH TABLE

INPUTS									OUTPUTS				
$\overline{EI}$	$\overline{0}$	$\overline{1}$	$\overline{2}$	$\overline{3}$	$\overline{4}$	$\overline{5}$	$\overline{6}$	$\overline{7}$	$\overline{GS}$	$\overline{A_0}$	$\overline{A_1}$	$\overline{A_2}$	$\overline{EO}$
H	X	X	X	X	X	X	X	X	H	H	H	H	H
L	H	H	H	H	H	H	H	H	H	H	H	H	L
L	X	X	X	X	X	X	X	L	L	L	L	L	H
L	X	X	X	X	X	X	L	H	L	H	L	L	H
L	X	X	X	X	X	L	H	H	L	L	H	L	H
L	X	X	X	X	L	H	H	H	L	H	H	L	H
L	X	X	X	L	H	H	H	H	L	L	L	H	H
L	X	X	L	H	H	H	H	H	L	H	L	H	H
L	X	L	H	H	H	H	H	H	L	L	H	H	H
L	L	H	H	H	H	H	H	H	L	H	H	H	H

H = HIGH VOLTAGE LEVEL
L = LOW VOLTAGE LEVEL
X = DON'T CARE (EITHER 1 OR 0)

Figure 8-16
Truth table for eight input priority encoder.

MULTIPLEXERS

A multiplexer is an electronic circuit that is used to select and route any one of a number of input signals to a single output. The simplest form of a multiplexer is a single pole multi-position switch. Figure 8-17 shows a rotary selector switch used as a multiplexer. Any one of six input signals can be connected to the output line by simply adjusting the position of this mechanical selector switch. Mechanical selector switches are widely used for a variety of multiplexing operations in electronic circuits. However, many applications require the multiplexer to operate at high speeds and be automatically selectable. Multiplexers of this type can be readily constructed with electronic components.

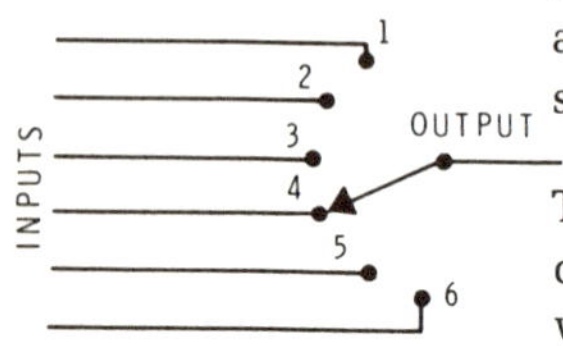

Figure 8-17
A rotary selector switch used as a multiplexer.

There are two basic types of electronic multiplexer circuits: analog and digital. The simple selector switch multiplexer in Figure 8-17 will work with either analog or digital signals. However when electronic multiplexers are constructed, they are primarily designed for either analog or digital applications. For analog applications relays and bipolar or MOSFET switches are widely used. For digital applications involving binary signals, a multiplexer can be simply constructed with standard logic gates. Our primary concern here of course is the digital multiplexer or binary data selector.

The circuit in Figure 8-18A is the simplest form of digital multiplexer. It has two input data sources and a single output. Either one of the input sources may be selected and fed to the output. The selection process takes place in AND gates 1 and 2. The flip-flop controls these two gates to determine which input is allowed to pass through OR gate 3 to the output. When this flip-flop is set, the Q output will be high enabling gate 1. The $\overline{Q}$ output will be low inhibiting gate 2. Data source 1 will therefore be allowed to pass through gate 1 and through the OR gate 3 to the output. Data source 2 will have no effect on the output state. Resetting the flip-flop reverses this condition. Gate 1 will be inhibited by Q thereby preventing data source 1 from affecting the output. However, data source 2 will be allowed to pass through gates 2 and 3 to the output. This circuit is equivalent to a single pole double throw switch as indicated in Figure 8-18B.

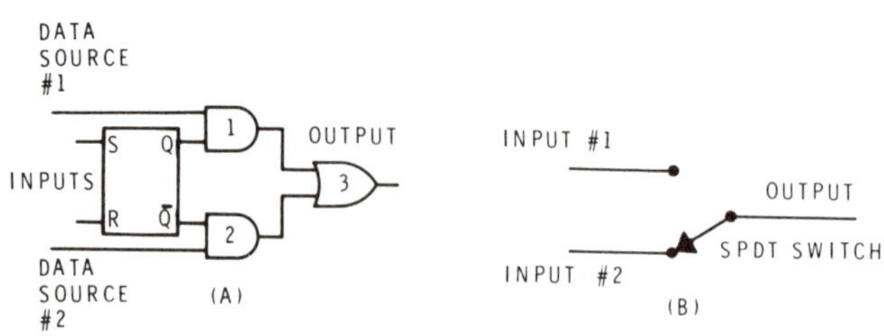

Figure 8-18
Two input digital multiplexer (A) and its mechanical equivalent (B). An SPDT switch.

An MSI functional circuit using this basic two input multiplexer is shown in Figure 8-19. Four 2 input multiplexers are combined to form a multiplexer for two four bit words. Word 1 has bits A1, B1, C1 and D1. Word 2 has bits A2, B2, C2 and D2. The enable (E) input controls the circuit. If E is high, the output of inverter 15 is low thereby inhibiting all of the AND gates and thus preventing either input word from appearing at the outputs. With E low, the circuit is enabled.

The select (S) input specifies which four bit input word appears at the output. When the select input is high, gates 2, 5, 8 and 11 will be enabled letting input word 1 appear at the output. If the S input is low, gates 1, 4, 7 and 10 will be enabled. This permits word 2 to be passed through to the output.

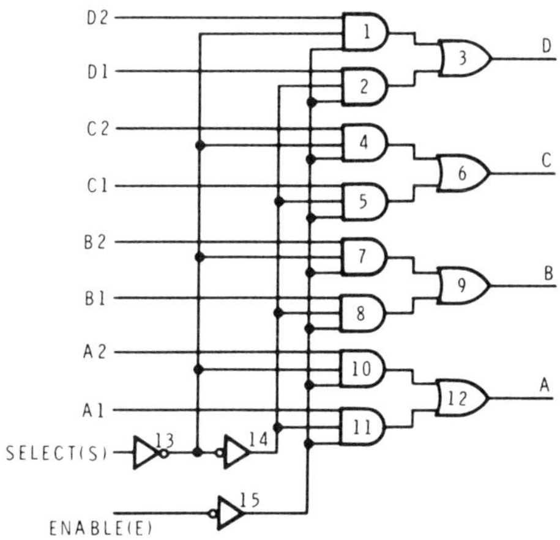

Figure 8-19

Quad two input multiplexer.

A four input multiplexer circuit is shown in Figure 8-20. Each input is applied to a NAND gate that is enabled or inhibited by a 1 of 4 decoder. The outputs of the NAND gates are ORed together in gate 5. As in other multiplexers, only one of the four inputs will be enabled and allowed to pass through to the output. The selection of the input is made by the decoder circuit. A two bit binary word AB is applied to the decoder. The decoder recognizes one of the four possible input codes and enables the appropriate gate. For example, when the two bit input word is 00 the $\overline{A}\overline{B}$ output line is high. This enables gate 1 and input 1 is allowed to pass through to the output. Input code 01 enables gate 2, input code 10 enables gate 3, and input code 11 enables gate 4.

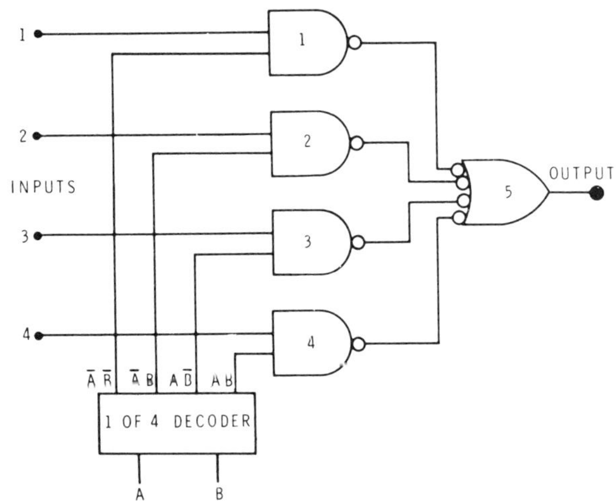

Figure 8-20
4 input multiplexer.

The simplest way to implement the 4 bit multiplexer is to combine both the decode and enable functions in the same gate. Such a circuit is shown in Figure 8-21. The arrangement is virtually identical to the 4 input multiplexer just discussed. However, additional inputs have been added to the input gates so that they also perform the decoding functions. The normal and complement outputs from the two bit binary input word AB are applied to the enable gates in the same way they would be applied to the decoder gates. If the binary input code 00 is applied, the $\overline{A}$ and $\overline{B}$ lines will be high. Gate 1 will be enabled and input number 1 will pass through gate 1 and gate 5 to the output. Gates 2, 3 and 4 will be inhibited at this time.

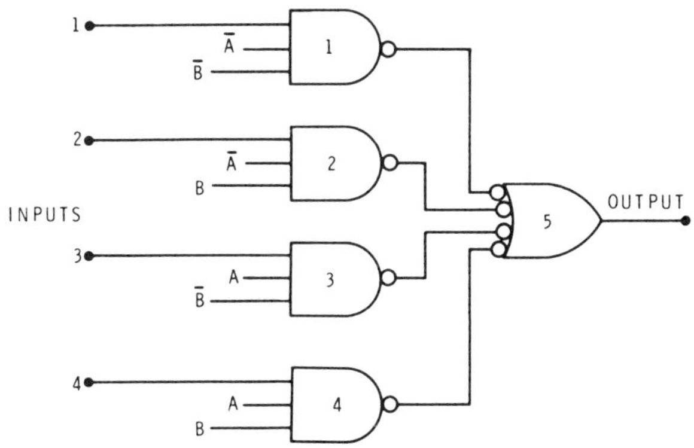

Figure 8-21

A 4 input multiplexer combining the decode and enable functions.

An eight input TTL binary multiplexer using this same technique is shown in Figure 8-22. Gates 1 through 8 enable or inhibit the eight data input lines D0 through D7. A three bit binary input word (ABC) enables one of the eight gates depending upon the input code. The six inverters at the data select inputs generate the normal and complement signals needed by the select gates. This three input word is an address code that designates which data input line is selected. If the binary input is 101, data input D5 is selected. The strobe enable line enables or inhibits all eight select gates. Both the normal (W) and complement (Y) output signals are available. Another name for the multiplexer is data selector.

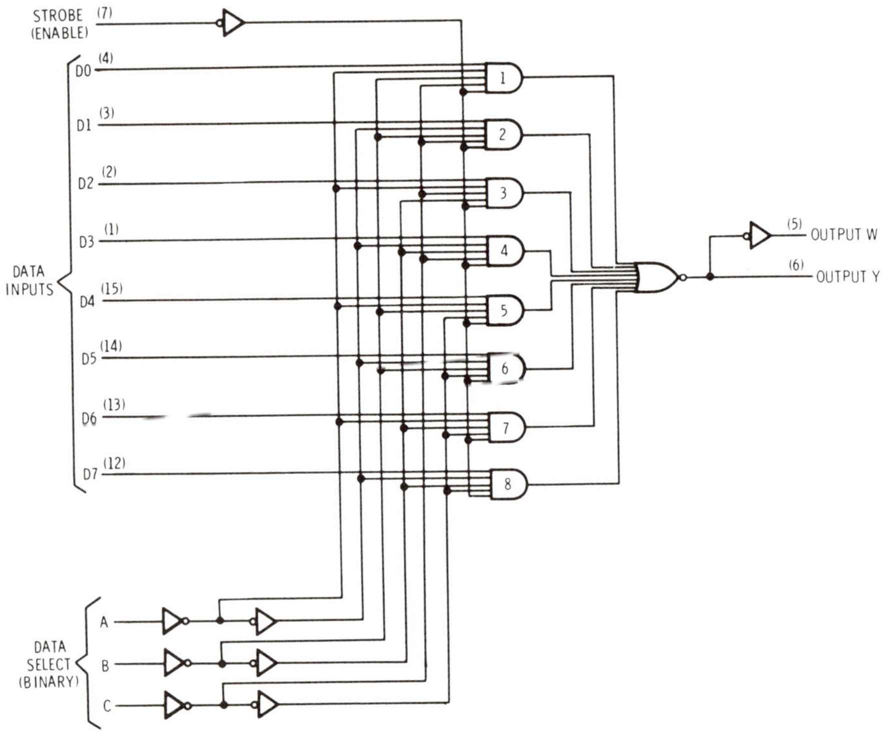

Figure 8-22
8 input multiplexer TTL IC type 74151.

Multiplexer Applications

Besides providing a convenient means of selecting one of several inputs to be connected to its single output, a multiplexer has several special applications which make it even more useful. Besides its data selector application, multiplexers are also used to provide parallel to serial data conversion, serial pattern generation and the simplified implementation of Boolean functions. Let's consider these important applications here.

Parallel to Serial Conversion. One of the most common applications of a multiplexer is parallel to serial data conversion. A parallel binary word is applied to the inputs of a multiplexer. Then by sequencing through the input enabling codes, the output of the multiplexer becomes a serial representation of the parallel input word. This function is illustrated in Figure 8-23. Here we show a four input multiplexer. (Multiplexer is often abbreviated MPX or MUX). The simple block diagram is often used to represent multiplexers in order to simplify their illustration. A two bit binary input word AB from a counter is used to select the desired input. Input word WXYZ is stored in a 4 bit storage register. The output of each of the flip-flops in the register is connected to one input of the multiplexer. As the two bit counter is incremented, the AB input select code is sequenced through its four states 00 through 11. The output (M) of the multiplexer is equal to the state of the flip-flop connected to the enabled input. This is illustrated by the truth table in Figure 8-23. By sequencing

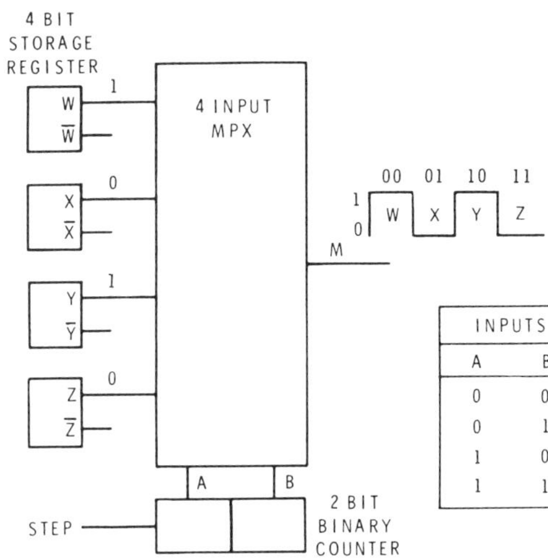

INPUTS		OUTPUTS
A	B	M
0	0	W
0	1	X
1	0	Y
1	1	Z

Figure 8-23

Four input multiplexer used as a parallel-to-serial converter.

through the four input states at a fixed rate, the parallel input word is converted to a serial output word. When the AB inputs are 00, the state of the W flip-flop appears at the multiplexer output. When the AB input state is 01, the state of the X flip-flop appears at the multiplexer output. Similarly, input select states 10 and 11 cause the states of flip-flops Y and Z respectively to appear at the multiplexer output. Depending upon how the inputs are connected to the register, the multiplexer can cause either the LSB or the MSB to occur first.

Serial Binary Word Generator. Another application of the multiplexer in digital circuits is the generation of a serial binary word. This application is virtually identical to the parallel to serial conversion technique just discussed. The primary difference is that for binary word generation, the serial word generated at the output of the multiplexer is generally a fixed value rather than one that can change as in the case of the parallel to serial converter. There are some occasions that require the generation of a single fixed serial word for some special function.

Figure 8-24 shows an eight input multiplexer used to generate a fixed serial binary output word. Notice that the eight inputs are connected to either +5 volts (binary 1) or ground (binary 0). The three bit input word ABC is used to select which of the inputs is routed to the output. By sequencing through the three bit input words from 000 through 111 with a binary counter, the binary states applied to inputs 1 through 8 are sequentially connected to the output. The binary word 10011010 is generated at the output. Each time the three bit input word is sequenced through the 000 to 111 state, this serial output word will be generated. Again, depending upon the application, the connections to the multiplexer input can be made such that either the MSB or LSB occurs at the output first. In this case the MSB appears at the output first. The truth table in Figure 8-24 completely defines the function of this circuit.

INPUTS			OUTPUT	
C	B	A	M	
0	0	0	0	1
0	0	1	1	0
0	1	0	2	0
0	1	1	3	1
1	0	0	4	1
1	0	1	5	0
1	1	0	6	1
1	1	1	7	0

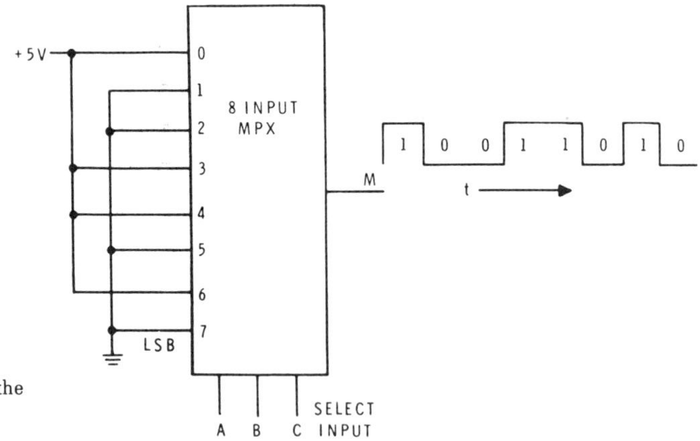

Figure 8-24
8 input multiplexer used to generate the serial binary word 10011010.

Boolean Function Generation. Multiplexers can greatly simplify the implementation of Boolean functions in the sum-of-products form. A close look at the multiplexer circuit in Figure 8-22 shows that it inherently implements the sum-of-products for all input combinations. The products $\overline{A}\,\overline{B}\,\overline{C}$ through ABC are developed by gates 1 through 8. By connecting a binary 1 or binary 0 to the appropriate data inputs, the products desired in the output can be selected.

For example, suppose that you wish to implement the Boolean function indicated below.

$$M = A\overline{B}\,\overline{C} + \overline{A}\,\overline{B}\,C + \overline{A}\,B\,C + A\,B\,C$$

By studying the logic diagram for the multiplexer in Figure 8-22, you can determine which gates generate each Boolean product. These are indicated in Table A for your convenience. Note that gate 1 generates the product $\overline{A}\,\overline{B}\,\overline{C}$. When a binary 1 is applied to the D0 input, a binary 1 will appear at the output if the data select inputs are 000. By applying a binary 0 to DØ, the 000 input state will be ignored and a binary 0 will appear at the output W.

Table A

Data Input	**Gate**	**Output**
DØ	1	$\overline{A}\,\overline{B}\,\overline{C}$
D1	2	$A\,\overline{B}\,\overline{C}$
D2	3	$\overline{A}\,B\,\overline{C}$
D3	4	$A\,B\,\overline{C}$
D4	5	$\overline{A}\,\overline{B}\,C$
D5	6	$A\,\overline{B}\,C$
D6	7	$\overline{A}\,B\,C$
D7	8	$A\,B\,C$

Therefore, you can see that the desired Boolean products can be selected by applying a binary 1 to the appropriate input associated with the gate generating that product. Those products you wish to delete from the output, you apply a binary 0 to the gate generating that product.

To generate the expression indicated earlier then, binary 1 states are applied to the D1, D4, D6 and D7 inputs. These are gates 2, 5, 7 and 8 in Figure 8-22. The complete Boolean function generator is shown in Figure 8-25.

Other more complex Boolean functions can also be implemented with multiplexers by connecting other input variables to the multiplexer inputs instead of a fixed binary 1 or binary 0 level. Four variable sum-of-products from inputs A, B, C, and D for example, can be implemented with an eight input multiplexer by connecting the D and $\overline{D}$ input states to selected multiplexer inputs to implement the desired function.

By using standard MSI multiplexer packages, the implementation of Boolean functions is greatly simplified. With this technique it is not necessary to interconnect multiple SSI logic gate packages to implement the desired function. This greatly reduces the number of integrated circuits used, the power consumption, size and the need for interconnection.

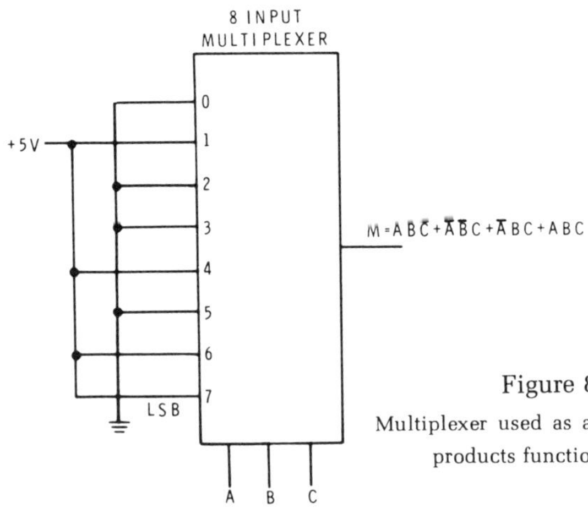

Figure 8-25

Multiplexer used as a Boolean sum-of-products function generator.

DEMULTIPLEXERS

A demultiplexer is a logic circuit that is basically the reverse of a multiplexer. Where the multiplexer has multiple inputs and a single output, the demultiplexer has a single input and multiple outputs. The input can be connected to any one of the multiple outputs. The demultiplexer is also known as a data distributor or data router.

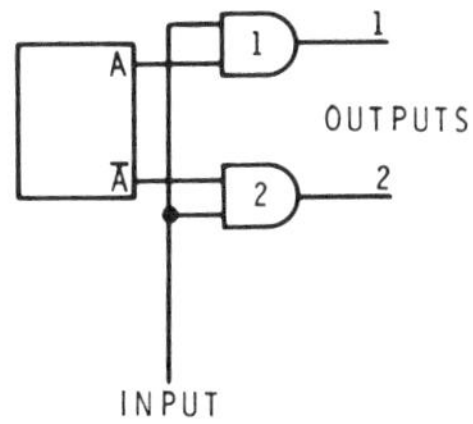

Figure 8-26
Two output demultiplexer.

A simple two-output demultiplexer circuit is shown in Figure 8-26. The single input is applied to both AND gates 1 and 2. The A flip-flop selects which gate is enabled. When the A flip-flop is set, gate 1 will be enabled and gate 2 will be inhibited. The input therefore will pass through gate 1 to output number 1. Resetting the flip-flop enables gate 2 and the input is passed to output number 2.

A four-output data distributor is shown in Figure 8-27. Here the single input is applied to four gates simultaneously. As in the multiplexer, additional inputs on the select gates are used for decoding. A two bit word AB from a counter is used to select which gate is enabled. If the two input binary word AB is 11, gate 4 will be enabled and the input will pass through gate 4. The other three gates will be inhibited at this time.

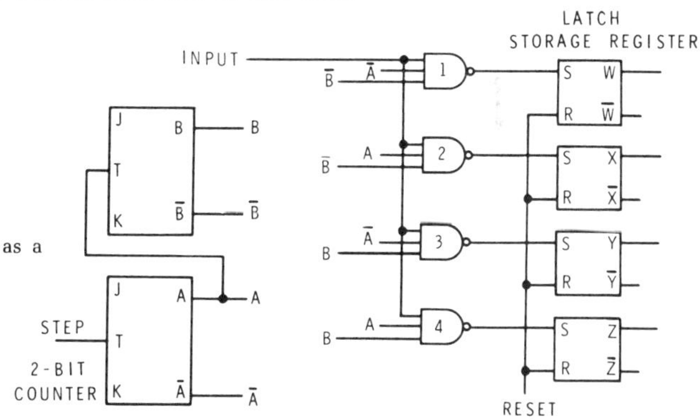

Figure 8-27
A four-output demultiplexer used as a serial to parallel converter.

The data distributor shown in Figure 8-27 is being used as a serial to parallel converter. This is one typical application of a demultiplexer circuit. A four bit serial word is applied to the input. As the input bits occur, the two bit counter is incremented. This causes the gates in the distributor to be enabled one at a time, sequentially from top to bottom. The step input to the two bit counter is in synchronism with the occurrence of the bits in the serial word.

The latch storage register with flip-flops WXYZ is initially reset prior to the application of the serial input. The flip-flops in the storage register are connected to the output of the data distributor and are sequentially set or left reset as the serial word occurs. Once each of the four gates has been enabled in sequence, the register contains the serial input word. Its outputs can then be observed simultaneously. The serial input word has been converted to a parallel output word.

Figure 8-28 shows the waveforms of the circuit in Figure 8-27. The input is the serial number 1101. The waveforms show the outputs of gates 1 through 4 and the flip-flop outputs WXYZ. The first bit of the serial input is a binary 1. It occurs during the A B input sequence. During this time, gate 1 is enabled and, since the input is a binary 1, its output will go low. This will set the W latch, causing the W output to go high. The $A\bar{B}$ input selection sequence is next. Note that this is synchronized with the next input, which is also a binary 1. This input state causes gate 2 to be enabled. Since the input signal is a binary 1 at this time, the output of gate 2 will go low, thereby setting the X flip-flop. The X output goes high as indicated. During the next input selection sequence $\bar{A}B$, the serial input word is 0. Gate 3 is enabled. The input is binary 0 at this time so the output of gate 3 remains high. This has no effect on the Y flip-flop so it remains reset. The AB input selection sequence is next. It occurs in synchronism with the next serial bit which is a binary 1. Gate 4 is enabled and, with the binary 1 input, its output is low. This sets the Z flip-flop, causing its output to go high. Looking at the states of the flip-flop after the fourth serial input bit has occurred, you can see that the parallel output 1101 is available. Note that all of the reset inputs to the latch flip-flops in the storage register are connected together to form a common reset line. Prior to the application of the serial input, a low signal is applied to the reset input to clear the register to the 0000 state.

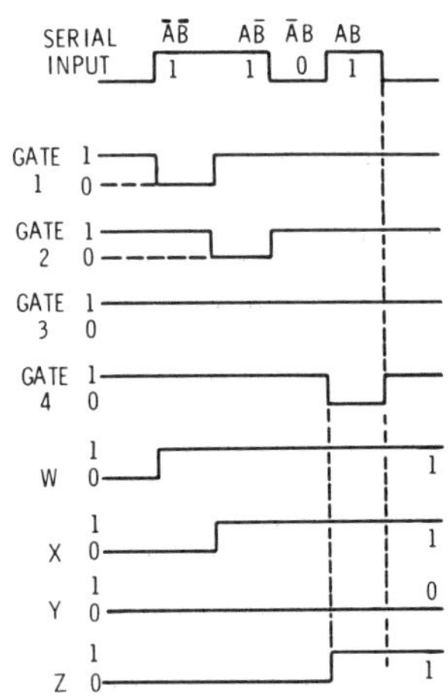

Figure 8-28

Waveforms of a serial to parallel conversion with a demultiplexer.

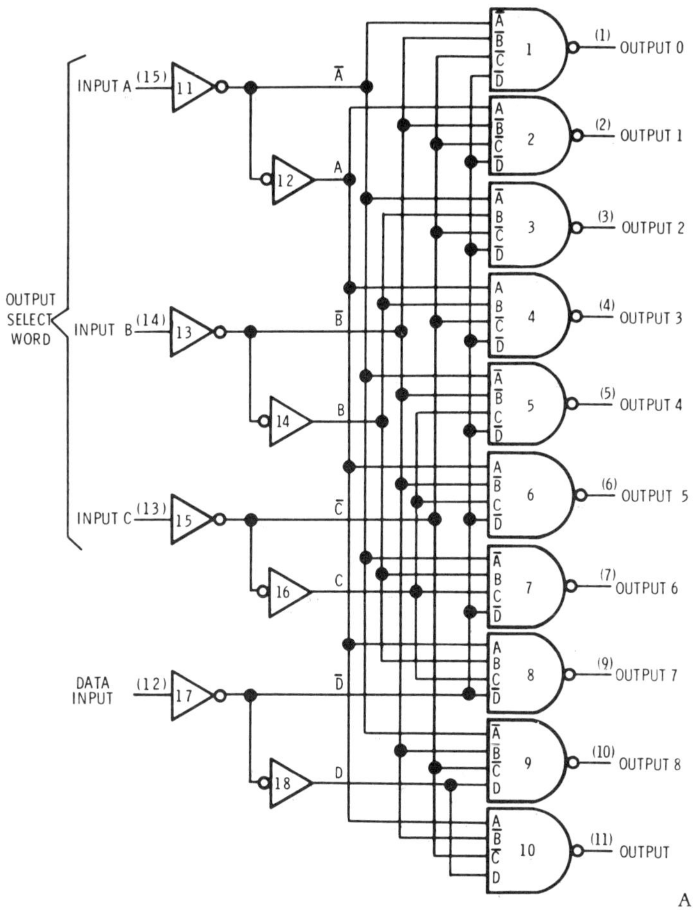

Figure 8-29
A 7442 decoder used as an 8 output data distributor.

A close look at the data distributor circuit in Figure 8-27 shows that it is essentially a decoder where the decode gates all have a common input. Because of this particular configuration, a standard MSI decoder circuit can often be used as a data distributor. Figure 8-42 shows how a 7442 BCD to decimal decoder can be used as an 8 output data distributor. When this circuit is used as a data distributor, inputs A, B and C are used to select the desired output. These three inputs will enable one of the gates 1 through 8. The data input is applied to the D input of the circuit. Note that data input is inverted by inverter 17 and then applied to gates 1 through 8. The data input will appear at the output of the gate selected by the three bit input word ABC. For example, if the input state is 000, gate 1 will be enabled. The data applied to the D input will appear at the output of gate number 1. In this application gates 9 and 10 of the decoder are not used.

EXCLUSIVE OR

One of the most widely used of all combinational logic circuits is the exclusive OR. It occurs so frequently in logic circuits that it is often considered to be one of the basic logic functions such as AND, OR and NOT. The exclusive OR is a two input combinational logic circuit that produces a binary 1 output when one, but not both, of its inputs is binary 1.

The standard OR logic circuit is generally referred to as an inclusive OR. The OR circuit produces a binary 1 output if any one or more of its inputs are binary 1. The exclusive OR produces a binary 1 output only if the two inputs are complementary. The table below compares the output for the standard inclusive OR and exclusive OR circuits. The inputs are A and B, the output is C.

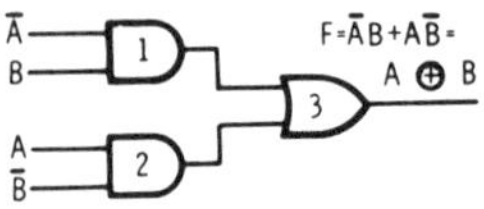

Figure 8-30
Basic exclusive OR logic circuit.

Inclusive OR

A	B	C
0	0	0
0	1	1
1	0	1
1	1	1

Exclusive OR

A	B	C
0	0	0
0	1	1
1	0	1
1	1	0

The exclusive OR logic function can be written as a Boolean expression. By using the technique you learned earlier, you can write the logic equation from the truth table. By observing the input conditions that produce binary 1 outputs, you can write the sum-of-products output. The exclusive OR function is indicated below.

$$C = \overline{A}\,B + A\,\overline{B}$$

A special symbol is used to designate the exclusive OR function in Boolean expressions. Like the plus sign represents OR and the dot represents the AND function, the symbol $\oplus$ represents the exclusive OR function. The exclusive OR of inputs A and B is expressed as indicated below.

$$C = A \oplus B = \overline{A}\,B + A\,\overline{B}$$

The exclusive OR function can be simply implemented with standard AND and OR gates as shown in Figure 8-30. The expression X-OR is often used as a short hand method for indicating the exclusive OR function.

Figure 8-31 shows several ways of implementing the exclusive OR function with NAND and NOR gates. The exclusive OR function implemented with NAND gates is illustrated in Figure 8-31A. The NOR implementation of the X-OR function is shown in Figure 8-31B.

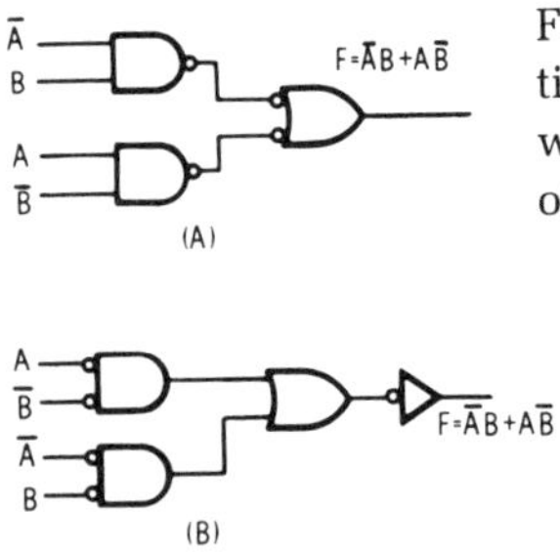

Figure 8-31
Implementation of the X-OR function with NAND gates (A) and NOR gates (B).

Figure 8-32 shows how the exclusive OR function can be performed using the wired OR connection. Here the open collector outputs of TTL or DTL gates can be connected together to produce the OR function required by the X-OR operation.

The exclusive OR circuits in Figures 8-30, 8-31, and 8-32 all assume that both the normal and complement versions of the A and B input signals are available. If they are not, then input inverters can be used to produce them. In some circuits this means extra components and a greater number of interconnections. The exclusive OR circuit in Figure 8-33 avoids this problem. Only the A and B input signals are required in order to generate the X-OR function at the output. This circuit can be readily constructed for example from a standard quad two input NAND gate such as the TTL7400.

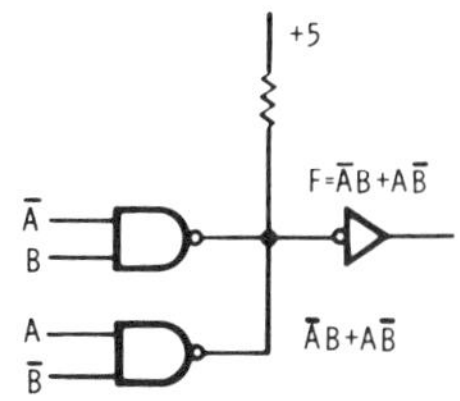

Figure 8-32
The X-OR function complemented with the wired-OR connection.

In order to avoid the necessity of drawing the exact logic diagram for each exclusive OR circuit used, the simplified symbol shown in Figure 8-34 is used. This symbol can be used to represent any of the exclusive OR circuits that we have described so far.

With modern integrated circuits, it is generally not necessary to actually construct exclusive OR circuits from individual logic gates. Instead exclusive OR circuits are available in MSI form. For example, the 7486 TTL integrated circuit contains four completely independent X-OR circuits.

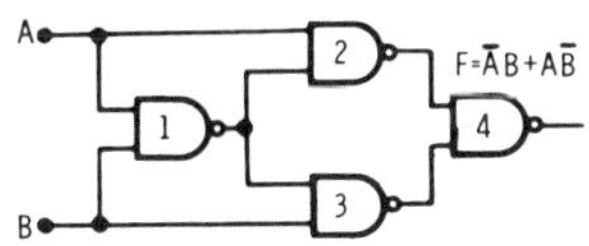

Figure 8-33
X-OR circuit not requiring complement inputs.

Figure 8-34
Standard symbol for an exclusive OR circuit.

Exclusive NOR

An often used version of the exclusive OR is the exclusive NOR (X-NOR) circuit. The truth table for this circuit is given below.

A	B	C
0	0	1
0	1	0
1	0	0
1	1	1

Note that the output of the equivalent circuit is a binary 1 when inputs A and B are equal. If both inputs are 0 or both inputs are 1, the output will be a binary 1. As a result, the exclusive NOR is sometimes referred to as an equivalence circuit or a comparator. Comparing this to the exclusive OR function you can see from the truth table that the output of the X-NOR is the complement of the X-OR.

The Boolean equation of the X-NOR circuit can be written from the truth table. It is indicated below.

$$C = \overline{A}\,\overline{B} + A\,B$$

Since the form of this equation is similar to that for the exclusive OR, (sum-of-products) the equivalence function can be implemented by using any one of the exclusive OR circuits given previously by simply rearranging the inputs. Alternately, all of the previously given exclusive OR circuits can also be used to perform the equivalence operation by leaving the inputs as designated and complementing the output. Figure 8-35 shows several methods of implementing the exclusive NOR function. The simplified symbol shown in Figure 8-36 is frequently used to indicate the X-NOR operation.

Applications of the Exclusive OR

As indicated earlier there are many applications for the exclusive OR logic circuit. There are many special combinational circuits that take advantage of the special characteristics of the exclusive OR. Let's take a look at some of the most widely used applications of the exclusive OR and the exclusive NOR circuits.

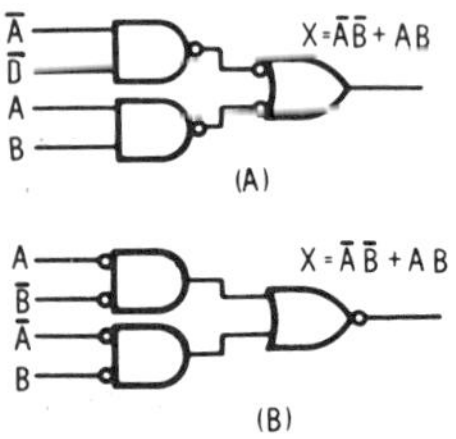

Figure 8-35
Methods of implementing of exclusive NOR or equivalence function with NAND gates (A) and NOR gates (B).

Binary Adder. A binary adder is a circuit that adds two binary numbers. The output of the adder is the sum of the two input numbers. A binary adder is the basic computational circuit used in digital computers, electronic calculators, microprocessors and other digital equipment employing mathematical operations.

The basic rules for a binary addition are very simple. These are indicated below.

```
  0    0    1    1
 +0   +1   +0   +1
 --   --   --   --
  0    1    1   10  ← carry
```

These rules indicate how two single bit numbers are added. Naturally, these rules can be extended to multibit numbers. Several examples of the addition of multibit numbers are shown below.

```
         1   carry   |         111  carries  |
  10     1010        |    7    0111          |    12    1100
 + 3    +0011        |  +11   +1011          |   +10   +1010
 ---    -----        |  ---   -----          |   ---   -----
  13     1101        |   18   10010          |    22   10110
```

Figure 8-36
Symbol for the exclusive NOR function.

A close look at the rules for binary addition indicated above show that if put in truth table form they would be identical to the logical function of an exclusive OR circuit. The exclusive OR inputs A and B are the two single bits to be added while exclusive OR output C is the single bit sum. As you can see, the exclusive OR is a binary adder. The only function not taken care of by the exclusive OR circuit is the carry function. When you're adding two binary 1 bits, a binary 1 carry will be generated. This carry operation can be simply implemented with an AND gate that will produce a binary 1 output only when both inputs are binary 1. Combining the AND gate and the exclusive OR we can develop a basic single bit binary adder circuit as shown in Figure 8-37. This circuit is generally referred to as a half adder.

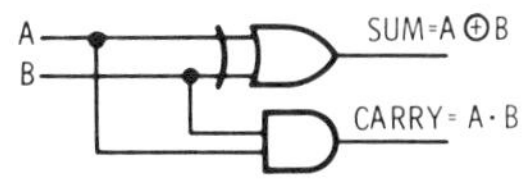

Figure 8-37
Half adder circuit.

To add multibit numbers, we must provide an adder circuit for each of the two corresponding bits to be added. However, the half adder circuit does not provide for a carry input from a lower order bit position. Therefore, an adder circuit must be developed that will add together the two input bits then add to that sum the carry from the next least significant bit position. Such a circuit combines two half adder circuits to form a full adder. This circuit is shown in Figure 8-38. The half adder made up of exclusive OR gate 4 and AND gate number 1 performs the addition of the two input bits A and B. The output of exclusive OR gate 4 is the sum of these two bits. To this sum is added the carry input (Ci) from the adjacent lower order bit position. The sum of bits A and B is added to the carry input in the half adder circuit made up of exclusive OR gate 5 and AND gate 2. The output of exclusive OR gate 5 is the correct sum. Note that because two half adders are used there will be two carry outputs. Since a carry can be generated from either the addition of the two inputs A and B or the addition of their sum and the carry, the two carry outputs are ORed together in gate 3 to produce a correct carry output (Co) that will feed the next most significant bit adder in a multibit adder.

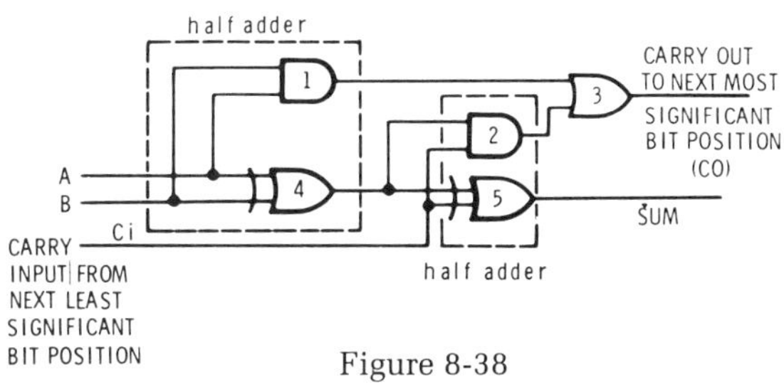

Figure 8-38
Full adder circuit.

Figure 8-39 shows a block diagram of an adder circuit used to produce the sum of two four bit binary numbers. The inputs are two four bit binary numbers A and B. Input number A is made up of bits A1, A2, A3 and A4. Input number B consists of bits B1, B2, B3, and B4. Each of the corresponding bits of the two numbers is added or summed in an adder circuit. Note that the least significant bits A1 and B1 are added in a half adder. Since there is no lesser significant bit, no carry input is required and a half adder circuit will suffice. All other bit positions require a full adder circuit to accommodate the carry input from the next lower order bit position. The output is a four bit parallel sum of the two input numbers with bits S1, S2, S3, and S4. The carry output of the most significant bit full adder also represents the fifth or most significant output bit in those situations where the four bit input numbers produce a five bit sum.

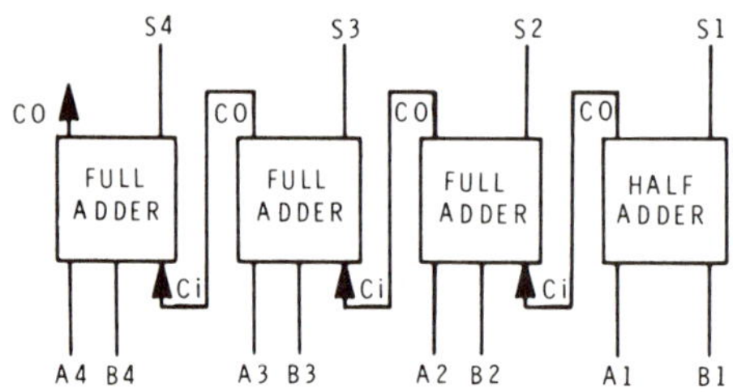

Figure 8-39
Four bit parallel adder.

While the adder circuits described here can be constructed of exclusive OR gates and other logic elements, it is generally unnecessary since single bit adders and four bit adders like those discussed here are available as complete MSI integrated circuits in a single easy to use package.

Parity Generator/Checker. A parity generator is a combinational logic circuit that generates a single output that indicates the presence or absence of a bit error in a binary word. In digital applications requiring the storage of binary data in an electronic memory or in the transmission of binary information from one location to another, there is the likelihood of an error being made. Because of electrical noise or circuit failure, a binary 1 bit may be stored or transmitted as a binary 0. A binary 0 bit could be stored, transmitted or received as a binary 1. In most electronic equipment it is desirable to know when such errors occur. A parity generator circuit performs this function.

The parity generator circuit looks at the binary word to be stored or transmitted and generates a single output known as a parity bit. This parity bit is then added to the other bits of the word and stored or transmitted with it. When the stored word is retrieved from memory for use or when a transmitted word has been received, a parity check operation is performed. The parity checker generates a parity bit from the received data and compares this bit with the parity bit stored or transmitted with the original information. If the two parity bits are identical, no error exists. A difference in parity bits designates an error.

The method of generating a parity bit is to observe the binary word to be stored or transmitted and determine the number of binary 1's in that word. A parity bit will be generated based on this information such that the total number of binary 1's in the word including the parity bit will be either odd or even. The table in Figure 8-40 shows all 16 possible combinations of four bit binary words. The odd and even parity bits for these words are designated in the adjacent columns. Note that for odd parity, a binary 1 or binary 0 parity bit is added to make the total number of bits in the word, including the parity bit, odd or even.

TABLE I

A	B	C	D	ODD PARITY	EVEN PARITY
0	0	0	0	1	0
0	0	0	1	0	1
0	0	1	0	0	1
0	0	1	1	1	0
0	1	0	0	0	1
0	1	0	1	1	0
0	1	1	0	1	0
0	1	1	1	0	1
1	0	0	0	0	1
1	0	0	1	1	0
1	0	1	0	1	0
1	0	1	1	0	1
1	1	0	0	1	0
1	1	0	1	0	1
1	1	1	0	0	1
1	1	1	1	1	0

Figure 8-40
Odd and even parity bits for a four bit word.

The basic circuit element used to generate the parity bit is the exclusive OR circuit. A look at the truth table for an exclusive OR circuit will reveal that it is basically an odd or even detector circuit. If the two inputs are equal or even, the exclusive OR output is a binary 0. But if the inputs are odd or complementary, the output is a binary 1. The exclusive OR gate then can be used to compare two binary bits and indicate whether they are odd or even, equal or unequal. An exclusive OR gate then is used to monitor each two bit group in a binary word. These exclusive OR outputs are then further compared with other exclusive OR circuits until a single output bit indicating odd or even is generated.

Figure 8-41 shows how exclusive OR gates are cascaded or pyramided to produce a parity generator circuit. A 4-bit binary number input to the parity generator circuit is stored in a register made up of flip-flops A, B, C and D. X-OR gate 1 monitors bits A and B while X-OR gate 2 monitors bits C and D. The outputs of these two X-OR circuits are then monitored by X-OR gate 3. The result is an even parity output bit. Inverter 4 generates the complement or the odd parity output. Using your knowledge of the exclusive OR circuit and the information in Table I, trace the various binary states from the flip-flop outputs to the output of the parity generator circuit to be sure that you fully understand its operation.

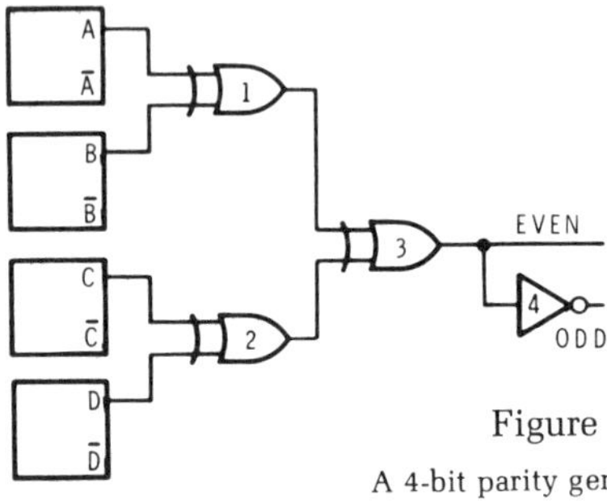

Figure 8-41
A 4-bit parity generator circuit.

A parity bit for any size binary word can be generated by simply using as many exclusive OR gates as necessary to monitor all input bits. Additional exclusive OR gates are then used to monitor the output states of the exclusive OR gates used in monitoring the input bits. This cascading or pyramiding of exclusive OR gates is continued until a single output bit is generated.

Once the parity bit has been generated, it is generally stored or transmitted along with the input word. When the binary word and its parity bit are read from memory or received at the remote location, it can be tested for bit errors in a parity checker circuit. A parity checker consists of a parity generator circuit identical to those just discussed. This parity generator looks at the stored or received word and again generates a parity bit. This bit is then compared to the parity bit stored or transmitted along with the word. This comparison takes place in another exclusive OR circuit. Figure 8-42 shows a parity checker circuit for a 4-bit binary word with parity bit.

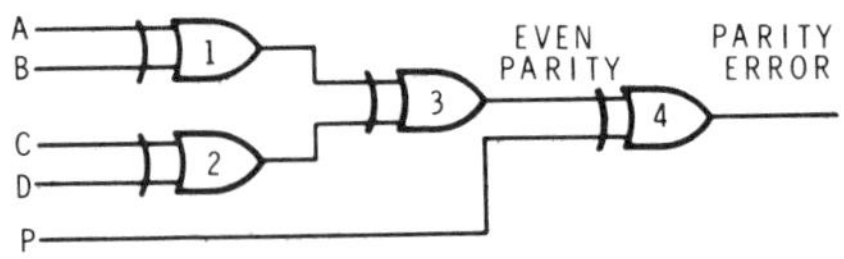

Figure 8-42
A 4-bit parity checker.

Note that the parity generator circuit consists of exclusive OR gates 1, 2, and 3 and is identical to the parity generator circuit in Figure 8-41. Exclusive OR gate 4 compares the output of the parity generator with the received parity bit P. In this circuit we are assuming the use of even parity. If the internally generated parity bit is the same as the received parity bit, the output of the exclusive OR circuit will be binary 0 indicating no parity error. However, if the two parity bits are different, the exclusive OR output will be binary 1 indicating a parity error.

The output of a parity checker circuit can then be used in a variety of ways to indicate the occurrence of a parity error. It can be used to turn on an indicator light indicating an error state. It can be used to initiate a series of logic operations that will either accept or reject the data depending on the error state. Or it may be desirable simply to count and record the number of parity errors that occur.

As you probably realize, a parity error detection method does not ensure complete freedom from or knowledge about all possible error conditions. The parity technique assumes that an error will occur in only one bit position of a word. If parity errors occur in two bit positions, it is possible for the word to be transmitted incorrectly while no parity error will be indicated. This situation rarely occurs since in most electronic storage and transmission systems the reliability is sufficient to eliminate the possibility of multibit errors. However, errors in a single bit position are common. The parity detection and checking process is a very reliable and useful indication of errors.

Even more sophisticated combinational logic circuits have been developed to detect when more than one bit error is produced. In systems requiring ultra high reliability and performance, such sophisticated circuits can be used to detect and even correct any bit error that occurs. In some high speed computers, multiple bit errors are automatically detected and corrected before the information is processed.

While parity generator and checker circuits can be constructed with individual exclusive OR gates, they are also available in MSI form. Figure 8-43 shows a typical commercial parity generator/checker MSI circuit. This circuit is capable of performing either the generation or checking function. As a generator, it can monitor up to nine input bits. The ninth bit is applied to either the odd or even input as required by the application. Both odd and even parity bit outputs are provided. In the checking function, this circuit will monitor an 8 bit word and generate an appropriate parity bit which is then compared with a received parity bit applied to either the odd or even input. The error indication appears at the odd or even output depending upon whether the odd or even parity convention is used.

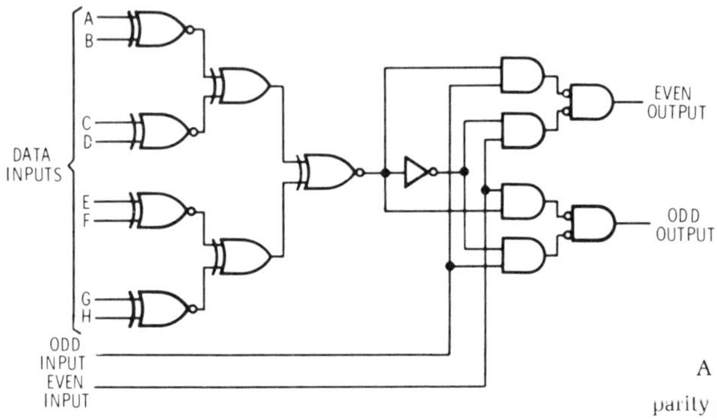

Figure 8-43
A commercial MSI parity generator/checker IC.

Binary Comparators. A binary comparator is a combinational logic circuit that looks at two parallel binary input words and generates a binary 1 output signal if the two numbers are equal. If the numbers or words are not the same, the output will be a binary 0.

As you saw earlier, the exclusive NOR circuit is essentially a single bit binary comparator. When the two inputs are alike, the output is binary 1. When the input bits are different, the output is binary 0. By using an exclusive NOR circuit for each pair of bits in the two numbers to be compared, a complete binary comparator circuit can be constructed.

Figure 8-45 shows a four bit binary comparator circuit. Word 1 with bits A1, A2, A3, and A4 are stored in the A register. The word to be compared is stored in the B register with bits B1, B2, B3, and B4. Each pair of bits is applied to an exclusive NOR circuit. The outputs of the exclusive NOR s are fed to a four input AND gate. When the two binary words are alike, the outputs of the exclusive NOR s will be binary 1. With all binary 1 inputs to the AND gate, the output will be binary 1 indicating the equality of the two words. If any one or more bits of the input words are different, the output of the related exclusive NOR circuit will be binary 0. Naturally, this will inhibit the AND gate and produce a binary 0 output which indicates inequality. Additional exclusive NOR gates and AND gates inputs can be added as required to compare any size binary number.

Commercially available MSI binary comparators are available thus eliminating the need to assemble such circuits from individual gates. Typically, these comparators are designed for comparing two four bit binary words. In addition to providing an output that indicates the equality of the two words, most MSI comparators also generate two additional output signals, one indicating when one word is greater than the other and another indicating when one word is less than another. Figure 8-44 shows a block diagram of such a comparator. If input word A has a binary value that is numerically larger than word B, the A greater than B output (A > B) will be binary 1. The A < B output will be binary 1 if A is less than B.

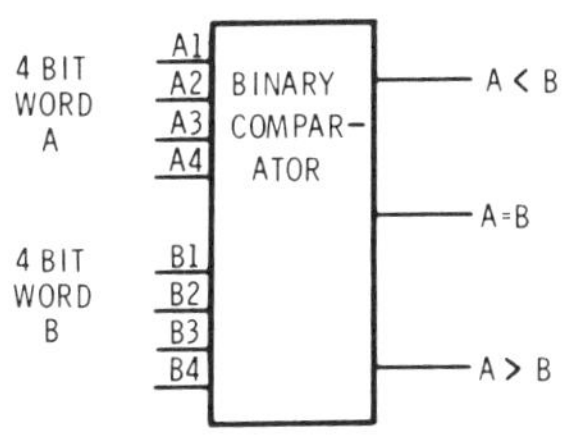

Figure 8-44

Four bit binary comparator.

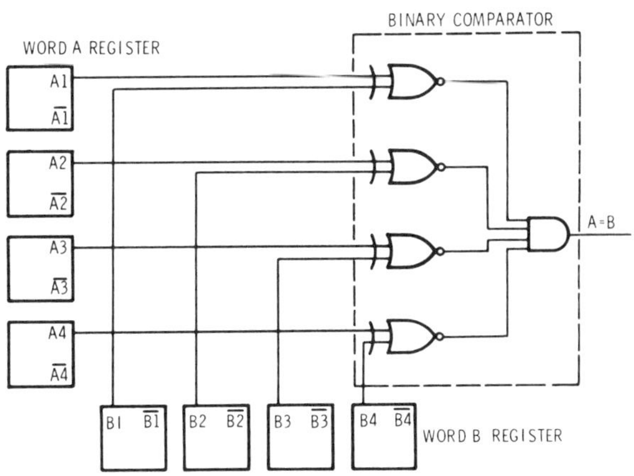

Figure 8-45

Typical 4-bit commercial MSI binary comparator circuit.

CODE CONVERTERS

A code converter is a combinational logic circuit that converts one type of binary code into another. There are many applications in digital systems where two or more different binary codes are used. There are applications where it is desirable to take advantage of the characteristics offered by the different types of codes. By using code converters, the various circuits can be made compatible. Some of the most commonly used code converters are listed below.

Binary to BCD
BCD to binary
Binary to Gray
Gray to Binary
8421 BCD to XS3
XS3 to 8421 BCD
ASCII to EBCDIC
EBCDIC to ASCII

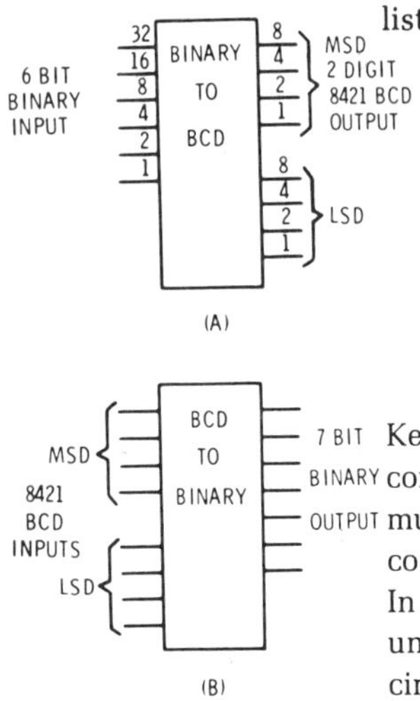

Figure 8-46

Binary to BCD (A) and BCD to binary (B) code converters.

Keep in mind that a combinational logic circuit can be constructed to convert any code into any other code. In a sense, any multibit input-multibit output combinational logic circuit can be considered a code converter. Any combination of input bits can be considered a binary code. In the same way any combination of output bits can also be considered a unique binary code. Using this broad definition of a code converter then circuits such as encoders and decoders could also be considered special forms of code converters. In general, you will find the term code converter used primarily to reference those special circuits indicated in the list above.

The most commonly used code converters are those for converting between binary and BCD. Block diagrams of such converters are shown in Figure 8-46. In Figure 8-46A, a 6 bit pure binary input number is converted into its two digit BCD equivalent output. Any 6 bit input number 000000 through 111111 (0 through 63 decimal) will be converted by the circuit into its BCD output equivalent. For example, the number 63 in binary is 111111. Its BCD output equivalent is 0110 0011.

A BCD to binary converter is shown in Figure 8-46B. A two digit BCD input word is converted into a 7 bit pure binary output equivalent. The input numbers can be anything from 00 through 99 or 0000 0000 through 1001 1001. If the BCD input equivalent of the number 99 is applied, the pure binary output number 1100011 will appear at the output. Both binary to BCD and BCD to binary converters are available as MSI combinational ICs, thereby eliminating the need to implement such circuits with gates.

Other frequently used code converters are binary to Gray and Gray to binary. There are many applications where the Gray or cyclical code must be used in order to minimize errors when changing from one state to another. While the Gray code is good for minimizing errors in generating certain types of data, the Gray code cannot be used in arithmetic operations. To permit arithmetic operations to be performed, Gray to binary code conversion is necessary.

Both Gray to binary and binary to Gray code converter circuits are shown in Figure 8-47. The Gray to binary circuit is shown in Figure 8-47A while the binary to Gray circuit is shown in Figure 8-47B. The most significant bit of both the Gray and binary words will be the same and therefore no code conversion is necessary. Note the use of exclusive OR circuits to perform the code conversion. The Gray and equivalent binary codes are shown in Figure 8-48.

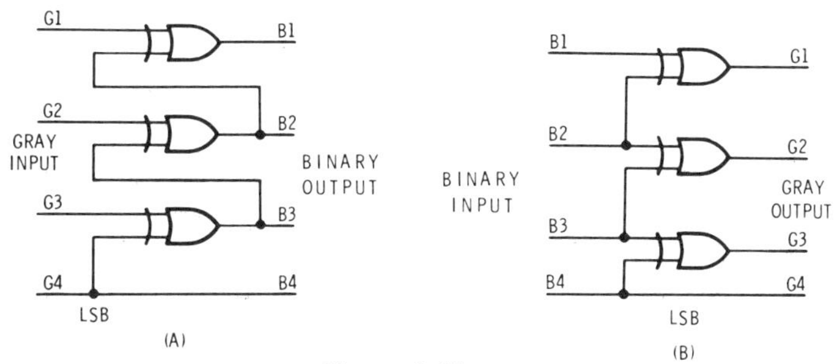

Figure 8-47

Gray to binary (A) and binary to Gray (B) code converters.

DECIMAL	BINARY				GRAY			
D	B4	B3	B2	B1	G4	G3	G2	G1
0	0	0	0	0	0	0	0	0
1	0	0	0	1	0	0	0	1
2	0	0	1	0	0	0	1	1
3	0	0	1	1	0	0	1	0
4	0	1	0	0	0	1	1	0
5	0	1	0	1	0	1	1	1
6	0	1	1	0	0	1	0	1
7	0	1	1	1	0	1	0	0
8	1	0	0	0	1	1	0	0
9	1	0	0	1	1	1	0	1
10	1	0	1	0	1	1	1	1
11	1	0	1	1	1	1	1	0
12	1	1	0	0	1	0	1	0
13	1	1	0	1	1	0	1	1
14	1	1	1	0	1	0	0	1
15	1	1	1	1	1	0	0	0

Figure 8-48

Binary and Gray codes.

While code conversion is most often accomplished with combinational logic circuits, many types of code converters use sequential circuits. Various combinations of flip-flops, counters and shift registers can be used to perform code conversion. A simple example is the serial Gray to binary code converter shown in Figure 8-49. Here a serial Gray code is applied to the JK inputs of a JK flip-flop, MSB first. The normal flip-flop output is a serial binary code.

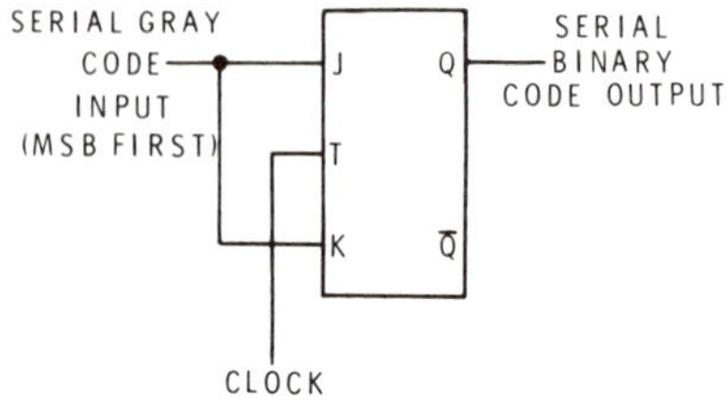

Figure 8-49
Serial Gray to binary sequential code converter.

READ ONLY MEMORIES

A read only memory (ROM) is an electronic circuit used to permanently store binary information. Practical read only memories are available for storing as many as 65,000 bits of data. Normally the ROM is organized to store equal length multibit words. For example a typical ROM might be capable of storing 512 eight bit words or 4096 bits.

The main feature of a read only memory is that the binary information contained in the memory is permanently stored there. The data is written into the memory when it is manufactured. The contents of the memory cannot usually be changed thereafter. This is in contrast to many electronic read/write memories that can both store and read out data. This type of memory is generally referred to as random access memory (RAM). You can think of it as many storage registers.

The general organization of both read/write memories and read only memories is basically the same. Both contain a number of memory locations where data can be stored. In the ROM, the data is stored there permanently and can be read out in any order. In the RAM, data may be written into or read out of any portion of the memory at any time. While the read/write memory is more flexible, it is also more expensive. It is this type of memory that is normally used as the main storage section of a digital computer. Our discussion here will center on the ROM which is a useful circuit in implementing digital systems.

ROM Operation

Figure 8-50 shows a general block diagram of a read only memory. It consists of three major sections: The address decoder, the memory storage elements, and the output circuits.

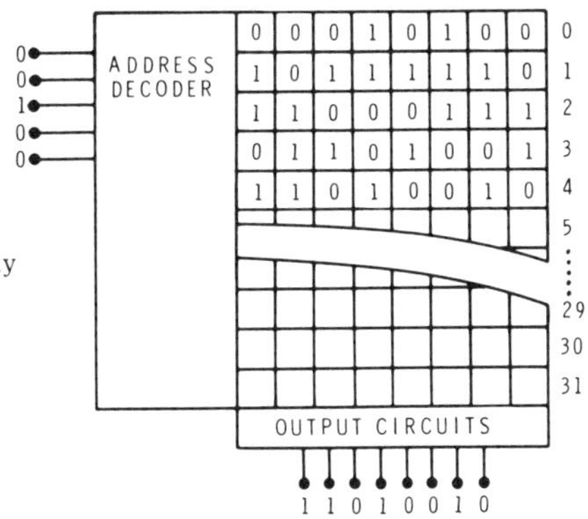

Figure 8-50
General block diagram of a read only memory.

The address decoder is like any binary decoder in that it accepts a multibit binary input word and decodes all possible input states. Only one of the decoder outputs will be activated. In the address decoder of Figure 8-50, there are five input bits meaning that a total of $2^5 = 32$ different states can be decoded. This five bit input word specifies one of 32 individual memory locations. This input word is generally referred to as the address.

The main body of the memory consists of electronic circuits or components that are used to store the binary data. These storage elements are arranged so that a specific number of multibit binary words may be stored. The organization in Figure 8-50 permits 32 eight bit words to be stoed. The memory locations are designated 0 through 31. Applying a five bit address code to the input will cause the contents of the addressed location to appear at the output. Note that if the address input code is 00100, the contents of memory location 4 appear at the output. All other memory locations are ignored at this time. The output circuits buffer the memory contents so that the data can be used in other logic circuits.

ROM Construction

There are many different ways to implement read only memories with electronic components. Any component or circuit capable of storing a binary 1 or binary 0 condition can be used. Magnetic cores and capacitors are examples of elements that have been used to store binary data in a ROM. Most modern read only memories, however, are semiconductor circuits. Both bipolar and MOS types are used. Since ROMs are capable of storing a significant amount of data they are generally classified as large scale integrated (LSI) circuits. Most ROMs, both bipolar and MOS types, are housed in standard dual in-line packages. Because of the wide variety of possible applications, ROMs are considered to be custom circuits. The user specifies the memory contents prior to the manufacture of the device. In this section, we investigate the most popular types of integrated circuit ROMs in use today.

Diode Matrix ROM. Figure 8-51 shows a read only memory constructed with a one of eight decoder and a diode matrix. The one of eight decoder accepts a three bit address input word and generates all possible decode output combinations. This means that the decoder will recognize the three bit input number applied to it and enable only one of the eight outputs. For example, if the binary input number is 011, the number 3 output line will go low. All other decoder output lines will be high at this time. The decoder is similar in operation to the type 7442 discussed earlier.

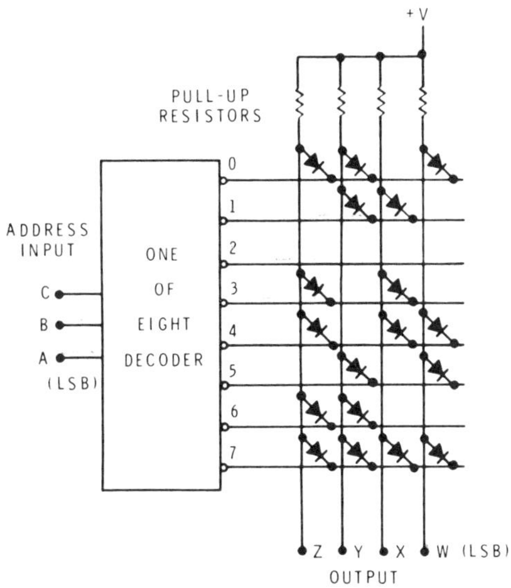

Figure 8-51
A ROM constructed from a decoder and diode matrix.

When the number 3 decoder output line goes low, it brings the cathode ends of the diodes connected to this line low. The diodes conduct through their associated pull-up resistors. This forces lines X and Z low. Lines W and Y are high at this time because of the pull-up resistors. Since all other decoder outputs are high, the other diodes in the circuit are cut off at this time. Observing the output lines ZYXW then you see the output code 0101. At address location 011 (3), the binary number 0101 is stored.

Consider the effect of applying the address 110 to the decoder input. This will bring decoder output 6 low causing output lines Y and Z to go low. Lines W and X will be high at this time. This means that the output number is 0011. The contents of memory location 110 (6) is the four bit number 0011.

A close look at the ROM in Figure 8-51 should reveal that the data is stored in the memory as either the presence or absence of a diode. In this circuit a diode connection between the decoder output and the output line causes a binary 0 to be read out when that address line is enabled. The absence of a diode causes a binary 1 to be read out. Another way to look at the read only memory is to consider each output line with its associated diodes and pull-up resistors as a diode OR gate. A low on any dode input causes the output to go low.

There are some commercial read only memory ICs designed and constructed exactly like that shown in Figure 8-51. Integrated circuit ROMs are constructed initially so that a diode is connected at each possible memory location. This means that all memory locations are initially programmed with binary 0's. To store data in the memory, an external pulse signal is applied to the output lines in such a manner as to reverse bias certain diodes and cause them to be destroyed. By destroying a diode and causing it to open, a binary 1 state is programmed. Such ROMs can be programmed by the manufacturer or the user. Read only memories that permit the user to store the data that he needs are called programmable read only memories (PROMs).

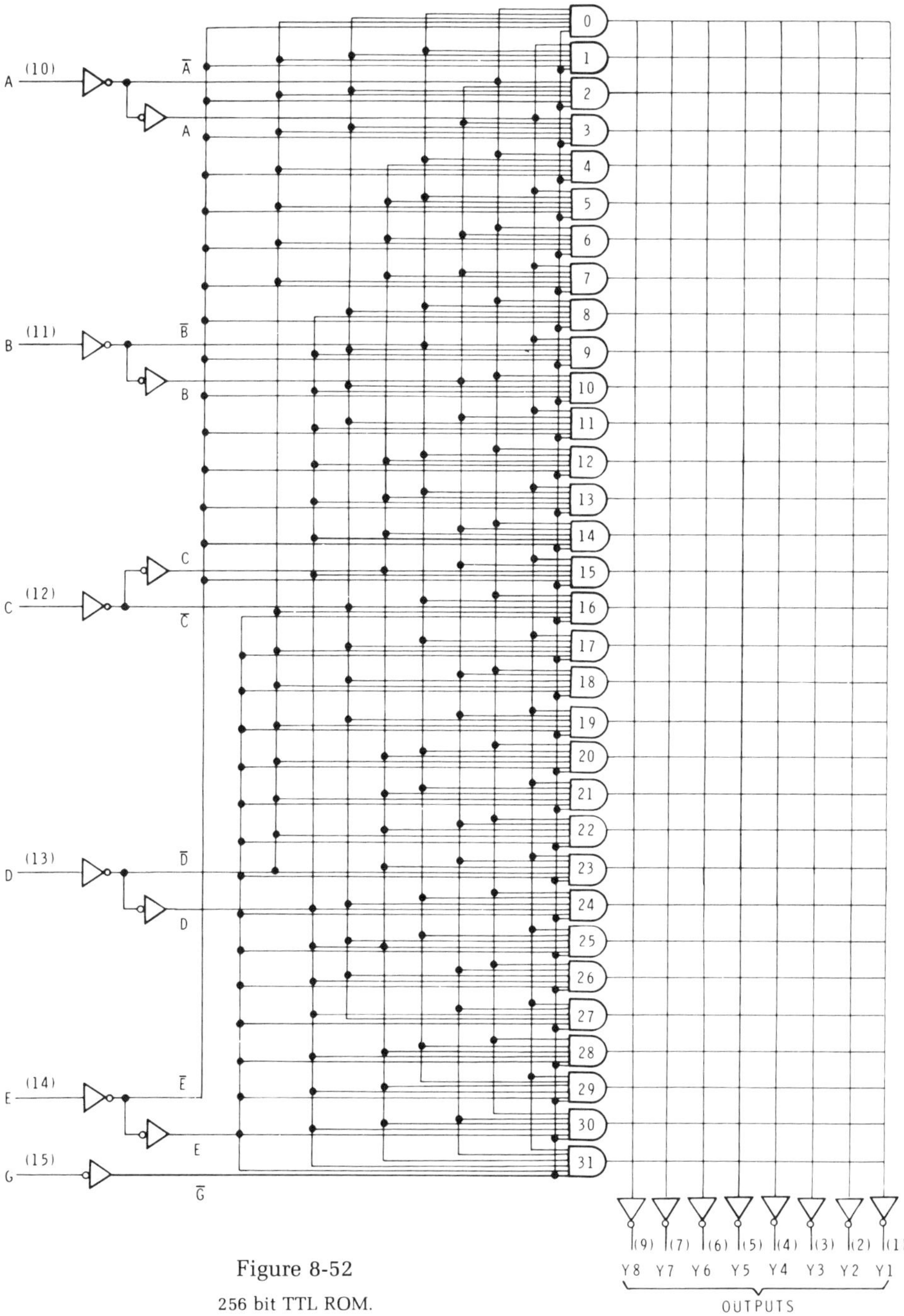

Figure 8-52
256 bit TTL ROM.

Bipolar ROM. A typical commercial bipolar read only memory is shown in Figure 8-52. This 256 bit ROM uses TTL circuitry. The memory is organized as 32 eight bit words. The address inputs labeled A through E are used to select one of the 32 words stored in the memory. Note that the circuitry is basically a 1 of 32 decoder.

Figure 8-53 shows the detailed circuitry of the ROM. Illustrated here is one of the 32 address decoding gates and the 8 output buffer circuits. The output of each decoding gate is a transistor with eight emitters. These emitters can be interconnected to the eight output buffers. The programming of the memory is done by either connecting or leaving open these emitter connections. If an emitter is connected to an output buffer, the output voltage will go low when that decoding gate is addressed. If the emitter is not connected, a high level voltage is read out of the associated buffer when the gate is addressed.

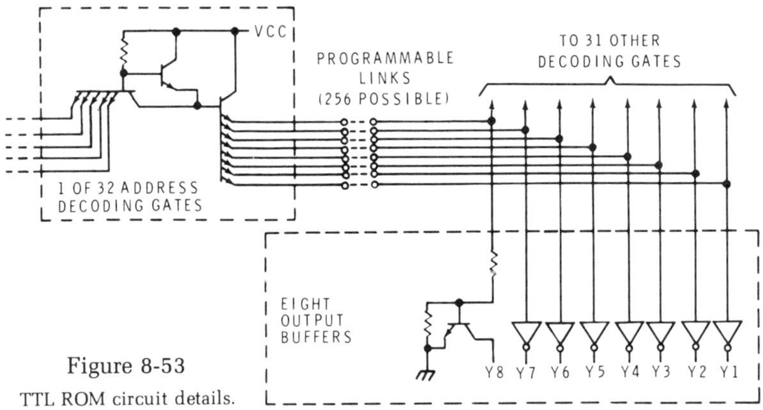

Figure 8-53
TTL ROM circuit details.

The decoding gate output emitters to be used are connected to the respective inputs of the eight output buffers when the integrated circuit is manufactured. The user specifies the memory contents, and the manufacturer produces special masks which will cause the interconnecting metalization of the integrated circuit to be properly arranged to store the desired data. Note that the output buffers have an open collector output. This permits the outputs to be wire-ORed with other similar memories so that the storage capability can be expanded. Three state output circuitry is used on some TTL ROMs. Input line G on the ROM in Figure 8-52 is used to enable or disable the circuitry so that this device can be combined with others similar to it in forming a memory with many more locations. This line is often referred to as a chip select line and is used as an extra address bit input in expanded memories.

Figure 8-54 shows two ways that a standard size ROM can be used to make larger memories. Figure 8-54A shows two ROMS connected to form a memory for 32 sixteen bit words. Each ROM can store 32 eight bit words as indicated by the designation 32 x 8 or 32 by 8. The five address lines are in parallel thus each ROM is enabled at the same time. One half of the 16 bit word is stored in the upper ROM and the other 8 bit segment in the lower ROM. Since the two ROMs are addressed simultaneously, both parts of the word will be read out at the same time. Input line G enables the memory when low.

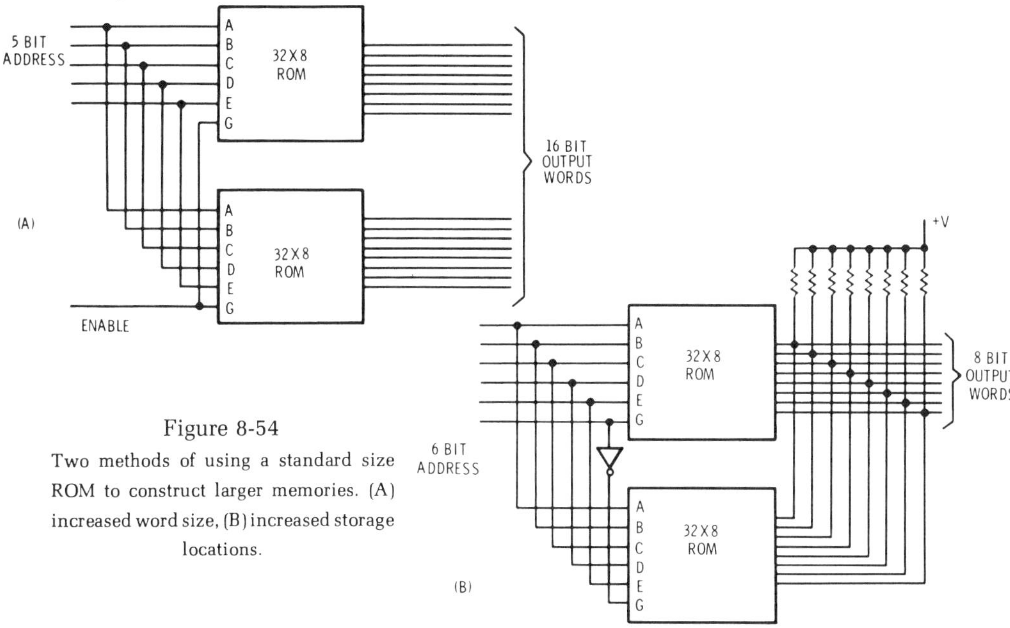

Figure 8-54

Two methods of using a standard size ROM to construct larger memories. (A) increased word size, (B) increased storage locations.

Figure 8-54B shows how the 32 x 8 ROMs can be used to form a memory for storing 64 eight bit words. Thirty-two of the words are in the upper ROM and the other thirty-two are in the lower ROM. The ROM outputs are wire-ORed. The five address lines A through E are in parallel. The chip enable lines (G) are used as a sixth address line. Remember, it takes six bits to address 64 words ($2^6 = 64$.). This sixth input line is the MSB of the address. The five lower order bits address both ROMs simultaneously. But only one of the two ROMs will be enabled by input G. The inverter keeps these lines complementary.

If the input address (GEDCBA) 101101 is applied, location 13 (01101) in each ROM will be addressed. However, input G is binary 1. This disables the upper ROM so all of its output lines are high. Input G to the lower ROM is low because of the inverter. Therefore, this ROM is enabled and the word at location 01101 is read out.

MOS ROMS. Many read only memories are implemented with metal oxide semiconductor integrated circuits. MOSFET circuitry lends itself well to the implementation of a read only memory. Because of the small size of most MOSFET circuits, many logic and memory elements can be constructed in a small space. This high density circuitry permits read only memories with a very high bit content to be readily manufactured. Many thousands of bits of data can be stored on a silicon chip approximately 1/10 inch square. Such MOS ROMs are low in cost and consume very little power.

The basic organization and structure of an MOS ROM is essentially the same as any read only memory. An address decoder selects the desired word. The presence or absence of a semiconductor device in a matrix network specifies a binary 1 or binary 0 stored in the addressed location. In MOS ROMs, the basic storage element is an enhancement mode MOSFET. The presence of an MOSFET programs a binary 1. The absence of such a device means a binary 0 has been programmed.

Figure 8-55 shows the basic internal structure of a typical PMOS ROM. P-type material is diffused into the substrate in long strips called bit lines as indicated. These P-type diffusions form the source and drain connections

Figure 8-55
Basic structure of a PMOS ROM.

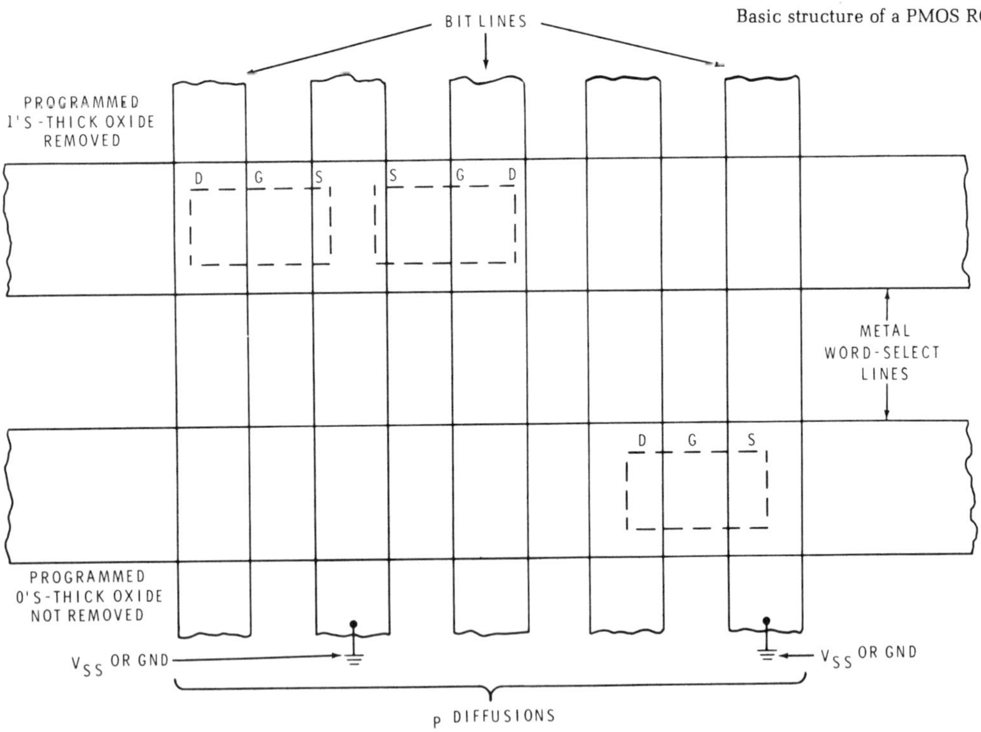

of the MOSFETs. Perpendicualr to the P diffusion areas are metal word select lines. These metal areas form the gate elements of the MOSFETs. Figure 8-55 shows several examples of how the MOSFETs are formed. The source (S), gate (G), and drain (D) of each MOSFET is identified. To program the memory, the MOSFETs formed by this structure are either enabled or disabled by appropriate masking operations during manufacturing. As indicated earlier, if the MOSFET is enabled, a binary 1 will be stored in that location. Disabling the MOSFET causes a binary 0 to be stored in the selected location.

In the MOSFET ROM structure, the metal word select lines are connected to the gates of the MOSFETs where binary 1s are stored. These metal word select lines are driven by the outputs of a decoder. The source terminals of the MOSFETs are connected either to ground or to the source supply voltage V_{SS}. The drain connections of the MOSFETs are designated as the bit lines. If one of the metal word select lines goes negative, the MOSFETs associated with that word will conduct and ground (or V_{SS}) will appear on the bit line.

Figure 8-56 shows the MOSFET ROM circuit. Q1 is a MOSFET formed by the process illustrated in Figure 8-55. The presence or absence of this transistor is a function of the masking process carried out during manufacturing. Note that the gate of Q1 is enabled by the output of decoder X. If the word select line is negative, Q1 will conduct and a binary 1 bit will appear on the bit line. However, this binary 1 may or may not reach the output of the ROM depending upon the state of Q2. Q2 and decoder Y are also used in selecting the desired output word.

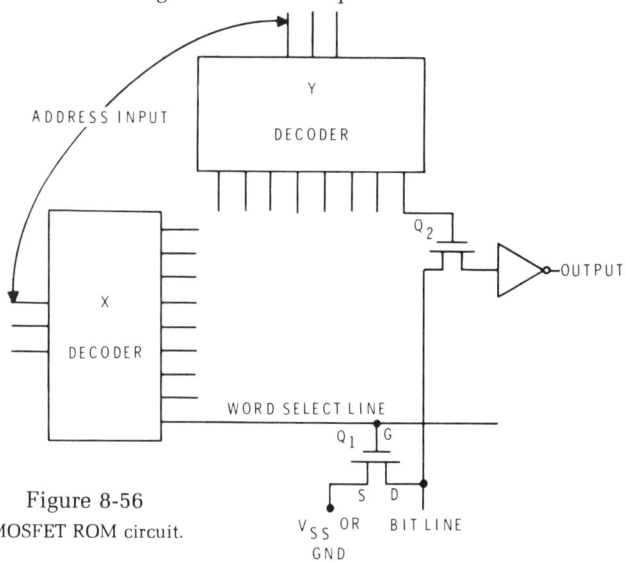

Figure 8-56
MOSFET ROM circuit.

Most MOS ROMs use an XY matrix decoding method. In Figure 8-56, two 1 of 8 decoders are illustrated. Two 3 bit words are used to address a particular word. The two 3 bit input numbers are simply treated as a single six bit address. Six bits define $2^6 = 64$ bit locations. By using two 1 of 8 decoders, a total of 64 words can be addressed. The word in memory is selected by enabling each decoder with the appropriate three bit word. If the Y decoder enables Q2, Q2 will conduct and connect the bit line to the output buffer. If the decoder does not enable Q2, the output on the bit line shown will not appear at the output despite the fact the word select line may have enabled Q1.

Access Time. Like any logic circuit, a ROM has propagation delay. This means that there is a finite time between the application of an input address and the appearance of data at the output. This propagation delay is referred to as access time. This is the time it takes to find a word in the ROM and read it out. For bi-polar ROMs, this access time is usually less than 100 nanoseconds and can be as low as 20 nanoseconds. For MOS ROMs, the access time is typically several hundred nanoseconds.

ROM Applications

The ROM is extremely versatile in implementing logic functions. An appropriately programmed ROM can often be used to replace many different types of combinational circuits. It is particularly useful in replacing complex logic functions with multiple inputs and outputs. The ROM offers the advantages of faster and easier design, lower cost, smaller size and often lower power consumption.

Combinational logic circuits generate output signals that are a function of

1. the input states
2. the types of gates used and
3. the particular unique interconnection of these gates.

The desired output states for a given set of inputs are produced by properly interconnecting the correct types of logic gates. This same logical function can be simulated by a ROM. The desired inputs are applied to the ROM address lines. These inputs specify a unique memory location. In this memory location is a binary bit pattern whose output states duplicate those produced by an equivalent combinational logic circuit. Instead of actually generating the desired output function with a logic circuit, we store the desired output states in the memory and read them out when the proper inputs appear on the address lines.

A ROM performs what is known as a table look up function. All of the memory locations can be considered to be entries in a large table of numbers. By applying an address to the ROM, we are in effect looking up one unique number in the table. In a sense, the ROM does not perform a logic operation. The desired output states for a given set of input conditions are simply stored in the memory.

The following examples will illustrate some of the many applications of a ROM.

Random Logic. A read only memory can be used to quickly and easily implement any random logic function involving multiple inputs and multiple outputs. To design such a combinational logic circuit with standard logic gates, you first develop a truth table that defines the operation to be performed. From the truth table the Boolean equations are then written. Boolean algebra is then used to minimize these equations. From the equations, the logic circuit is developed and then implemented with standard NAND gates, NOR gates and inverters.

When a read only memory is used, the only design step is the implementation of the truth table. The truth table defines the inputs and outputs. This is all of the information that is necessary to develop a read only memory that will perform the desired logic function. The input logic states are assigned as addresses of a read only memory. In the memory locations corresponding to the addresses are stored binary words that cause the output lines to assume the desired states with the given input address. By using a ROM you can go from truth table to finished logic circuit in one simple step. Design time is reduced considerably.

If the function to be implemented involves only a few inputs and a few outputs, such a circuit is best implemented by conventional means with logic gates. However, if the number of inputs and outputs is four or more, the use of a ROM becomes practical. Since a ROM costs more than standard SSI logic circuits, it is not practical or economical to use a ROM where very simple functions must be implemented. Four inputs and four outputs are generally regarded as the decision point between a read only memory vs. a conventional logic circuit.

Code Conversion. As we indicated earlier, code conversion refers to any multi-input/multi-output combinational logic circuit. A code converter is nothing more than a special application of such a logic circuit. Since read only memories can readily replace multi-input/multi-output combinational logic circuits, a ROM provides a simple and low cost means of code conversion.

To use a ROM as a code converter, the input code is made equal to the binary address code in the ROM. In the memory location specified by the input or address code is the desired output code. No complex logic functions must be implemented to achieve this result. The desired output codes are simply stored in the memory locations and are read out when the equivalent input code is applied. All of the most commonly used code conversion processes mentioned earlier have been implemented with read only memories.

Arithmetic Operations. Arithmetic operations are some of the most difficult functions to implement with digital circuitry. Simple combinational logic circuits have been developed to perform additions and subtractions. Various algorithms have been developed for using addition and subtraction along with other digital operations such as shifting to perform multiplication and division. More complex mathematical functions such as the trigonometric and logarithmic functions are even more difficult to implement. The read only memory provides a very simple and direct method of implementing the more complex arithmetic operations.

The multiplication of two binary numbers requires a significant amount of logic circuitry. While there are numerous methods for carrying out multiplication, all of them require an extensive amount of circuitry. Multiplication can be performed with a read only memory without the need for complex circuitry or high cost. In addition, the ROM can provide this function at lower cost, in a smaller space and at a significantly higher speed.

The truth table shown in Figure 8-57 illustrates the concept of multiplying two binary numbers using a ROM. To simplify the explanation we will use only two bit binary numbers. Multiplying two binary numbers produces a product whose length is twice that of one of the input numbers. The two bit binary numbers serving as the multiplier and multiplicand will form a four bit product. The two input numbers are grouped so that they form a single four bit binary input number which serves as the address input to the ROM. At the address formed by the input numbers, the correct four bit product corresponding to the two bit numbers is stored. When any combination of the two bit input numbers appears on the ROM address lines, the correct product is read out. For example when one input is 10 (2) and the other is 11 (3), the input address formed is 1011. At this location is the number 0110 (6) which is the product of 2 and 3. By using a larger ROM, numbers requiring more bits can be used.

INPUTS		OUTPUTS
0 0	0 0	0 0 0 0
0 0	0 1	0 0 0 0
0 0	1 0	0 0 0 0
0 0	1 1	0 0 0 0
0 1	0 0	0 0 0 0
0 1	0 1	0 0 0 1
0 1	1 0	0 0 1 0
0 1	1 1	0 0 1 1
1 0	0 0	0 0 0 0
1 0	0 1	0 0 1 0
1 0	1 0	0 1 0 0
1 0	1 1	0 1 1 0
1 1	0 0	0 0 0 0
1 1	0 1	0 0 1 1
1 1	1 0	0 1 1 0
1 1	1 1	1 0 0 1

Figure 8-57
Truth table for 2 bit ROM multiplier.

A ROM is particularly useful in handling complex mathematical operations such as the trigonometric and logarithmic functions. Instead of having the digital circuitry actually compute the sine, cosine or tangent of a number, the ROM simply stores the trigonometric functions corresponding to the angles. In the same way, a ROM can be used to store the logarithms of specified input numbers. In these applications, the ROM is virtually a log table or trig function table. The desired angle or input number is applied to or assigned a binary address that is applied to the ROM. At the address representing the desired input angle or number is stored the correct trigonometric function or logarithm.

Microprogramming

Microprogramming is a technique originally developed to systematize the automatic control logic in a digital computer. The heart of a microprogrammed control unit is a read only memory. The read only memory is combined with other logic elements to perform sequential logic operations. This combination is called a microprogrammed controller.

Most sequential operations are carried out by counters and shift registers in combination with combinational logic circuits. These circuits are used to generate a sequence of timing pulses that will control the operations in other parts of the digital system. The signals generated may increment counters, cause data transfers to take place between registers, enable or inhibit various logic gates, select a multiplexer channel or permit a decoding operation to take place. All of these operations will be timed so that they occur in the correct sequence to perform the desired operation.

In large digital systems the control logic circuitry can become very complex. This is particularly true of the control logic in a digital computer. By the use of microprogrammed controller, the entire network of sequential and combinational logic circuits can be replaced by a very simple circuit containing a read only memory. Figure 8-58 shows several methods of implementing sequential logic functions with a ROM.

In Figure 8-58A, the ROM is driven by a binary counter. A periodic clock signal increments the four bit binary counter. The counter output is used as the ROM address input. The address decoder is assumed to be part of the ROM itself. The ROM output consists of parallel 8 bit words. Since there are four input bits, the ROM therefore contains sixteen 8 bit words. As the binary counter is incremented, a sequence of 8 bit words appears at the ROM output. The words stored in the ROM are programmed such that the binary states appearing at the ROM outputs will cause the desired logic operations to take place in the correct sequence. Here the eight output lines can be used for a variety of control purposes. The states of these outputs are strictly a function of the bits stored in the ROM. The rate of change of the ROM outputs is a function of the frequeny of the clock pulses stepping the binary counter.

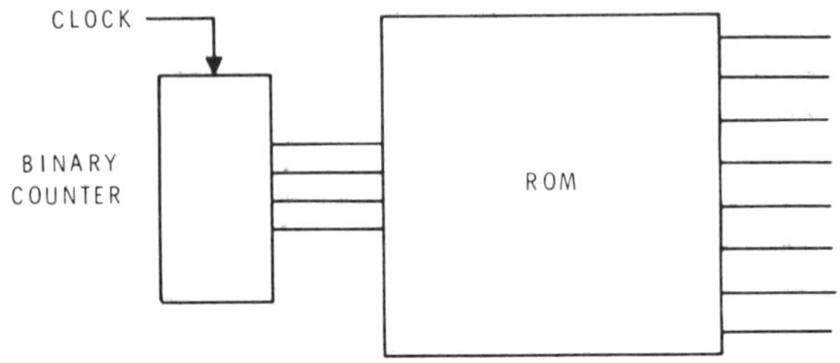

(A)

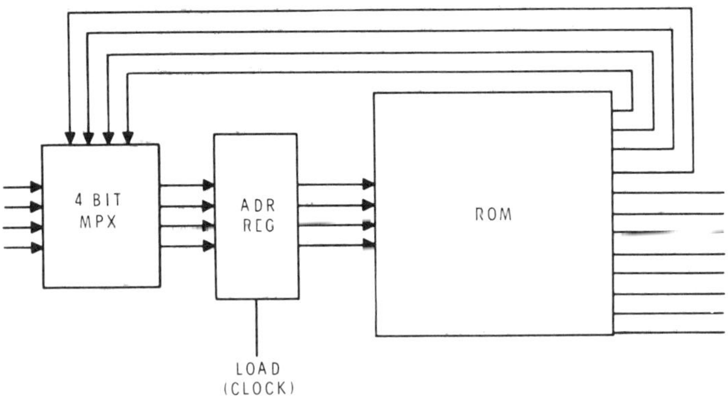

(B)

Figure 8-58

Microprogrammed controllers using a ROM.

A more sophisticated version of this same circuit is shown in Figure 8-58B. Again, a ROM is the main circuit element with the desired output states specified by the ROM contents. Note, however, that four of the ROM output bits are fed back around to the inputs of a 4-bit multiplexer. Another group of four input bits is applied to the multiplexer as well. The multiplexer can select either of the two 4-bit sources as an address and feed them to an address register. The address register in turn selects a specific ROM word. In this circuit, the four bits define sixteen words in the ROM. The output is a 12-bit word, eight bits for controlling external operations and four bits which are used to determine the next address of the word in the ROM.

To operate this circuit, a four bit starting address is applied to the multiplexer from an external source. This is applied to the address register. One specific word in the ROM is addressed and its outputs appear. At this point the state of the multiplexer is changed so that the next input to the address register will come from the four bits in the current ROM output word. This permits the ROM to select the next word that should appear at the output. By repeatedly loading the address register with a clock signal, the ROM addresses are sequenced and a desired pattern of output pulses is produced. With this arrangement, the ROM words addressed can be either sequential or in any desired order. The four bit address output from the ROM could specify the next location in sequence. Alternately, it can select any other word in memory. All of this is determined beforehand in the design of the circuitry. Once the proper sequence of operations is determined, the contents of the ROM can be specified.

The term micro programming is applicable to these circuits in the sense that the words stored in the ROM make up a specific program for carrying out a specific function. Each word in the ROM is referred to as a microinstruction. The bits of that word appearing at the ROM outputs cause certain operations to occur; in other words, that word instructs the external circuitry in the function to be performed. Each word or microinstruction stored in the memory makes up a microprogram. The microprogram defines the complete operation to be initiated. The circuits given in Figure 8-58 are only a few of many different ways that microprogrammed operations can be implemented.

PROGRAMMABLE LOGIC ARRAYS

A programmable logic array (PLA) is an integrated circuit logic network that can be used to perform many different types of combinational logic functions. It offers the digital designer an alternative to the use of combinational logic circuits made with standard SSI and MSI ICs or read only memories. For many applications the PLA offers a significant improvement in performance over both conventional logic circuit implementation and read only memories.

Basically the PLA is a bipolar or MOS logic network that can be programmed during manufacturing to produce a wide variety of combinational logic functions. It is capable of translating any input code into any output code. The circuit is designed to generate a large number of sums of partial products.

Figure 8-59 shows a general logic diagram of a PLA. The multiple binary inputs (I_1 through I_{14}) are applied to inverters which are used to generate both the normal and complement versions of the input signals. The inverter outputs may then be interconnected to one of many AND gates. In one particular commercial PLA, a total of 96 twelve input AND gates are provided. These AND gates generate the product terms of the input variables. Up to 14 different input variables may be handled by this particular PLA. The products or partial products of the inputs formed by

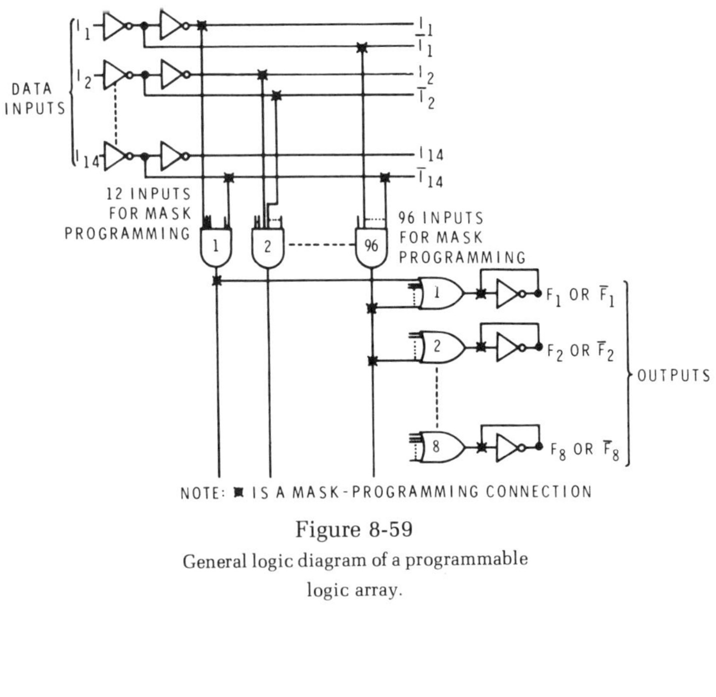

Figure 8-59

General logic diagram of a programmable logic array.

the AND gates are then connected to OR gates to form the output sums (F1 through F8). The selection of which input variables are applied to which AND gate outputs are connected to the eight available output OR gates is determined during the manufacturing of the device. By properly designing the mask that specifies the interconnections of the logic gates on the chip, a huge number of circuit configurations are possible. Note also in Figure 8-59 that even the use of an inverter on the outputs of the OR gates is programmable.

To design a logic circuit with a PLA involves basically the same procedure as developing the logic for any combinational logic circuit or selecting an ROM. The procedure generally starts with the truth table that defines the output states for each combination of input states. The output equations are written in sum-of-products form. The required product terms are listed and these are converted into the appropriate mask programming instructions for making the IC.

The PLA is particularly valuable in implementing large complex combinational logic circuits. Simple functions are readily implemented with SSI logic gates. More complex functions can be handled by one of the many available MSI functions. But when there are many input variables and many output variables, the use of standard SSI and MSI packages also becomes complex and cumbersome. The PLA can be used to generate the desired complex function and house it in a single integrated circuit package.

The PLA also offers numerous advantages over the read only memory for implementing some complex logic functions. Of course, a ROM can be used to handle logic functions involving any number of inputs or any number of outputs. However, the read only memory becomes very inefficient when all possible combinations of the input variables are not used. For example, a four input logic circuit has 16 possible different combinations. The design application may only call for the use of nine of these. The four bits on input to a ROM to be used in implementing the function define 16 memory locations, seven of which would not be used. Despite the fact the seven locations would not be used they are still present in the device and are essentially wasted. However, by using a PLA, the same logic function can be implemented more economically. PLAs offer the digital designer another option in implementing combinational logic circuits. For large, complex logic functions involving four or more inputs and outputs, it offers advantages over SSI and MSI combinational circuits and ROMs for some applications.

Unit 9

DIGITAL DESIGN

INTRODUCTION

In this unit, you are going to learn how to design digital circuits. You will apply your knowledge of digital circuit operation and application to develop a digital circuit to solve a particular problem or meet a particular need. First, we will show you how to establish a design criterion for the equipment under consideration. Then we will consider procedures for designing both combinational and sequential logic circuits. These step-by-step procedures will lead you from a problem definition to a completed electronic circuit. The emphasis will be on implementing your design with modern integrated circuits. A variety of design examples will be given to show you some of the many applications for digital circuits.

DESIGN CRITERIA

The first step in designing a digital circuit is to define what the circuit must accomplish. You can do this by outlining the circuit specifications. These details will accurately specify the purpose of the circuit and the desired performance. The remaining design steps will convert this set of specifications into a practical working circuit that meets the design objectives. In designing the circuit there must be some standard for evaluating your design. In other words, there must be some criterion for determining whether you have adequately met your design objectives. In digital equipment design, as in the design of almost any type of electronic equipment, the primary design criterion is to achieve maximum performance for the lowest cost. This broad general criterion is made up of many parts which define what we mean by maximum performance and lowest cost. Together these make up the design criteria for the equipment. Let's see what each of these important considerations mean.

Maximum Performance

The term maximum performance can have a variety of meanings depending upon the circuit or equipment being designed. The definition of maximum performance therefore is a direct function of the application. Some of the factors that make up maximum performance include operating speed, accuracy, size, power consumption, reliability, and the number of unique features.

If we are going to market this equipment then we want to make it as desirable as possible in terms of cost, performance, and unique features. We want it to be marketable and competitive. These are only a few of the factors that define what maximum performance is. Each of these must be defined by itself to match the specific application. As you can see, what we want to do is to develop the best piece of equipment possible for our investment in design and production time and money.

Lowest Cost

Our design objective is to achieve the maximum performance possible for the lowest cost. The cost means both materials and time. A low cost design will have fewer parts. Fewer parts means a smaller printed circuit board on which to mount them and lower cost. The fewer the number of parts in a design the greater the reliability.

Another important factor is design and production time. The cost of any product includes the time required to design and produce it. The faster and easier the circuit is to design, the less it costs. In addition, the fewer the number of hours required to construct and test that unit the lower its cost.

Trade-Offs

A design that achieves maximum performance for lowest cost is an efficient design. The time and money spent in developing the unit is minimal while we achieve the benefit of maximum performance. By letting our criterion of maximum performance for lowest cost be our primary design goal, you must realize that in practical situations trade offs are generally necessary. This means it is not always possible to achieve the very highest of performance for the lower cost. As a general rule, high performance costs more money. If very high performance standards must be met, then we must accept the fact that the penality we will pay is higher cost. For example, to achieve the highest operating speed, higher cost digital integrated circuits must be used. This higher speed also generally requires a higher power consumption. Therefore, to achieve the highest possible speed, we are sacrificing cost and power consumption. Increased accuracy generally requires higher quality circuits and in some cases a greater quantity of circuitry. Again, higher costs, higher power consumption will be sacrificed. Increasing the number of components also increases the production time and reduces the reliability. High performance designs in addition, are generally more sophisticated and complex. This, in turn, means a greater design time. Therefore, an attempt to achieve maximum performance will invariably increase cost. For that reason you must be ready to trade off higher performance and features in order to obtain a lower cost. The design procedure is basically one of juggling the performance requirements and the costs to achieve a desired performance level for the lowest possible cost. You are seeking an efficient, middle-of-the-road solution.

There are many situations where it is desirable to select high performance as the single, most desirable design characteristic. It may be absolutely necessary to achieve the desired level of performance regardless of the cost. In another application, just the opposite may be true. Instead of optimizing your design for high performance you might want to optimize it for lowest possible cost. In order to achieve the low cost, high performance and features will naturally be sacrificed.

While optimizing your design for maximum performance or lowest cost is sometimes necessary, most design projects will be in that middle ground where your job will be to balance performance and cost to achieve an acceptable level in both.

COMBINATIONAL LOGIC CIRCUIT DESIGN

In this section we are going to give you a step-by-step procedure that can be used to design virtually any combinational logic circuit. The procedure will lead you from the basic problem definition to the completed circuitry using modern integrated circuits. The steps in this design procedure are:

1. Problem definition
2. Truth table development
3. Writing logic equations
4. Minimizing the logic equations
5. Selecting the circuitry and implementing the design

Each of these steps will be described in detail. Design examples will be used to illustrate these steps. In addition, you will learn an important technique for minimizing logic equations. This technique involves the use of a Karnaugh map. This is a method of putting the logic equations into a graphical form that permits rapid circuit minimization without the use of Boolean algebra. Finally,a variety of design problems will be given to permit you to practice this procedure yourself. Now, let's take a look at the steps in the design procedure.

Problem Definition

The first step in the design of any combinational logic circuit is complete problem definition. This means that you must thoroughly identify all functions of the circuit. You will know from the specific application what operations the circuit must perform. Your initial job will be to outline them completely.

The best and most thorough method of problem definition is to write out a complete description of the application and the desired functions to be accomplished. While it may seem unnecessary to put this information in writing, by doing so it forces you to completely identify and explain what is going to take place. In this description you will identify the types and number of input signals to the circuit. You will also identify the type and number of outputs to be produced by the circuit. Make your circuit description as complete as possible. The exact form of the description is not important. It can be a descriptive narrative in paragraph form. Alternately it can simply be a list of inputs, outputs and functions to be performed.

Once you have completed the functional description of your circuit, make up a table of specifications. This table of specifications will duplicate some of the information contained in your circuit description. However, the information will be more concisely stated. The specifications will list number and type of inputs, number and type of outputs, desired operating speed, desired power consumption, a cost objective, a size and weight objective, types of integrated circuits to be used, interface requirements for both the inputs and outputs including logic level specifications, and any other information pertinent to the operation of the circuit. Regardless whether your design is to be a circuit within a larger system or a complete digital system itself; the functional description, problem definition and complete set of specifications will give you all of the information necessary to complete the design.

The process of writing down the circuit description and its specifications is a valuable excercise. It forces you to think through the problem and to identify it as carefully as possible. In preparing this information you will discover many things that you may not have thought of initially. Problem definition is more than just a busy work exercise. It is a vital first part of the design process and the success or failure of the design can be traced almost directly to the thoroughness of this problem definition.

Another benefit of complete problem definition and specification outlining is that you will have a complete set of documentation for your circuit or system that can be used later in preparing instructional manuals for the equipment, engineering reports, journal articles, and other requirements for this information.

To illustrate this concept of problem definition, and the other steps of a design procedure, we are going to take a typical design example and follow it through each of the design steps. A circuit example will be a simple one to start with in order to help you easily grasp the concepts involved. Later, several more detailed examples will be given. In addition, practice problems are provided to permit you to practice these steps yourself.

Example Problem. The design objective is to develop a combinational logic circuit that will monitor a four bit binary word and generate a binary 1 output signal when any one of the six invalid four bit states in the 8421 BCD code occur.

Specifications. The detailed specifications for the circuit described above are as follows:

1. Four bit parallel binary input word
2. Signal output that will be a binary 1 state when any one of the six invalid BCD code numbers occur.

3. Use TTL integrated circuitry with standard TTL logic levels of binary 0 = +0.4 volts and binary 1 = +3.5 volts.
4. Speed of operation: Propagation delay of this circuit shall be less than 100 nanoseconds. In other words, the output will become binary 1 in 100 nanoseconds or less from the time the input code is any one of the six invalid BCD values.
5. Minimize cost, size and power consumption. It is desirable to have the entire circuit contained within a single dual inline package IC.

Truth Table Development

The next step in the design procedure is to convert your problem description into a truth table. The truth table as you will recall is a chart that completely identifies all possible input combinations and the corresponding logic output states. The truth table completely defines the operation of the circuit. The truth table can be developed directly from your problem description and specifications.

The first step in developing the truth table is to determine the number of inputs to the logic circuit. This, of course, is a function of the application and this information should have been defined in the problem description. The number of inputs will determine the maximum number of input states that can occur with this number of variables. The total number of states that can occur is equal to 2^n where n is the number of inputs. For example, if you have defined four circuit inputs, there are a total possible number of $2^4 = 16$ different states that can occur. Depending upon the application, all or possibly only some of the total number of possible states may occur. In the problem description these should be identified.

Begin the construction of the truth table by writing down all possible binary input states. You can do this by simply listing the binary numbers from zero through the maximum upper limit which is a function of the number of input states. For four inputs and a maximum of 16 possible states, you will simply list all four bit binary numbers 0000 through 1111 in sequence. This completely defines all possible input states.

In columns adjacent to the input states in the table, record the output variables specified by the problem. In these columns, identify the desired output states for each input combination. If some of the input states are not used, identify them as being invalid or "don't care" states even though they are not used or needed.

Figure 9-1 shows the truth table for the BCD invalid code detector. The input is a parallel four bit word. We label each of the input bits with a letter for identification. Remember that any input lettering or numbering scheme can be used to suit the application. Short mnemonic names designating the signal or its function can be used. Alpha-numeric combinations can also be used. For this application, the letters A, B, C, and D are adequate. The output of the circuit is designated F. This is the logic signal that will be a binary 1 if an invalid code is detected.

INPUTS	OUTPUT
A B C D	F
0 0 0 0	0
0 0 0 1	0
0 0 1 0	0
0 0 1 1	0
0 1 0 0	0
0 1 0 1	0
0 1 1 0	0
0 1 1 1	0
1 0 0 0	0
1 0 0 1	0
1 0 1 0	1
1 0 1 1	1
1 1 0 0	1
1 1 0 1	1
1 1 1 0	1
1 1 1 1	1

Figure 9-1
Truth table for BCD invalid code detector.

Notice in the truth table that all sixteen possible combinations of four bits are listed. The first ten states 0000 through 1001 are the valid 8421 BCD codes. Since these are valid, the output F will be binary 0. For the inputs 1010 through 1111, the output F is binary 1 signaling an invalid code. There are no unused or "don't care" states. The truth table completely defines the operation of the circuit.

While our example here has only a single binary output, other combinational logic circuits may have multiple outputs. In this case the truth table will define these as well. A separate column for each output will be provided in the truth table.

Develop the Logic Equations

The next step in the design process is to write the Boolean logic equations from the truth table. This will put the logic function into a form where it can be manipulated with Boolean algebra. This will allow you to reduce the logic equation using Boolean techniques and thus minimize the amount of circuitry required to implement it. For some applications it may not be necessary to minimize the equation with Boolean algebra. Instead the equation will simply be used as a guide in implementing the function depending upon the types of circuits to be used.

To write the logic equation from the truth table, you observe the outputs column in the truth table and write a product term of the inputs for each output where a binary 1 state occurs. The result will be a sum-of-products logic equation. This process will lead to a single Boolean equation for each output.

Observing the BCD invalid code detector truth table in Figure 9-1 you can write the output equation.

$$F = A\overline{B}C\overline{D} + A\overline{B}CD + AB\overline{C}\overline{D} + AB\overline{C}D + ABC\overline{D} + ABCD$$

At this point it is possible to implement the Boolean equation directly with logic circuits. AND and OR gates can be combined to perform this function. However, in most cases it is desirable to use Boolean algebra or other means to reduce the equation to a simpler form. This can have the result of minimizing the number of gates and integrated circuit packages used in a design. Minimization generally leads to lower cost, smaller size, and reduced power consumption.

Circuit Minimization

By using the Boolean algebra techniques described in a previous unit, the logic equations you developed from the truth table can be reduced.

Shown below is the step-by-step procedure for using Boolean algebra to minimize the logic equation developed in the previous step. Keep in mind that this is only one of several approaches that can be used. Depending upon how you group the various logic terms, the individual steps may be different. In any case, your resulting reduced logic equation should be the same.

$$F = A\overline{B}C\overline{D} + A\overline{B}CD + AB\overline{C}\overline{D} + AB\overline{C}D + ABC\overline{D} + ABCD$$

Reduce by factoring

$$F = A\overline{B}C(\overline{D} + D) + AB\overline{C}(\overline{D} + D) + ABC(\overline{D} + D)$$

Reduce by Law of Complements and Law of Intersection

$$F = A\overline{B}C + AB\overline{C} + ABC$$

Reduce by factoring

$$F = A\overline{B}C + AB(\overline{C} + C)$$

Reduce by Law of Complements and Law of Intersection

$$F = A\overline{B}C + AB$$

Reduce by factoring

$$F = A(\overline{B}C + B)$$

Reduce by the Law of Absorption $\overline{B}C + B = B + C$

$$F = A(B+C)$$

Expand by multiplying

$$F = AB + AC$$

As you can see a significant reduction in the equation takes place. It is obvious that it requires much less circuitry to implement the reduced version than it does the original equation derived from the truth table. This minimization step is very important to the design of the circuit.

The use of Boolean algebra is time consuming and burdensome. For some logic equations, a reduction can take place quickly without a lot of work. However, for large complex equations, the minimization process can require a substantial amount of time. You may have to rearrange the equation and regroup the terms several times before you arrive at a minimum result. In addition, the Boolean algebra procedure described in an earlier unit does not always lead to the optimum minimization. Because of the subtlety of logic circuits, some methods of circuit reduction do not show themselves in the equation reduction process. For that reason, Boolean algebra has its limitations. Other forms of minimization have been developed to provide the maximum amount of minimization possible and to do so quickly and conveniently. One of these techniques is known as Karnaugh maps.

Karnaugh Maps

A Karnaugh map is a graphical method of minimizing logic equations. The equations describing a digital logic function can be broken up and arranged in such a way that they form a map or illustration that permits rapid reduction or simplification. The Karnaugh map is an alternative to the use of Boolean algebra for minimizing logic expressions. In fact, the Karnaugh map is preferred over Boolean algebra because it makes the reduction process faster, easier, and more effective. This technique completely eliminates the need for using Boolean algebra and allows you to translate the logic function directly from the truth table into a Karnaugh map that then leads to the simplified form. With this technique, it is not always necessary to write the equations from the truth table first.

Karnaugh maps effectively replace Boolean logic equations in the sum-of-products form. For design purposes, these equations are derived from the truth table as we described earlier. Each of the product terms in the equation is referred to as a minterm. Each minterm is the product of the various input variables which are called literals. In the truth table describing the logic function, all possible input combinations are listed. For example, for a 2-input logic circuit, there are four possible input combinations. They are 00, 01, 10, and 11. If the input literals are given the names A and B, then the minterms are $\overline{A}\ \overline{B}$, $\overline{A}\ B$, $A\ \overline{B}$, and $A\ B$.

Instead of writing the product term with respect to the literals, they are often simply expressed as the letter *m* with a subscript equal to the decimal value of the binary number representing that minterm. For example, the product term $\overline{A}\ \overline{B}$ represents input states of 00. The minterm designation then would be m_0. The product term A B represents the input states 11. The decimal equivalent of this number is 3, therefore, the minterm designation would be m_3. Figure 9-2 lists the product terms for a two input logic circuit, their binary and decimal equivalents and the minterm designation.

DECIMAL	BINARY A	BINARY B	PRODUCT TERM	MINTERM DESIGNATION
0	0	0	$\overline{A}\ \overline{B}$	m_0
1	0	1	$\overline{A}$ B	m_1
2	1	0	A $\overline{B}$	m_2
3	1	1	A B	m_3

Figure 9-2
Minterm designations for logical products of two literals.

A Karnaugh map takes this information and translates it into graphical form. Figure 9-3 shows a Karnaugh map for a two input logic circuit. Since there are two input variables, there are four possible product terms. Each product term is represented by a cell or square in the map.

To show how the map and the equation are related, let's take several examples of converting from the equation to the map and from a map to the equation.

Consider the Boolean equation: $C = \overline{A}\,B + A\,\overline{B}$

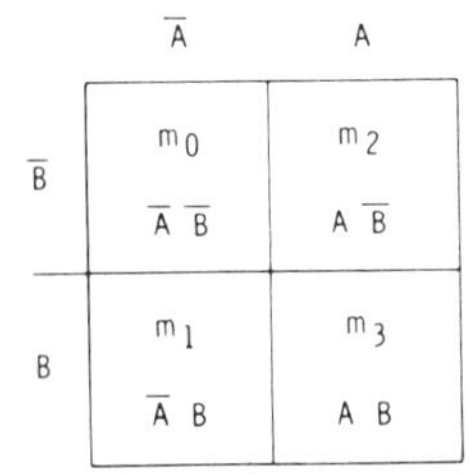

Figure 9-3
Two-variable Karnaugh map.

This is the equation for an exclusive OR circuit. Note that it contains two product terms of the two variables A and B. To plot this equation on the map, we simply place a binary 1 in those squares representing the product terms in the equation. This is shown in Figure 9-4. The minterm designations are not generally included within the squares on the map. Instead, the squares themselves can be identified by referring to the designations above and to the left of the map. At the top of the map are the designations $\overline{A}$ and A which refer to the two vertical columns. At the left of the map are the designations $\overline{B}$ and B which designate the two rows of squares in the map. The product term corresponding to a square is identified by simply reading upward and to the left for the letter designation defining that term and using these letters to form the product term.

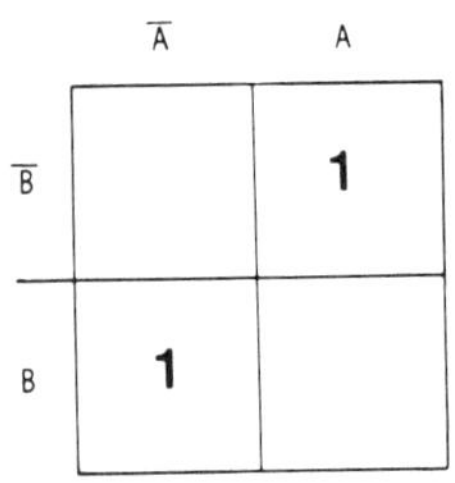

Figure 9-4
Karnaugh map for the exclusive OR function $A\overline{B} + \overline{A}B$.

In this example, we translated a known equation into Karnaugh map form. Keep in mind that the map can also be developed directly from the truth table. The output column of the truth table is observed and those input states corresponding to binary 1 outputs are translated into product terms that can then be plotted on the map.

As an example of translating from a Karnaugh map into the equivalent Boolean equation, consider the map shown in Figure 9-5. To write the output expression corresponding to this map, you develop a minterm for each square containing a binary 1. These minterms are then ORed together to form a sum-of-products Boolean equation. The map in Figure 9-5 designates the exclusive NOR function expressed by the equation:

$$C = \overline{A}\,\overline{B} + A\,B$$

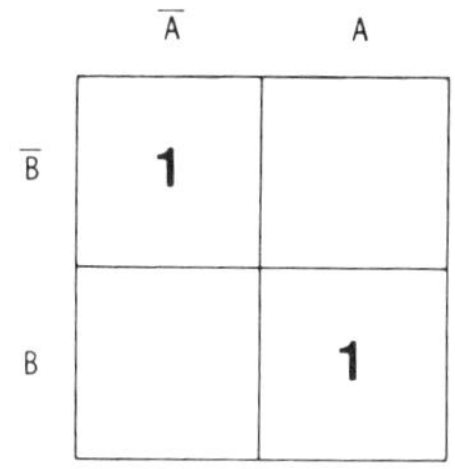

Figure 9-5
Karnaugh map for exclusive OR function $\overline{A}\overline{B}$ + AB.

We can also write this equation using the minterm designations:

$$C = m_o + m_3$$

Figure 9-6 shows the Karnaugh map for a three variable logic circuit. Since there are three input variables, there are eight possible input combinations. Each input state is represented by a square in the map. The minterm for each square in the map is designated. The relationships between the product terms, their binary and decimal equivalents and the minterm designation are given in Figure 9-7. Note that the columns and rows in the map are designated by the letters corresponding to the inputs.

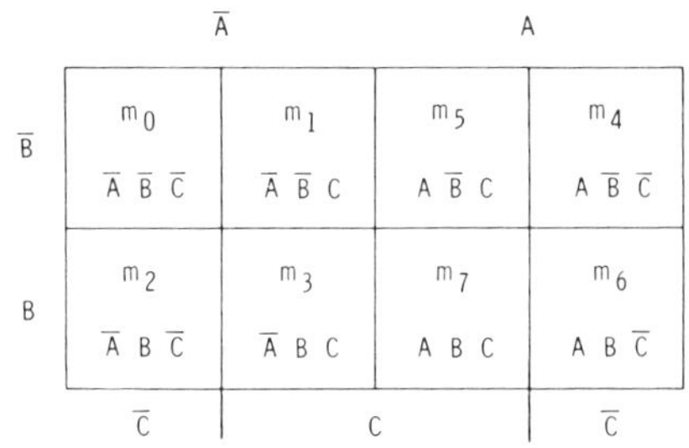

Figure 9-6
Three—variable Karnaugh map.

DECIMAL	BINARY A B C	PRODUCT TERM	MINTERM DESIGNATION
0	0 0 0	$\overline{A}\,\overline{B}\,\overline{C}$	m_0
1	0 0 1	$\overline{A}\,\overline{B}\,C$	m_1
2	0 1 0	$\overline{A}\,B\,\overline{C}$	m_2
3	0 1 1	$\overline{A}\,B\,C$	m_3
4	1 0 0	$A\,\overline{B}\,\overline{C}$	m_4
5	1 0 1	$A\,\overline{B}\,C$	m_5
6	1 1 0	$A\,B\,\overline{C}$	m_6
7	1 1 1	$A\,B\,C$	m_7

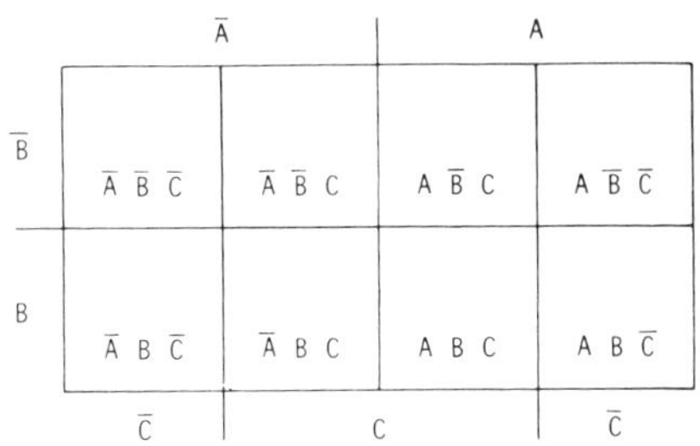

Figure 9-7
Minterm designations for logical products of three literals.

The two right-hand vertical columns are designated A, the two left-hand vertical columns are designated $\bar{A}$. Note that these vertical columns are designated in a different form by the input variable C. The two center vertical columns represent C while the two outside columns represent $\bar{C}$. The two horizontal rows of four squares are designated $\bar{B}$ and B. The minterm represented by each square can be determined by simply writing a product term made up of the three letters designating that square in its row and column position. It takes three input designations to define the coordinates of a square.

The method of recording a given Boolean equation in the Karnaugh map for three variables is similar to that for two variables. Consider the equation:

$$X = A\,\bar{B}\,C + A\,\bar{B}\,\bar{C} + \bar{A}\,B\,C$$

Each three-variable term is designated by a binary 1 in the appropriate square. See Figure 9-8.

The Karnaugh map in Figure 9-9 shows how an equation can be written from the map. The minterm represented by each square where a binary 1 appears is summed (logically ORed) with the other terms to produce the equation:

$$M = \bar{A}\,\bar{B}\,\bar{C} + \bar{A}\,B\,\bar{C} + A\,\bar{B}\,\bar{C} + A\,B\,C$$

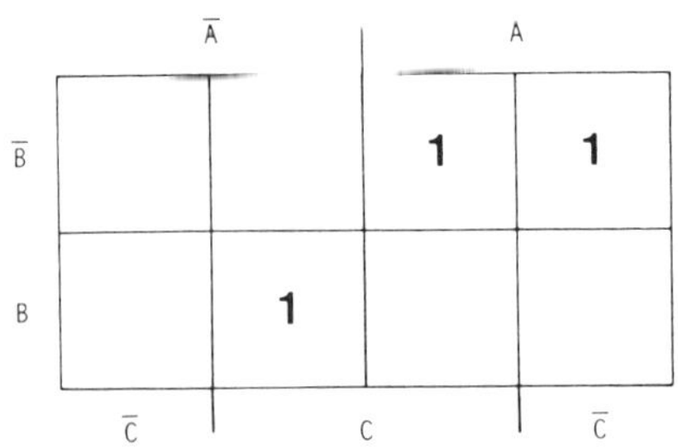

Figure 9-8
Karnaugh map for the equation
$X = A\bar{B}C + A\bar{B}\bar{C} + \bar{A}BC$

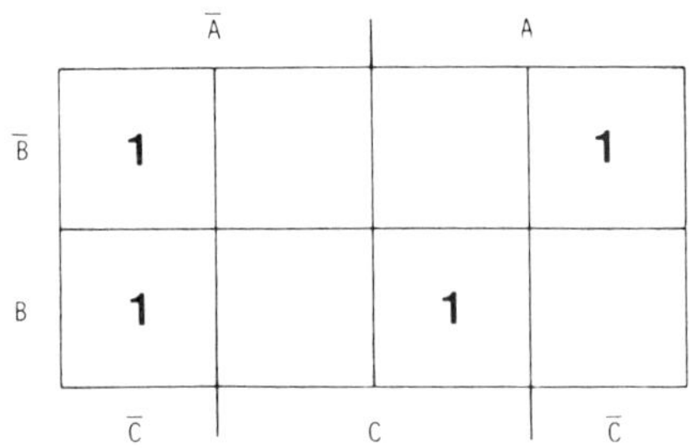

Figure 9-9
Karnaugh map of the equation
$M = \bar{A}\bar{B}\bar{C} + \bar{A}B\bar{C} + A\bar{B}\bar{C} + ABC$

In minterm form the equations are:

$$M = m_o + m_2 + m_4 + m_7$$

These same concepts can be applied to logic expressions involving four variables. The table in Figure 9-10 shows all sixteen possible combinations of four-input variables. All product terms with their binary and decimal equivalents are given.

DECIMAL	BINARY A B C D	PRODUCT TERM	MINTERM DESIGNATION
0	0 0 0 0	$\bar{A}\bar{B}\bar{C}\bar{D}$	m_0
1	0 0 0 1	$\bar{A}\bar{B}\bar{C}D$	m_1
2	0 0 1 0	$\bar{A}\bar{B}C\bar{D}$	m_2
3	0 0 1 1	$\bar{A}\bar{B}CD$	m_3
4	0 1 0 0	$\bar{A}B\bar{C}\bar{D}$	m_4
5	0 1 0 1	$\bar{A}B\bar{C}D$	m_5
6	0 1 1 0	$\bar{A}BC\bar{D}$	m_6
7	0 1 1 1	$\bar{A}BCD$	m_7
8	1 0 0 0	$A\bar{B}\bar{C}\bar{D}$	m_8
9	1 0 0 1	$A\bar{B}\bar{C}D$	m_9
10	1 0 1 0	$A\bar{B}C\bar{D}$	m_{10}
11	1 0 1 1	$A\bar{B}CD$	m_{11}
12	1 1 0 0	$AB\bar{C}\bar{D}$	m_{12}
13	1 1 0 1	$AB\bar{C}D$	m_{13}
14	1 1 1 0	$ABC\bar{D}$	m_{14}
15	1 1 1 1	$ABCD$	m_{15}

Figure 9-10
Minterm designations for all possible combinations of four literals.

A four variable Karnaugh map is shown in Figure 9-11. Each square in the map represents one of the four-variable minterms. The process of recording a given sum-of-products equation on the map is similar to the process described earlier for two- and three-input variables. In addition, the

	$\bar{A}$	$\bar{A}$	A	A	
$\bar{B}$	m_0 $\bar{A}\bar{B}\bar{C}\bar{D}$	m_2 $\bar{A}\bar{B}C\bar{D}$	m_{10} $A\bar{B}C\bar{D}$	m_8 $A\bar{B}\bar{C}\bar{D}$	$\bar{D}$
$\bar{B}$	m_1 $\bar{A}\bar{B}\bar{C}D$	m_3 $\bar{A}\bar{B}CD$	m_{11} $A\bar{B}CD$	m_9 $A\bar{B}\bar{C}D$	D
B	m_5 $\bar{A}B\bar{C}D$	m_7 $\bar{A}BCD$	m_{15} $ABCD$	m_{13} $AB\bar{C}D$	D
B	m_4 $\bar{A}B\bar{C}\bar{D}$	m_6 $\bar{A}BC\bar{D}$	m_{14} $ABC\bar{D}$	m_{12} $AB\bar{C}\bar{D}$	$\bar{D}$
	$\bar{C}$	C	C	$\bar{C}$	

Figure 9-11
Four-variable Karnaugh map.

procedure for writing the equation from the map is also similar to that described before. As an example, the map shown in Figure 9-12 represents the equation below:

$$X = \bar{A}\,\bar{B}\,\bar{C}\,\bar{D} + \bar{A}\,\bar{B}\,C\,D + \bar{A}\,B\,\bar{C}\,\bar{D} + A\,\bar{B}\,\bar{C}\,\bar{D} + A\,\bar{B}\,C\,D$$

$$X = m_0 + m_3 + m_4 + m_8 + m_{11}$$

Match each term in the equation to the appropriate square in the map to be sure that you understand how the two are related.

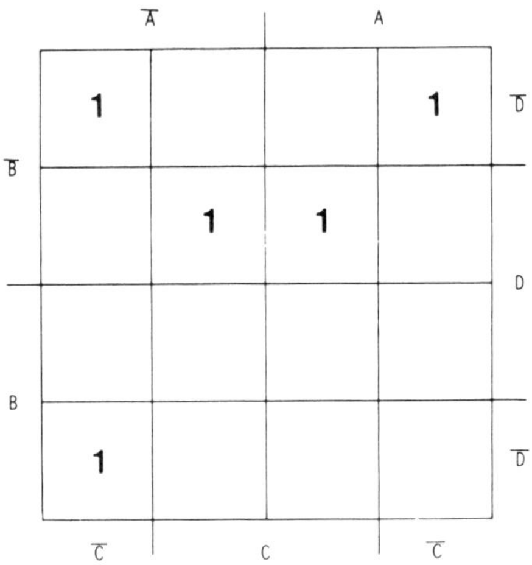

Figure 9-12
Karnaugh map for the equation $X = \bar{A}\bar{B}\bar{C}\bar{D} + \bar{A}\bar{B}CD + \bar{A}B\bar{C}\bar{D} + A\bar{B}CD + A\bar{B}\bar{C}\bar{D}$

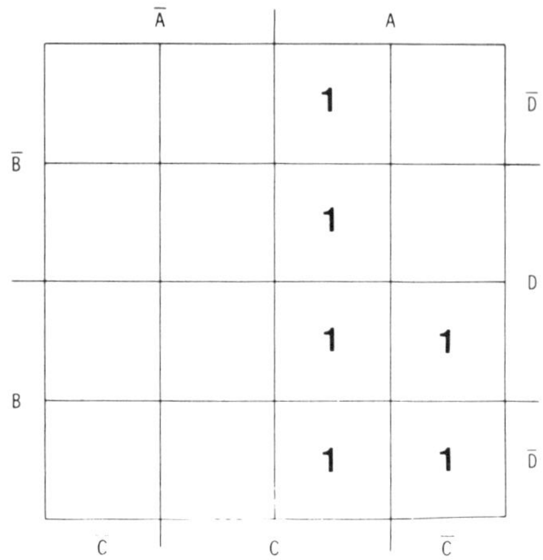

Figure 9-13
Karnaugh map for the BCD invalid code detector.

In designing the BCD invalid state detector circuit, you wrote the Boolean equations from the truth table. These are:

$$F = A\,\bar{B}\,C\,\bar{D} + A\,\bar{B}\,C\,D + A\,B\,\bar{C}\,\bar{D} + A\,B\,\bar{C}\,D + A\,B\,C\,\bar{D} + A\,B\,C\,D$$

$$F = m_{10} + m_{11} + m_{12} + m_{13} + m_{14} + m_{15}$$

These can be placed on a Karnaugh map as show in Figure 9-13.

Now that you know how Boolean equations are plotted on Karnaugh maps and how to read a Karnaugh map and translate it into a Boolean equation, you are ready to see how these maps can be used for circuit minimization.

The reduction of logic equations with Boolean algebra is largely by use of the law of complements ($A + \overline{A} = 1$). Putting the logical function in sum-of-product equation form, the minterms can then be grouped to permit the factoring out of common variables. This procedure generally produces law of complement expressions for one of the input variables. Thus, one of the input variables is eliminated from the group of minterms from which it was factored thereby simplifying the expression. Nearly all of the simplification that results from the use of Boolean algebra comes from the use of factoring and the reduction by the law of complements. The Karnaugh map effectively implements this technique in a graphical form.

Refer to the four-variable Karnaugh map in Figure 9-14. If you will study this map carefully, you will see that adjacent cells or adjacent minterms differ by only one of the input variables. In other words, only one variable will change in moving from one adjacent cell to the next, horizontally or

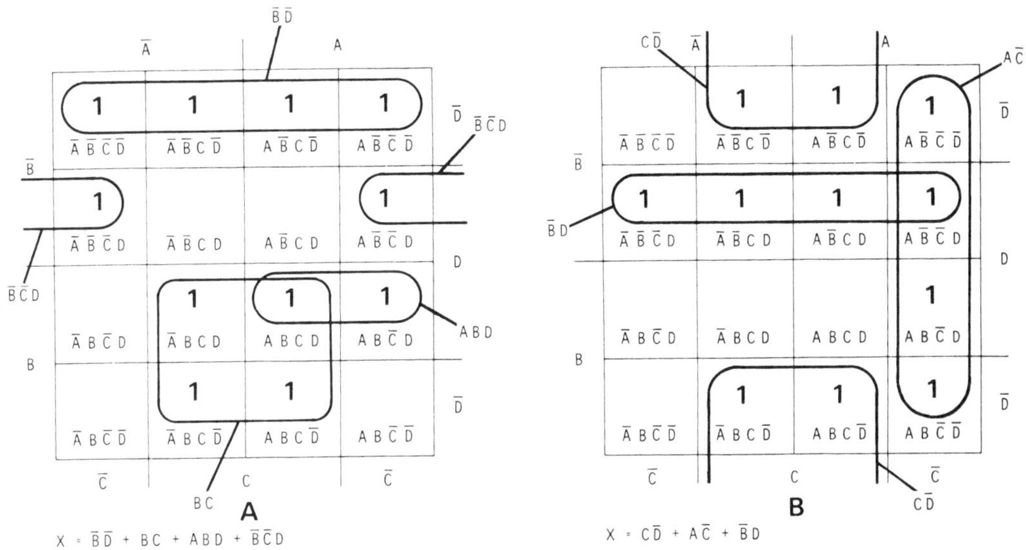

Figure 9-14
Four-variable Karnaugh map.

vertically. For example, consider the two vertical minterms identified by binary 1's in the map of Figure 9-15. In moving from the upper cell to the lower cell, only the B variable changes, from $\overline{B}$ to B. The A and C variables do not change. These two adjacent minterms specify a simplification that can be made. You can see this by writing the Boolean equation of the minterms recorded. Assume that the sum-of-product equations of these two minterms is equal to the function Y. The equation from the map in Figure 9-15, then is $Y = A \overline{B} C + A B C$.

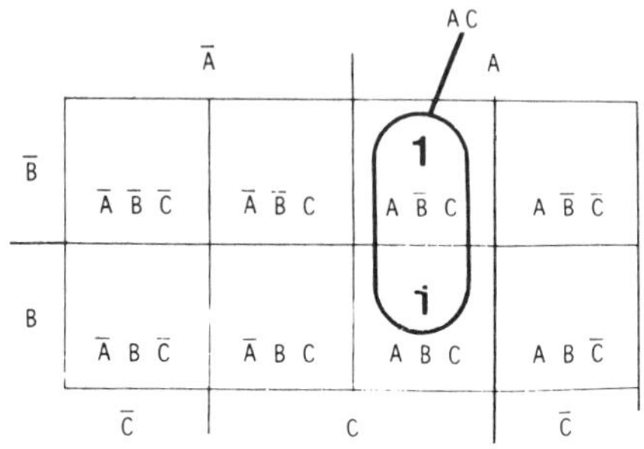

Figure 9-15
Adjacent cells in a Karnaugh map differ by only one literal.

Now, use Boolean algebra to simplify this expression. This is done as shown below.

$$Y = A \overline{B} C + A B C$$
$$Y = A C (\overline{B} + B)$$
$$Y = A C (1)$$
$$Y = AC$$

Note that AC was factored out of each minterm leaving an expression equal to $(\overline{B} + B)$. Since this is equal to 1, the expression is considerably simplified. The B literal simply drops out leaving the expression Y = AC. Logic equation minimization with a Karnaugh map is based on this concept.

The basic procedure for reducing a logic equation by a Karnaugh map is to first map the expression by putting a binary 1 in each cell representing the minterms in the sum-of-products logic equation. Then, horizontal and vertical adjacencies in groups of two or four are identified. We then note which variables change from one cell to the next in each set of grouped adjacent terms. These inputs then drop out of the expression. The remaining input terms are regrouped in sum-of-product forms to produce the simplified expression. An example will illustrate this process.

Consider the logic expression:

$$Y = \overline{A}\,\overline{B}\,\overline{C} + \overline{A}\,\overline{B}\,C + \overline{A}\,B\,\overline{C} + A\,\overline{B}\,C$$
$$Y = m_0 + m_1 + m_2 + m_5$$

To simplify this logic expression we first record the minterms on a Karnaugh map. Since there are three-input variables, an eight-cell Karnaugh map will be used. Binary 1's are entered in those cells representing the minterms in the equation. This is indicated in Figure 9-16.

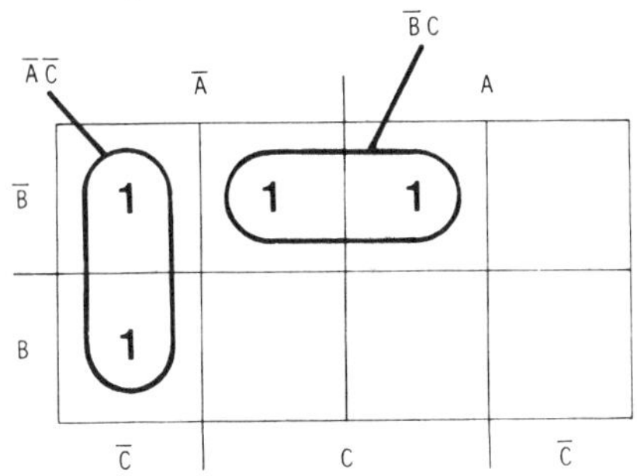

Figure 9-16
Using a Karnaugh map to minimize the expression $Y = m_0 + m_1 + m_2 + m_5$

Next, adjacent minterms are grouped by some power of 2 (2, 4, 8, etc.). Each group of two or four minterms is identified by a circle enclosing the binary 1's on the map as indicated in Figure 9-16.

Each circled group is then observed to determine which variable changes in moving from one adjacent cell to the next in the group. In the vertical group of Figure 9-16, the variable B changes in moving from the upper cell to the lower cell. This indicates then that the B term drops out leaving only the $\overline{A}$ and $\overline{C}$ terms. Therefore, this group of two adjacent variables represent the logic expression $\overline{A}\ \overline{C}$.

Observing the horizontal grouping of two variables in Figure 9-16, we see that the variable that changes in moving from one cell to the next is the A variable. The A term therefore drops out leaving the $\overline{B}$ and C terms. Once the variable that changes has been identified, a new shorter minterm is developed from the variables that have not changed. A product term of the variables that do not change is formed, in this case $\overline{B}$ C. These shorter product terms for each group are then summed (ORed) to produce the reduced expression. Therefore, the reduced expression from the map in Figure 9-16 is:

$$Y = \overline{A}\,\overline{C} + \overline{B}\,C$$

The ability to use a Karnaugh map to produce a minimum equation reduction results from being able to properly group the minterms and recognize all of the adjacencies or combinations of adjacencies.

Figure 9-17 shows two additional examples of mapping logic equations, grouping minterms, and generating the reduced expression. In Figure 9-17A, the Boolean equation is:

$$X = A\,\overline{B}\,C + A\,\overline{B}\,\overline{C} + A\,B\,\overline{C}$$

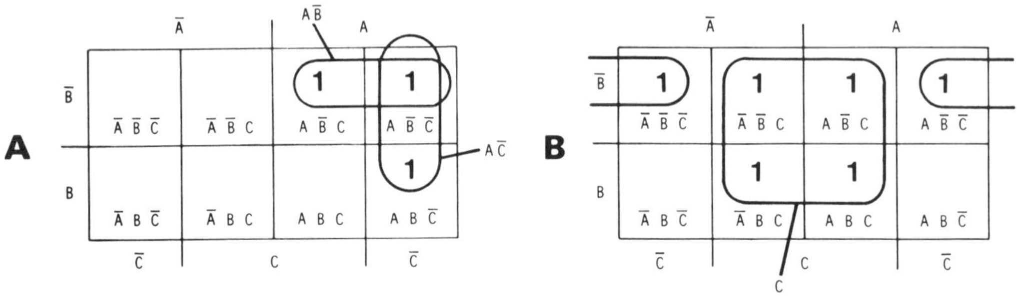

Figure 9-17
Examples of reducing three variable logic equations with maps.

The three minterms are recorded in the appropriate cells in the map by marking 1s in the appropriate cells. Adjacent minterms are then grouped to identify the redundant input terms. Note that the minterm $A\,\overline{B}\,\overline{C}$ is used twice. Any given minterm may be used as many times as needed to form adjacent groups of two or four.

To determine the redundant input variables, you note which variable changes when moving from one cell to the next within the groups you formed. The horizontal group of two identifies the change of variable $\overline{C}$. In moving from one cell to the next the A and $\overline{B}$ terms do not change but the C term does. This means that the C term drops out. We then form a new product term made up of the variables that did not change, in this case $A\,\overline{B}$.

Next, we observed the vertical group of two to determine the redundant variable. In this case the redundant variable is B since in moving from one cell to the next within that vertical group the B term changes from B to $\overline{B}$. The A and $\overline{C}$ terms do not change therefore they represent the new product term for use in the minimized expression. The new product terms are then logically summed or ORed to produce the output expression:

$$X = A\overline{B} + A\,\overline{C}$$

As you can see from Figure 9-17A the original and reduced expressions are considerably different. The reduced expression is far more economical in the use of circuitry.

Figure 9-17B illustrates another example of minimizing a three-variable Boolean equation. This equation is:

$$X = \overline{A}\,\overline{B}\,\overline{C} + \overline{A}\,\overline{B}\,C + \overline{A}\,B\,C + A\,\overline{B}\,\overline{C} + A\,\overline{B}\,C + A\,B\,C$$
$$X = m_0 + m_1 + m_3 + m_4 + m_5 + m_7$$

Groups of two or four adjacent functions are then formed as shown. It is desirable to form the largest grouping possible with the minterms recorded on the map. The larger the grouping, the greater the reduction that will take place. Note in the group of four that, in moving from one adjacent cell to the next, the A and B terms change. In moving vertically, the B terms change. Moving horizontally, the A terms change. The only input variable that does not change from one of these four cells to the next is the C input term. This means that A and B inputs are redundant and can be factored out of the equation and dropped. This 4-bit grouping then results in a substantial minimization and simply represents the input variable C. What this grouping tells us is that the output expression will be affected only by the variable C regardless of the A and B input states.

The grouping formed by the minterms $\overline{A}\,\overline{B}\,\overline{C}$ and $A\,\overline{B}\,\overline{C}$ illustrates a special case of grouping and shows the unique characteristics of a Karnaugh map. Note that between these two minterms, only the A variable changes. This means that these two cells can be considered adjacent. Another way of looking at this is to assume that the left-hand edge of the Karnaugh map is adjacent to the right-hand edge. Think of the map as being a cylinder where the right-hand and left-hand edges are curved around to form the cylinder. This adjacency then results in the expression $\overline{B}\,\overline{C}$. These terms are common to both of these cells with the $\overline{A}$ term changing and, therefore, dropping out of the expression. The reduced output expression then is:

$$X = \overline{B}\,\overline{C} + C$$

The power of the Karnaugh map is evident from these examples. With a little practice in mapping and grouping the minterms, you can quickly reduce logic expressions to their minimum form. The map provides a visual means of recognizing patterns in the minterm groupings so that redundancies in the input variables can be easily recognized and eliminated, thereby leaving only the essential input terms to implement the function.

The benefit of the Karnaugh map in speeding up and simplifying logic equation reduction becomes more evident as more input variables are used. Four-input variables produce sixteen different input states. These can be combined in a variety of ways to form logic equations. Expressions involving minterms of four variables or more are difficult to work with by using standard Boolean algebra techniques. But by mapping them, you automatically group the related minterms so that the redundancies can be readily identified.

Figures 9-18A through 9-18D show several examples of the use of four-variable Karnaugh maps. The reduced equations for each example are given. Study the various groupings of minterms to be sure you understand how the reduced expression is obtained for each group. As you study each example, keep in mind these important facts:

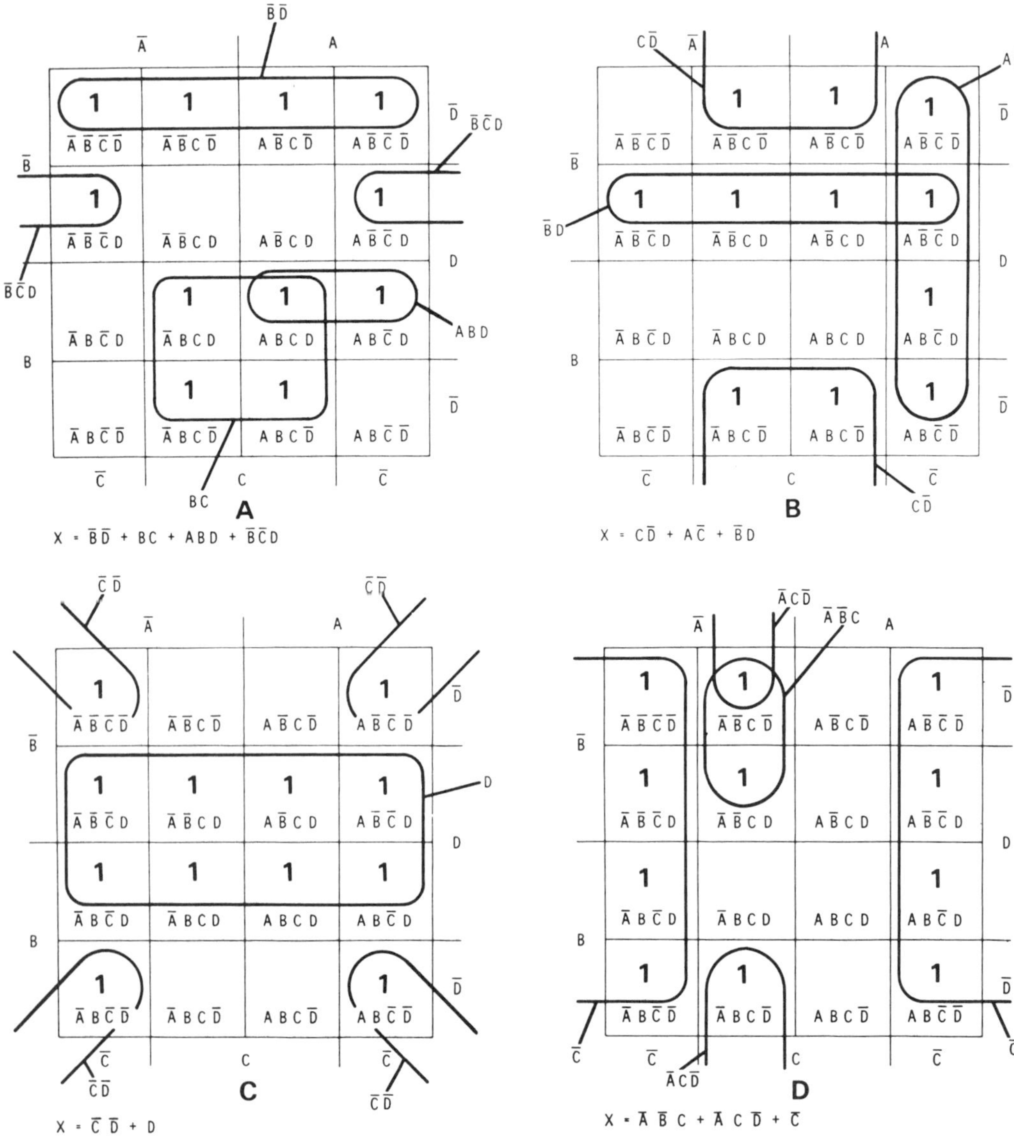

Figure 9-18
Examples of Karnaugh map usage.

1. Adjacent minterms are grouped by twos, fours, eights, and higher powers of 2 as required. The groupings may be horizontal or vertical and may involve adjacent terms that do not "appear" adjacent in the map. In a sixteen cell, four-variable Karnaugh map, adjacencies or redundancies can occur in the cells in the left and right most columns. Assume that the map is formed into a cylinder where the left and right edges are made adjacent. In the same way a cylinder can be formed by causing the upper and lower edges of the map to be made adjacent. The term "redundant" simply means that in moving from one cell to an adjacent cell only one of the input variables change.
2. Try to use each minterm in a group of two or four. After you have made your initial groupings, go back and study them to be sure that you have not overlooked various combinations of two, for example, which would be made into a single logic grouping of four. The larger the number of minterms enclosed within a loop the greater the reduction that results. There will be some occasions where a minterm cannot be included in a group of two, four or eight. In such cases no reduction is possible and the minterm must be treated by itself.
3. In each group of two or four variables, simply move from one cell to the next noting which variables change. The variables that change are redundant and can be dropped out of the minterm. Form a new product expression using the variables that do not change.
4. The minimized expression is formed by producing a sum-of-products expression made up of the reduced product expressions resulting from each group of minterms.

Keeping in mind the rules and characteristics just discussed, the examples in Figure 9-18 should be self-explanatory. There is one special case, however, that you may not easily recognize. In Figure 9-18C the minterm in each corner of the map is marked. Since minterms on the left and right columns and in the upper and lower rows are considered to be adjacent, the minterms in each corner of the map can be considered as a single group of four. This is more easily seen by determining which variables change in moving from one corner cell to the next. As you can see the A and B variables change in moving between these four cells. The $\overline{C}$ and $\overline{D}$ variables, however, are common to these cells. This permits a reduction of this group to the simple two-variable term $\overline{C}\,\overline{D}$. Note also in this map the group of eight minterms in the center of the map. In moving from one adjacent cell to the next you will see that the A, B, and C variables do change at one point or another. The only variable that does not change is the D variable. Therefore, this group of variables simply represents the input variable D.

Another special case that may occur is when all squares in the map are marked. In this case the function represented by the map is simply a binary 1.

Summary of Karnaugh Map Usage. The list below is a summary of the rules and procedures for using Karnaugh maps in reducing logic equations.

1. Study the truth table or the logic equations for the function to be minimized. Determine the number of input variables and construct a Karnaugh map containing a number of cells equal to two raised to a power equal to the number of input variables.
2. Map the minterms directly from the truth table. If you constructed a truth table as part of your design procedure there is no need to translate the truth table into a Boolean equation first. If you have an equation instead of a truth table, plot the minterms in the map from the equation itself.
3. Group the minterms in the map in units of two, four, and eight terms. Try to include each minterm in the largest group possible to ensure a minimum solution. Each minterm should be used at least once and can be used as many times as necessary to form the groups to produce a minimum result. Identify each group of two, four or eight terms by enclosing them within a loop or circle.
4. Note the input variables that change in moving from one minterm to the next in each group. The variables that change are redundant and drop from the expression. Another way of looking at this is to observe the variables within each group that remain the same in moving from one adjacent cell to the next within that group. Use these variables to form a product expression that will appear in the reduced equation.
5. Once you have determined a product expression for each group of minterms on the map, write the final output expression by ORing together the product terms developed for each group.

In Figure 9-19 we show the Karnaugh map used in reducing the Boolean expression for our BCD invalid code detector. Note that we can identify two groups of four variables. The reduced equation is much simpler than the equation we wrote from the truth table given earlier.

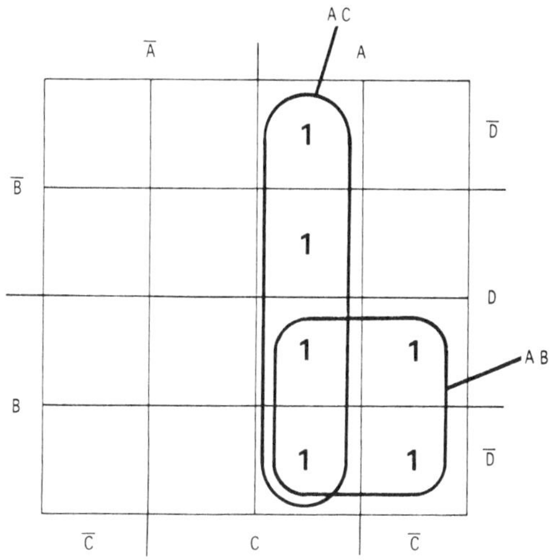

Figure 9-19
Karnaugh map for reducing the equation for a BCD invalid code detector.

Don't Care" States. There are some design situations where all combinations of the input variables will not occur. For example, you may have identified the need for four input variables and your design calls for the use of only seven of the sixteen possible input states. You can usually determine by the application which input combinations can never occur. In other situations there are various combinations of input states which will not effect the operation of the circuit and therefore you do not care whether they occur or not. It is useful to identify these "don't care" states. In most applications you should have no difficulty in determining what these "don't care" states are. Such states are of value in minimizing the logic expression through the use of a Karnaugh map. The "don't care" states are plotted on the Karnaugh map along with the minterms specified by the truth table or the equation. In most cases they will aid in the reduction of the circuitry required to implement the desired function.

To illustrate this concept, assume that your design calls for four-input variables and the output function is indicated by the equation below.

$$M = \overline{A}\,\overline{B}\,C\,D + \overline{A}\,\overline{B}\,C\,\overline{D} + A\,\overline{B}\,C\,D + A\,B\,C\,\overline{D} + A\,B\,C\,D$$

Figure 9-20A shows how this function is plotted on a Karnaugh map. The variables are grouped to reduce the amount of circuitry required to implement the function. This greatly simplifies the function as you can see by the reduced equation below.

$$M = \overline{A}\,\overline{B}\,C + \overline{B}\,C\,D + A\,B\,C$$

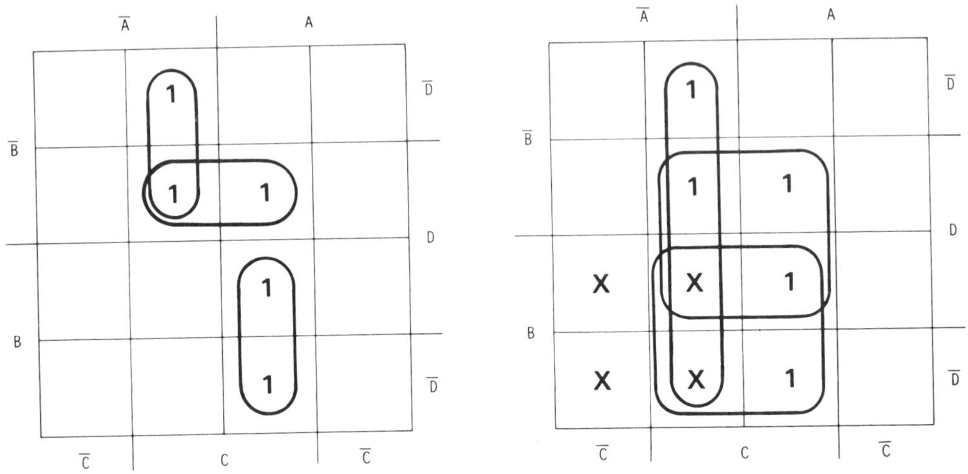

$M = \overline{A}\,\overline{B}\,C\,D + \overline{A}\,\overline{B}\,C\,\overline{D} + A\,\overline{B}\,C\,D + A\,B\,C\,D + A\,B\,C\,\overline{D}$

$M = \overline{A}\,\overline{B}\,C + \overline{B}\,C\,D + A\,B\,C$ **A**

$M = \overline{A}\,C + C\,D + B\,C$ **B**

Figure 9-20
Example of the use of "don't care" states in minimizing logic functions.

In your design process assume you determine that there are four "don't care" states. These are: $\overline{A}\,B\,\overline{C}\,\overline{D}$, $\overline{A}\,B\,\overline{C}\,D$, $\overline{A}\,B\,C\,D$, and $\overline{A}\,B\,C\,\overline{D}$. These four "don't care" states can then be plotted on the map as shown in Figure 9-20B. The "don't care" states are indicated by X's instead of binary 1's used to designate the required minterms. You can now use the "don't care" states along with the designated minterms to produce further circuit reductions. The X's can be grouped along with the 1s to form larger loops. The more minterms that you can include within a group, the greater the resulting minimization. As you can see, three groups of four can be formed. This resulting equation is:

$$M = \overline{A}\,C + C\,D + B\,C$$

This equation is far simpler than either the original equation or the first reduced version. When designing a combinational logic circuit, don't forget to make every attempt to identify these "don't care" states since significant reductions in circuit size and complexity can result.

Implementing the Logic Equations

You are now ready to select the circuitry to implement your design. Remember that your basic goal is to perform the desired function for the lowest possible cost. This means you will select the lowest price available circuitry. You will attempt to minimize the number of components in the design. This not only lowers the cost, but will also reduce power consumption, size and weight and increase reliability.

In all new equipment design situations you will be using integrated circuits. There are very few if any applications where benefits can be derived by using discrete component circuitry. Therefore, our discussion here is limited to selecting the types of integrated circuits appropriate to your design.

There are four practical ways to implement a combinational logic function with integrated circuits. They are:

1. SSI
2. MSI
3. ROM
4. PLA

Each of these approaches has its own benefits and limitations. In the sections to follow we will discuss each of these methods of implementation. We will use the BCD invalid code detector circuit as an example in evaluating each of these methods.

SSI Implementation. The most direct method of implementing your logic equations is to use SSI logic gates. Then by working from the equation derived from the truth table or the minimized version from the Karnaugh map, implement the circuit with available NAND and NOR gates. This literal approach is best employed when the function to be implemented is simple. For larger more complex functions some of the other techniques should be used.

To illustrate the use of SSI circuits in implementing our BCD invalid code detector, consider the original Boolean equation:

$$F = A\,\overline{B}\,C\,\overline{D} + A\,\overline{B}\,C\,D + A\,B\,\overline{C}\,\overline{D} + A\,B\,\overline{C}\,D + A\,B\,C\,\overline{D} + A\,B\,C\,D$$

This expression is readily implemented with SSI logic circuits as shown in Figure 9-21A. Each product term requires a four-input gate. Dual four-input TTL gates such as the 7420 can be used. One gate will be used for each minterm expression in the equation. All of these product terms will then be ORed together to produce the output function F. And since there are six terms, a six-input OR gate is required. An eight-input TTL gate such as the 7430 can be used for this purpose. Two of the inputs will not be used and can be simply connected to one of the other inputs. Note that the four-input variables must come from a source where both the normal and complement signals are available. If the complements are not available then they can be generated with inverters as shown in Figure 9-21B. A standard TTL circuit is the 7404 hex inverter containing six inverter circuits. Only four of these are needed in this application. As you can see, it requires a minimum of four and possibly five TTL integrated circuit packages to implement this function. These ICs must be mounted on a printed circuit board and the interconnection pattern on the circuit board must be designed. This will take a substantial amount of time, and the printed circuit board required to hold these circuits will be fairly large.

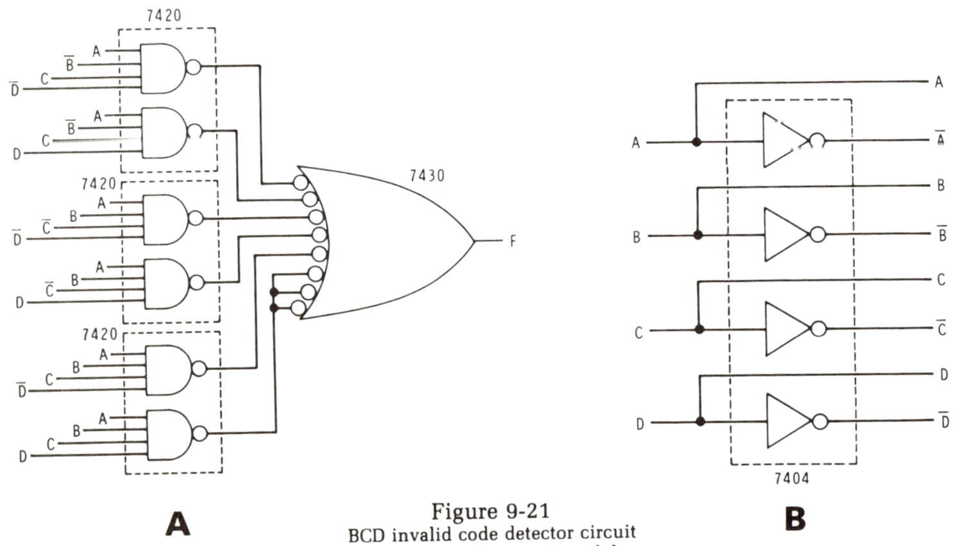

Figure 9-21
BCD invalid code detector circuit using SSI implementation of the equation.

Of course it is ridiculous to implement our BCD invalid code detector circuit as we have indicated. You have already shown that by the use of the Karnaugh map you can reduce the original equation to the simplified expression:

$$F = A\,C + A\,B$$

Figure 9-22 shows how this equation can be implemented with SSI circuits. Only three two-input logic gates are required. This means that a standard quad two-input NAND gate such as the TTL 7400 can be used to implement this expression. As a result we have implemented our detector circuit with a single integrated circuit. This reduces the package count, power consumption and printed circuit board design time as well as size and weight. Note that the input variable D is not even required in implementing this function. Yet, this simple circuit will generate the same truth table as the more complex circuit in Figure 9-21. The value of minimization by the use of a Karnaugh map is evident in this example.

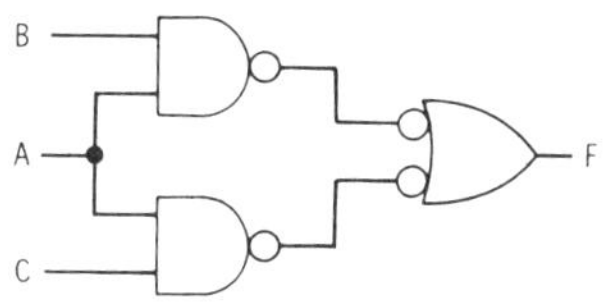

Figure 9-22
BCD invalid code detector circuit.
Minimum SSI circuit.

MSI Implementation. There are a variety of MSI functional circuits that can be used to implement combinational logic circuit designs. While these MSI circuits are designed to perform common combinational logic functions, they can often be adapted to perform other functions. Their use can result in a simplified, low cost method of implementing a logic expression. It is an alternative which should be thoroughly considered when designing digital circuits.

The two most useful MSI circuits for implementing logic equations are the decoder and the multiplexer or data selector. A decoder or 1 of N detector circuit accepts a number of logic inputs and recognizes all possible combinations of the input states. This is done by using AND or

NAND gates to detect each of the possible input conditions. Figure 9-23A shows a one-of-sixteen decoder circuit. It features four inputs which are then decoded by NAND gates to produce 16 outputs. A typical commercial version of this circuit is the 74154 TTL decoder. It is housed in a 24-pin dual in-line package. You can see from the figure that this decoder is a minterm generator. All 16 possible states are generated within the single IC thereby eliminating the need to interconnect external SSI gates and inverters. Note that in this circuit active low outputs are generated. To obtain active high outputs, inverters can be used on each output or the decode gates can be combined with other logic gates to produce the proper logic levels. When inputs E1 and E2 are low, all

A

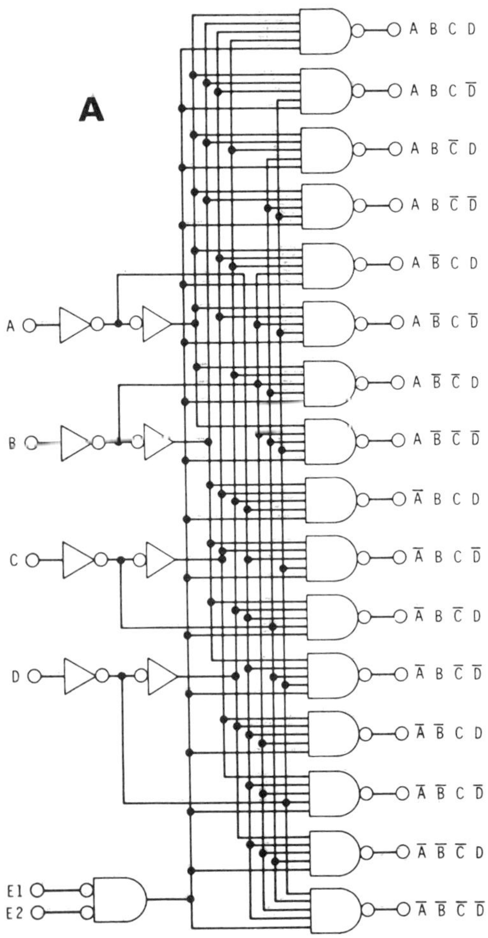

Figure 9-23
A MSI decoder (A) logic diagram and (B) block diagram.

B

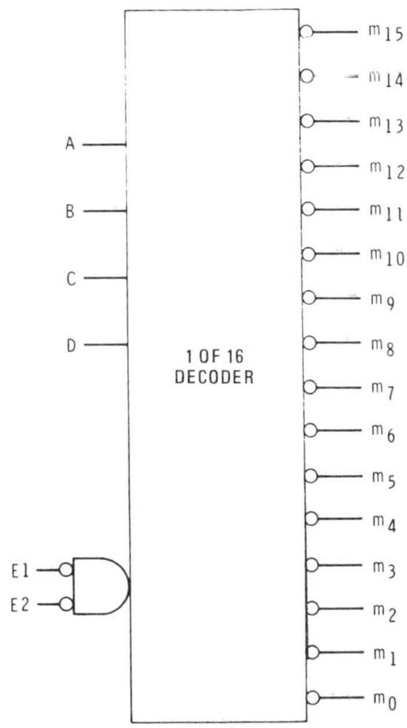

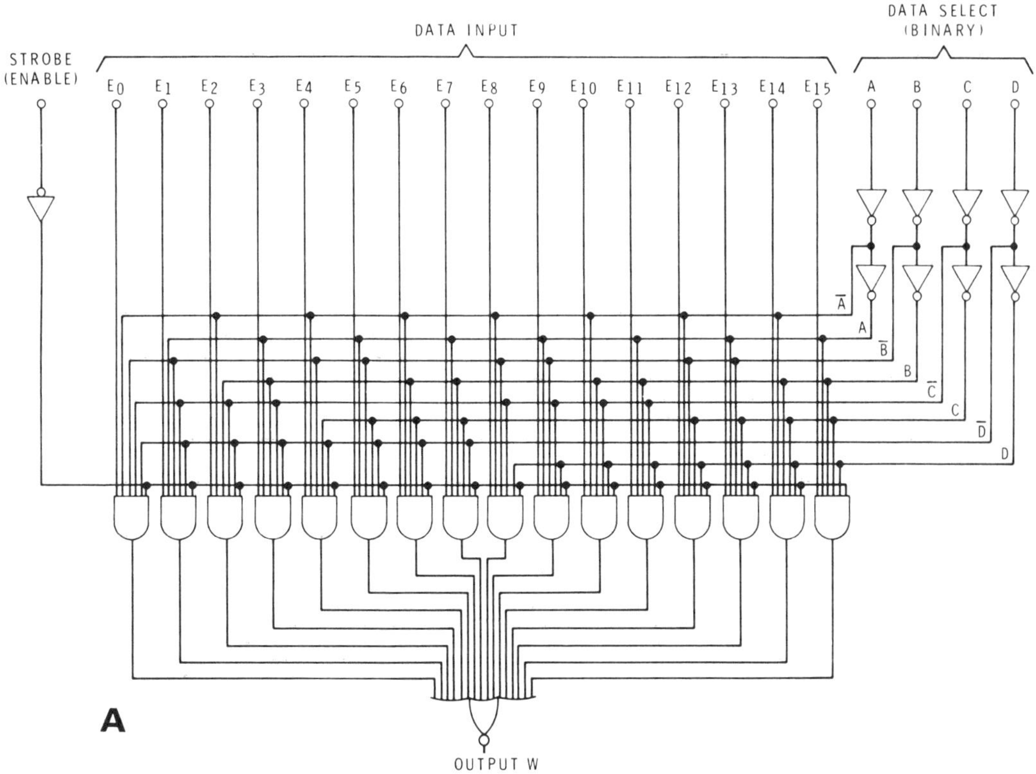

A

$$W = S(\bar{A}\bar{B}\bar{C}\bar{D}E_0 + A\bar{B}\bar{C}\bar{D}E_1 + \bar{A}B\bar{C}\bar{D}E_2 + AB\bar{C}\bar{D}E_3 + \bar{A}\bar{B}C\bar{D}E_4 + A\bar{B}C\bar{D}E_5 + \bar{A}BC\bar{D}E_6 + ABC\bar{D}E_7 + \bar{A}\bar{B}\bar{C}DE_8 + A\bar{B}\bar{C}DE_9 + \bar{A}B\bar{C}DE_{10} + AB\bar{C}DE_{11} + \bar{A}\bar{B}CDE_{12} + A\bar{B}CDE_{13} + \bar{A}BCDE_{14} + ABCDE_{15})$$

B

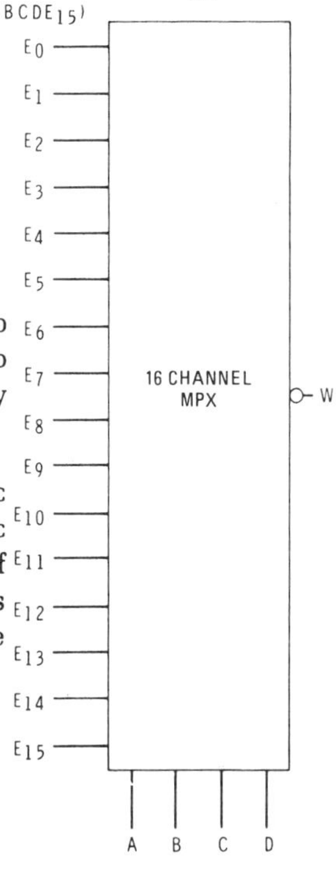

Figure 9-24
7415 TTL MSI multiplexer or data selector used to implement combinational logic circuits. (A) logic diagram, (B) block diagram.

sixteen gates are enabled. Figure 9-23B shows the block diagram used to represent this MSI decoder. Only the inputs and outputs are shown to simplify the drawing of the circuit. The outputs here are designated by the minterm numbers.

Another MSI circuit widely used for implementing combinational logic functions is the multiplexer or data selector. Figure 9-24 shows a logic diagram of a typical sixteen channel multiplexer circuit. Any one of sixteen input signals may be routed to the single output. Each input is applied to an AND gate that is enabled by the four-bit input code. The input code selects the input that is routed to the output.

Investigation of the data selector circuit in Figure 9-24 shows that all four-variable minterms are generated by the AND gates in the circuit. These are ORed together to produce a single output. The data selector circuit itself then implements a logic equation that is a sum of all possible minterms. An additional fifth input variable E can be accomodated by connecting it to one or more of the sixteen input lines. For example, any one of the sixteen four-bit minterms can be added to the output by applying a binary 1 to the appropriate data input. If a minterm is not needed in the output, the associated input line can be connected to binary 0. A fifth input code bit can be implemented by connecting it or its complement to the appropriate data inputs. A single strobe or enable line(s) is also used to enable or inhibit the entire circuit. Note that the output is active low. An inverter or other logic gate can be used to provide the complement if needed.

We can readily illustrate the use of MSI decoders and data selectors by showing how our BCD invalid code detector circuit can be implemented with them.

Figure 9-25 shows the 74154 one-of-sixteen MSI decoder used to generate the BCD invalid code detection function. Instead of working with the simplified Boolean equation for this function we work with the complete equation derived from the truth table. Each of the six four-variable minterms are defined. These minterms are generated by the decoder. The proper decoder outputs are then fed to a TTL NOR gate to produce the desired output function. Note that with this method of implementation two integrated circuits are required. Although we are able to implement this function more economically than by the previously described brute force SSI method, the result does not produce a minimum package count or the lowest cost design. MSI circuits are considerably more expensive than SSI circuits. In addition, this implementation requires two integrated circuits. The MSI implementation shown in Figure 9-25 is not the most desirable approach. In most cases it will not produce the most efficient design.

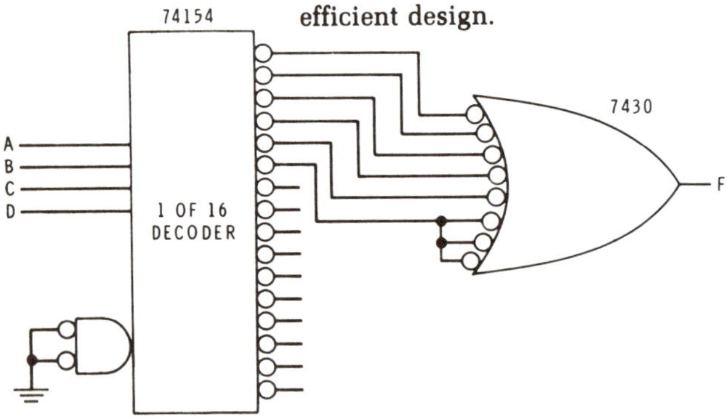

Figure 9-25
BCD invalid code detector function implemented with a 1-of-16 MSI decoder.

Figure 9-26 shows how the BCD invalid code detection function can be implemented with a sixteen channel data selector. A 74150 multiplexer circuit is used. Note that binary 1 (+5 volts) is applied to the six higher order inputs thereby enabling the gates which generate the proper minterms. All other inputs are connected to binary 0 to inhibit the remaining minterms from being applied to the output. Again we work from the expanded version of the logic equation rather than the simplified version. The implementation does result in the use of a single integrated circuit. But again this is an MSI device that is larger and more expensive than the simple SSI circuit we developed earlier. Note also that the circuit in Figure 9-26 has an active low output. Whenever we detect one of the six invalid BCD codes, the output of this circuit ($\overline{F}$) will go low instead of high as we indicated earlier. For many applications this is no disadvantage since a low output is just as valid an indication of the incorrect code as a high output. Nevertheless, our initial requirements stated that the output must be high. This may require the use of an external inverter thereby adding an additional IC package and further increasing cost, power consumption, and waste of space.

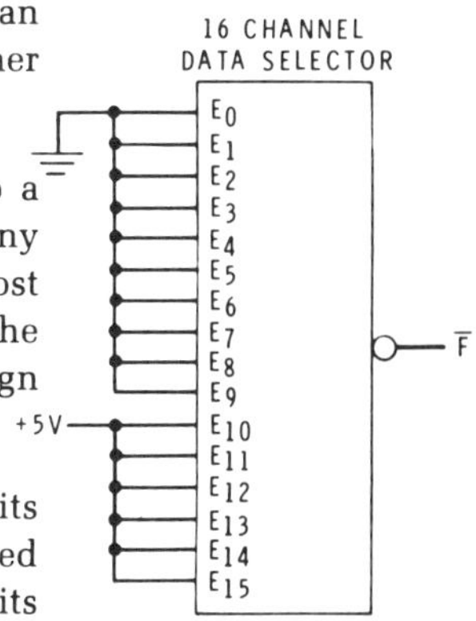

Figure 9-26
BCD invalid code detector circuit implemented with a MSI data selector.

While the use of MSI decoders and data selectors did not lead to a minimum implementation of our example problem, there are many situations where these devices will result in the minimum, lowest cost design. Every design will be different and you must evaluate each of the four basic alternatives before you find the one that meets your design criteria.

ROM Implementation. One of the easiest ways to design digital circuits is to use a read only memory. Virtually no design time or effort is required to use such a device. However, there are restrictions or limitations on its use. ROMs are generally large scale integrated circuits and must be custom manufacturered to your specifications. Therefore, they are expensive. In order to justify their use, the logical functions being implemented must require that degree of sophistication.

Here are some guidelines for determining whether a ROM should be used to implement a given logic function:

1. ROMs are used primarily for multiple input and multiple output logic circuits. The use of a ROM becomes practical and economically feasible only when the number of inputs and number of outputs are equal to or exceed four. Logical circuits having fewer inputs and outputs are generally more economically implemented with SSI or MSI logic circuits. Because of this restriction, a ROM is not applicable to our BCD invalid code detector. While the circuit requires four inputs, it has only a single output. Naturally, a ROM could be employed but more of its capabilities would be wasted than used.

2. ROMs are best employed where all possible input combinations are specified by the logic design. For example, in a four input variable circuit a ROM is economical only if all or most of the input states are used.

If the circuit you are designing has four or more inputs and outputs you should consider the use of a ROM. Evaluate your design by studying the truth table to see if it meets the criterion indicated in the two steps above. If it does, the implementation of the ROM can be taken directly from the truth table itself as the input signals specify the ROM address states while the output states specify the memory contents at each of the address locations. No further design procedure is required.

PLA Implementation. The fourth and final alternative available to the digital designer in implementing combinational logic circuits is the programmable logic array. This LSI device is used primarily for implementing large complex logic functions. Such devices do not become practical or economically feasible until the complexity of the design reaches a very high level. PLAs are used primarily for multiple input multiple output circuits. If your design calls for five or fewer inputs and outputs, a PLA will not result in a minimum design. The cost involved in programming the circuit to produce the desired function during manufacturing will make the cost unusually high. In such cases MSI circuit implementation should be employed. If the number of inputs and outputs exceeds five or six, then PLAs should be considered. Like any of the other alternatives, the PLA implementation should be evaluated carefully from a cost standpoint. Size, power consumption, and reliability should also be considered as usual. For both PLAs and ROMs, high volume of usage will greatly reduce the cost and make these alternatives more practical. For our BCD invalid code detector, the PLA would certainly not be applicable.

To design a logic circuit using a PLA, you use the procedure outlined previously. The design is first tabulated in a truth table. From the truth table the logic equations are written. Then by using Boolean algebra or Karnaugh maps, the equations are then minimized. The minimized equations are given to the manufacturer who will in turn design the mask that will properly interconnect the gates within the PLA device. The result will be a custom integrated circuit for your design.

All four of these design alternatives for implementing digital circuits depend directly upon the available commercial integrated circuits. Your ability to meet your design objectives is a function of the type, cost, performance, flexibility, and quality of the integrated circuits available to you. For that reason it is imperative that you become familiar with the literature of all of the manufacturers of digital integrated circuits. Order their catalogs and ask for individual device data sheets. Study these to determine what circuits are available, what their specifications are and how they can be used. In addition, most integrated circuit manufacturers supply application notes which describe the ways in which their component can be used. Most manufacturers also offer engineering and design assistance to help their customers in selecting the correct device. There are also many manufacturers that do custom design work. If the commercially available devices do not meet your applications then it is possible that special custom devices can be designed and implemented for you. We simply cannot overemphasize the importance of being familiar with and working with the manufacturers of the digital integrated circuits that you will use in your designs.

Multiple Output Combinational Circuits

The applications that we have considered involve circuits with a single output. Multiple input states are monitored and a single logic signal is developed to indicate when specific states occur. There are many applications, however, that require multiple outputs as well as multiple inputs. All of the design procedures that we have described so far are fully applicable to combinational circuits with multiple outputs. Only minor variations are necessary to achieve a correct design.

The methods of defining the problem and stating the design objectives are similar. You will completely specify the type and number of inputs and the type and number of outputs.

Your problem statement is then converted into a truth table that will completely define the operation of the circuit. The number of inputs will determine the total number of states that can exist. Then, instead of defining a single output based on these inputs, you will define all of the outputs required by the application. Simply, this means creating a separate column in your truth table for each circuit output. In each column a binary 1 is recorded adjacent to the set of input conditions necessary to produce that output. Don't forget to note the states that won't occur or states that have no meaning for this application. These "don't care" states will greatly aid in reducing the amount of circuitry required. Once the truth table is complete you will have thoroughly defined the circuit to be designed.

Next, you will observe the output columns in the truth table and write a separate Boolean equation for each. Use a Karnaugh map to minimize these output equations. This will result in a minimized or reduced output equation for each of the outputs required by the circuit. It is these minimized equations that you will implement in your final design.

When choosing the integrated circuits to implement your design there are several important points to consider. First, depending upon the complexity of the circuit ROMs and PLAs should be given first consideration. These will generally result in the simplest and smallest circuits. MSI logic circuits should then be considered if ROMs and PLAs are too complex and sophisticated for the application. For many common functions, a standard MSI circuit may already exist making the design unnecessary. Finally, SSI circuits should be considered for multiple output circuits of minimum complexity.

When implementing the multiple output function with SSI circuits, it is a good idea to study the minimized output equations derived from the Karnaugh maps to determine if common product terms exist. If the same product happens to occur in two or more of the output expressions then it is only necessary to generate this product once. This will further reduce the amount of circuitry required.

Design Examples

We have now described the procedure for designing combinational logic circuits. Virtually any logic design problem can be handled with these procedures. However, because of the wide range of applications there will be many variations. The only way to illustrate the use of this procedure is to provide information on several different types of applications. Your own ability to design digital circuits will come from practice. The design examples in this section will help to give you the experience necessary to achieve competence. The primary purpose of the examples in this section is to illustrate the many ways in which the procedures described can be used. Additional practice problems are given in the Self Test Review following this section.

INPUTS	OUTPUT
A B C D	F
0 0 0 0	0
0 0 0 1	0
0 0 1 0	0
0 0 1 1	1
0 1 0 0	0
0 1 0 1	1
0 1 1 0	1
0 1 1 1	0
1 0 0 0	0
1 0 0 1	1
1 0 1 0	1
1 0 1 1	0
1 1 0 0	1
1 1 0 1	0
1 1 1 0	0
1 1 1 1	0

Figure 9-27
Truth table for Design Example #1:
Two-of-four detector

Design Example #1. Design a two-of-four input detector circuit. The circuit has four inputs A, B, C, and D, and we want a binary 1 output condition F to occur whenever only two of these inputs are binary 1. Develop the truth table for this circuit, write the output equation, minimize it, and select a method for implementing it.

Solution to Design Example #1. The truth table for this circuit is shown in Figure 9-27. With four inputs there are sixteen possible combinations that can occur. Our design requirements stated that we wanted the output F to be binary 1 when only two of the inputs were binary 1. By observing the binary states of each of the sixteen possible input conditions, you can quickly identify those where only two of the inputs are binary 1. These states are indicated by a binary 1 in the F output column.

You can go directly from the truth table to a Karnaugh map to attempt simplification of this logic function. However, it is usually a good idea to write the logic equation from the truth table first. This step only takes a short time and helps you to visualize the function better. Writing the equation from the truth table gives us:

$$F = \overline{A}\,\overline{B}\,C\,D + \overline{A}\,B\,\overline{C}\,D + \overline{A}\,B\,C\,\overline{D} + A\,\overline{B}\,\overline{C}\,D + A\,\overline{B}\,C\,\overline{D} + A\,B\,\overline{C}\,\overline{D}$$

$$F = m_3 + m_5 + m_6 + m_9 + m_{10} + m_{12}$$

Next, using the logic equation or the truth table, plot the function on a Karnaugh map. This is done in Figure 9-28. A binary 1 is marked in those cells identified by the minterms specified by the truth table and the equation.

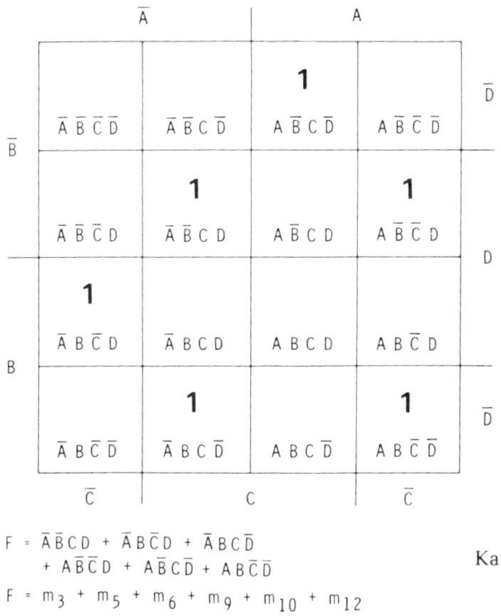

Figure 9-28
Karnaugh map for Design Example #1.

Observing the Karnaugh map should immediately tell you that absolutely no simplification of this logic function is possible. As you can see, the variables are widely spaced and separated. There are no two minterms that can be grouped together. Since no simplification is possible, the logic equation must be implemented directly.

An initial consideration of the four methods of implementing the logic function will quickly rule out the use of the ROM and the PLA. Since only a single output is required, the circuit implementation will be either by SSI logic elements or MSI functional devices. Your job is to evaluate these alternatives and select the best form of implementation for your design.

There are several ways that we can implement our two-of-four detector circuit. We can use SSI logic gates and implement the equation directly as shown in Figure 9-29. Here TTL SSI gates are used. Type 7420 dual four-input gates are used to form products of the inputs. A 7430 eight-input gate is used to produce the output sum. Depending upon the source of the inputs, the 7404 hex inverter IC may be needed to generate the complements of the input signals. With this circuit a total of five integrated circuits are required. While the cost of such circuits is extremely low (approximately 15 cents each in large quantities) they do take up a lot of space. A significant amount of time is required to lay out a printed circuit board to interconnect these devices. Therefore, it is desirable to investigate the methods of implementing this circuit with MSI functional devices.

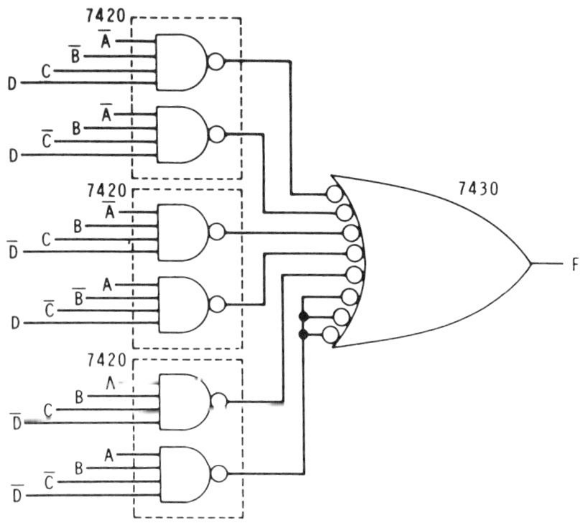

A

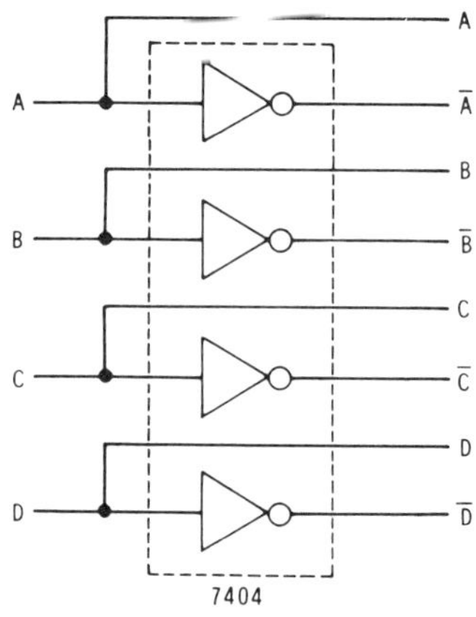

B

Figure 9-29
Two-of-four detector circuit.

Figure 9-30 shows how the two-of-four detector circuit can be implemented using a 74154 one-of-sixteen decoder and a 7430 eight-input gate. The one-of-sixteen decoder is used as a minterm generator and the appropriate outputs are ORed together in the 7430 gate. The size of the circuit is somewhat less than the SSI implementation mentioned earlier. The layout is simpler and the two circuits occupy much less space.

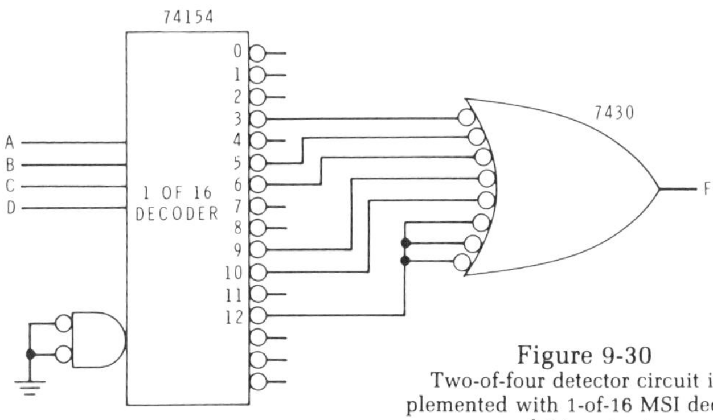

Figure 9-30
Two-of-four detector circuit implemented with 1-of-16 MSI decoder and 8-input gate.

A third alternative is to use an MSI data selector. The two-of-four detector circuit can be implemented with a 74151 multiplexer as shown in Figure 9-31. This single 16-pin dual-in-line IC seems to offer the most promising method of implementing the circuit.

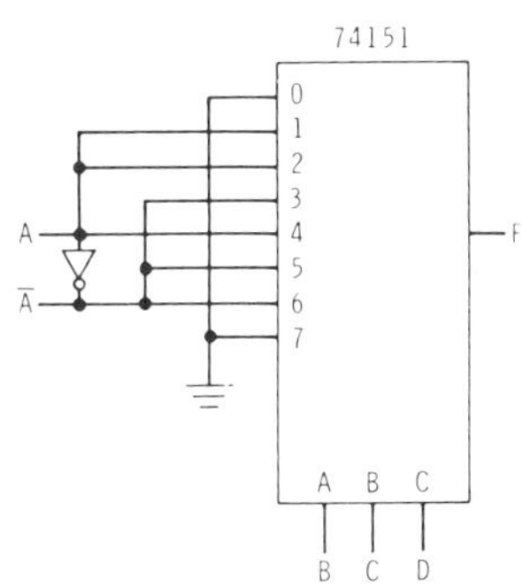

Figure 9-31
Two-of-four detector circuit implemented with MSI data selector.

The 74151 data selector or multiplexer circuit was described in detail in an earlier unit. The logic diagram of this circuit is repeated in Figure 9-32. Each of the gates in the multiplexer are enabled by the A, B, and C inputs. The B, C, and D inputs of our logic circuit will be applied to these lines. These are the least significant bits of the four-bit words. To see how the circuit works, consider the decimal value of these three least significant bits alone and analyze the truth table to determine which inputs on the multiplexer will be used. You should find that six of the sixteen possible states of these three inputs are used. The decimal values of the BCD input for each output F are 3, 5, 6, 1, 2 and 4. The unused input states are $\overline{B}$, $\overline{C}$, $\overline{D}$ and B, C, D, or 0 and 7. These correspond to the 0 and 7 inputs on the multiplexer. Since they will not be used, these two inputs are connected

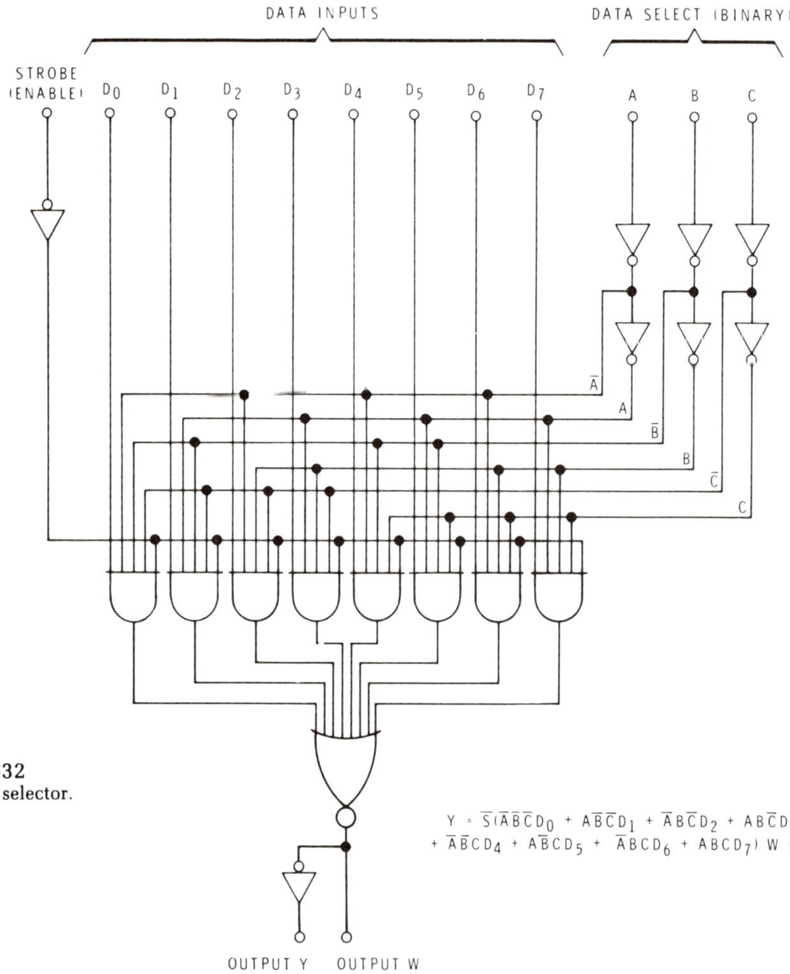

Figure 9-32
74151 TTL data selector.

$$Y = \overline{S}(\overline{A}\overline{B}\overline{C}D_0 + A\overline{B}\overline{C}D_1 + \overline{A}B\overline{C}D_2 + AB\overline{C}D_3 + \overline{A}\overline{B}CD_4 + A\overline{B}CD_5 + \overline{A}BCD_6 + ABCD_7)\quad W = \overline{Y}$$

to ground or binary 0 thereby disabling them. To the other six inputs we connect either the A or $\overline{A}$ signal in order to form the appropriate four-bit product terms. Depending upon the source of the A input, an external inverter may or may not be required as shown in Figure 9-31.

In evaluating our alternatives now we can see that the simplest and easiest to use is the 74151 multiplexer. It results in a single integrated circuit and design time and PC board layout time are at a minimum. However, this MSI device is more expensive compared to the SSI devices used and shown in Figure 9-29. The MSI multiplexer costs approximately twice as much as all of the integrated circuits in the SSI version. For that reason from a cost standpoint we may be inclined to select the SSI implementation method. However, keep in mind that the PC board layout time for the SSI circuit will be significant. In most cases it will be great enough to offset the extra cost of the 74151. For that reason the multiplexer implementation of the circuit probably offers the best solution to this particular design problem.

Design Example #2. This next example of combinational logic circuit design is more complex but is also more representative of the types of circuits that you will be designing. The techniques for designing multi-input multi-output logic circuits are demonstrated here.

Design a code converter circuit that will change the 8421 BCD code into the four-bit excess 3 code. Parallel inputs and outputs are required. (NOTE: Since the 8421 BCD input code is used, the six invalid states are considered as "don't care" states.)

Solution to Design Example #2. The first step in the design procedure is to develop a truth table. Since the inputs are the 8421 BCD code, four input lines are required. These are labeled A, B, C, and D. The excess 3 code also has four bits. These will be labeled W, X, Y, and Z. The truth table for this circuit is shown in Figure 9-33.

The next step is to write the Boolean equations from the truth table. Since there are four outputs from the circuit, you will develop an output equation for each. This is done by observing the positions of binary 1's in each output column. Then you write a sum-of-products expression involving the related minterms. The output equations for this circuit are:

$$W = \overline{A}\,B\,\overline{C}\,D + \overline{A}\,B\,C\,\overline{D} + \overline{A}\,B\,C\,D + A\,\overline{B}\,\overline{C}\,\overline{D} + A\,\overline{B}\,\overline{C}\,D$$
$$X = \overline{A}\,\overline{B}\,\overline{C}\,D + \overline{A}\,\overline{B}\,C\,\overline{D} + \overline{A}\,\overline{B}\,C\,D + \overline{A}\,B\,\overline{C}\,\overline{D} + A\,\overline{B}\,\overline{C}\,D$$
$$Y = \overline{A}\,\overline{B}\,\overline{C}\,\overline{D} + \overline{A}\,\overline{B}\,C\,D + \overline{A}\,B\,\overline{C}\,\overline{D} + \overline{A}\,B\,C\,D + A\,\overline{B}\,\overline{C}\,\overline{D}$$
$$Z = \overline{A}\,\overline{B}\,\overline{C}\,\overline{D} + \overline{A}\,\overline{B}\,C\,\overline{D} + \overline{A}\,B\,\overline{C}\,\overline{D} + \overline{A}\,B\,C\,\overline{D} + A\,\overline{B}\,\overline{C}\,\overline{D}$$

INPUTS 8421 BCD A B C D	OUTPUTS X S 3 W X Y Z
0 0 0 0	0 0 1 1
0 0 0 1	0 1 0 0
0 0 1 0	0 1 0 1
0 0 1 1	0 1 1 0
0 1 0 0	0 1 1 1
0 1 0 1	1 0 0 0
0 1 1 0	1 0 0 1
0 1 1 1	1 0 1 0
1 0 0 0	1 0 1 1
1 0 0 1	1 1 0 0
1 0 1 0	DON'T CARE
1 0 1 1	
1 1 0 0	
1 1 0 1	
1 1 1 0	
1 1 1 1	

Figure 9-33
Truth table of 8421 BCD to XS3 code converter circuit.

The next step is to map the output equations. A sixteen-cell Karnaugh map is used for each output. You can map each output function directly

Figure 9-34
Karnaugh maps for BCD to XS3 code converter.

from the truth table or from the equations you derived from the table. Don't forget to mark the "don't care" states with X's. Combine these X's with the binary 1s on the map to help in reducing the equations. Finally, minimize the equations by grouping the variables on the map and from those groupings write the reduced logic equations. Figure 9-34 shows the four output maps and the reduced equations.

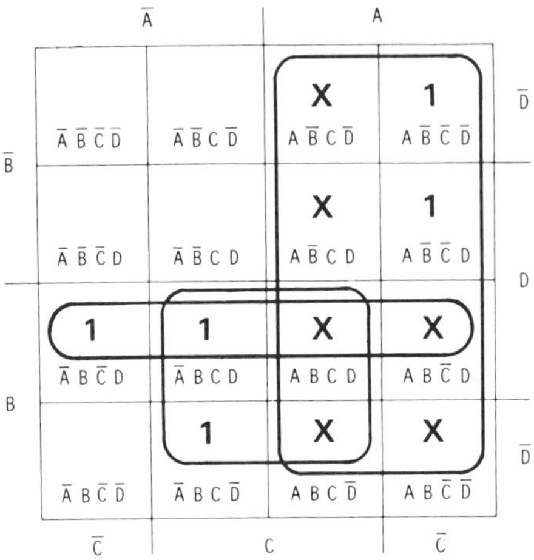

A

$W = \bar{A}B\bar{C}D + \bar{A}BC\bar{D} + \bar{A}BCD + A\bar{B}\bar{C}\bar{D} + A\bar{B}\bar{C}D$
$W = m_5 + m_6 + m_7 + m_8 + m_9$
$W = A + BD + BC$

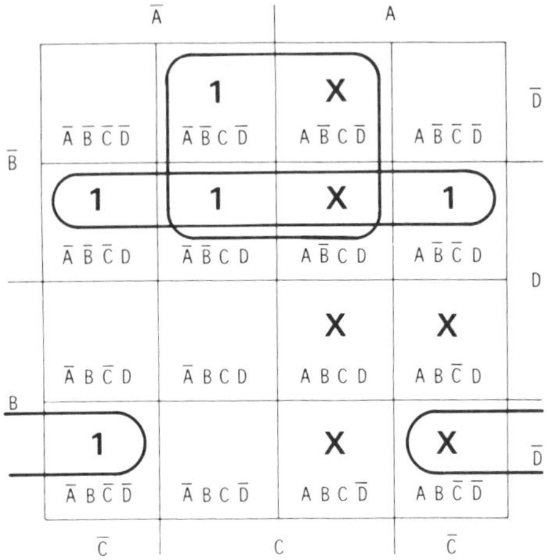

B

$X = \bar{A}\bar{B}\bar{C}D + \bar{A}\bar{B}C\bar{D} + \bar{A}\bar{B}CD + \bar{A}B\bar{C}\bar{D} + A\bar{B}\bar{C}D$
$X = m_1 + m_2 + m_3 + m_4 + m_9$
$X = \bar{B}D + \bar{B}C + B\bar{C}\bar{D}$

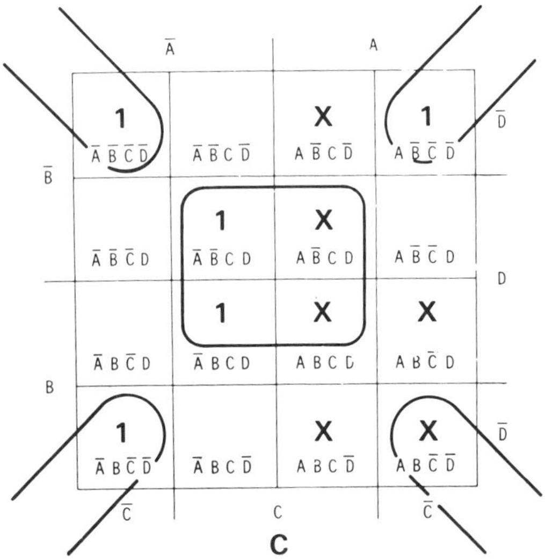

C

$Y = \bar{A}\bar{B}\bar{C}\bar{D} + \bar{A}\bar{B}CD + \bar{A}B\bar{C}\bar{D} + \bar{A}BCD + A\bar{B}\bar{C}\bar{D}$
$Y = CD + \bar{C}\bar{D}$

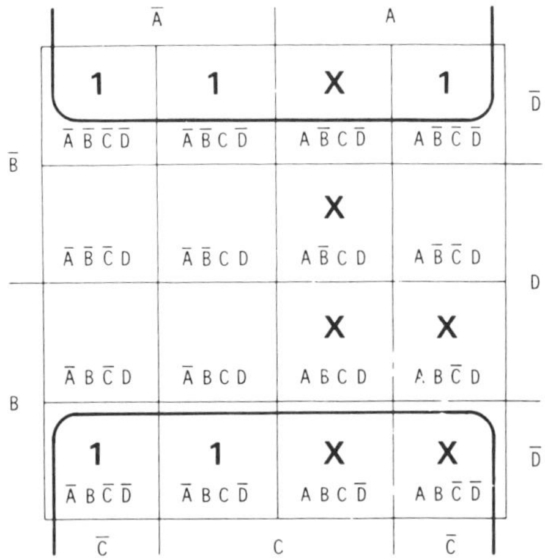

D

$Z = \bar{A}\bar{B}\bar{C}\bar{D} + \bar{A}\bar{B}C\bar{D} + \bar{A}B\bar{C}\bar{D} + \bar{A}BC\bar{D} + A\bar{B}\bar{C}\bar{D}$
$Z = \bar{D}$

As with any combinational logic circuit, there are several ways in which it can be implemented with hardware. Consider the various techniques described earlier and apply them to this problem to determine the optimum method of implementation.

Figure 9-35 shows the circuit implemented with SSI logic circuits. This circuit implements the minimized equations from the Karnaugh maps. Only four SSI packages are required. Assuming the use of 7400 TTL circuits, the following circuits are required:

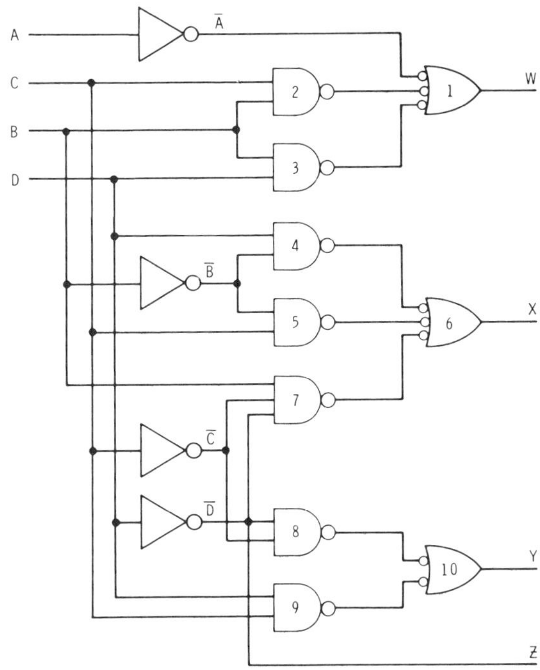

Figure 9-35
BCD to XS3 code converter implemented with SSI circuits.

1—7410 triple 3-input gate (gates 1, 6, and 7)
2—7400 quad 2-input gates (gates 2, 3, 4, 5, and gates 8, 9, and 10)
1—7404 hex inverter

The circuit is simple and straight forward. Trace the input lines and compare each of the circuits with the logic equations to be sure you see how the circuit is implemented.

Figure 9-36 shows how the BCD to XS3 code converter can be implemented with MSI data selectors. Type 74150 TTL MSI data selectors are used to implement the output equations for W, X, and Y. These sixteen-input multiplexers are driven by the four-line 8421 BCD input. The input lines of the multiplexers corresponding to the minterms appearing in the output equations are connected to +5 volts to enable them. The unused inputs are connected to ground to disable them. Output Z is implemented with an inverter connected to the D input. Note that this method of implementation requires four IC packages, three of which are 24 pin MSI devices. This method of implementation is larger and more expensive than the SSI implementation described earlier. It is not an efficient method of implementation.

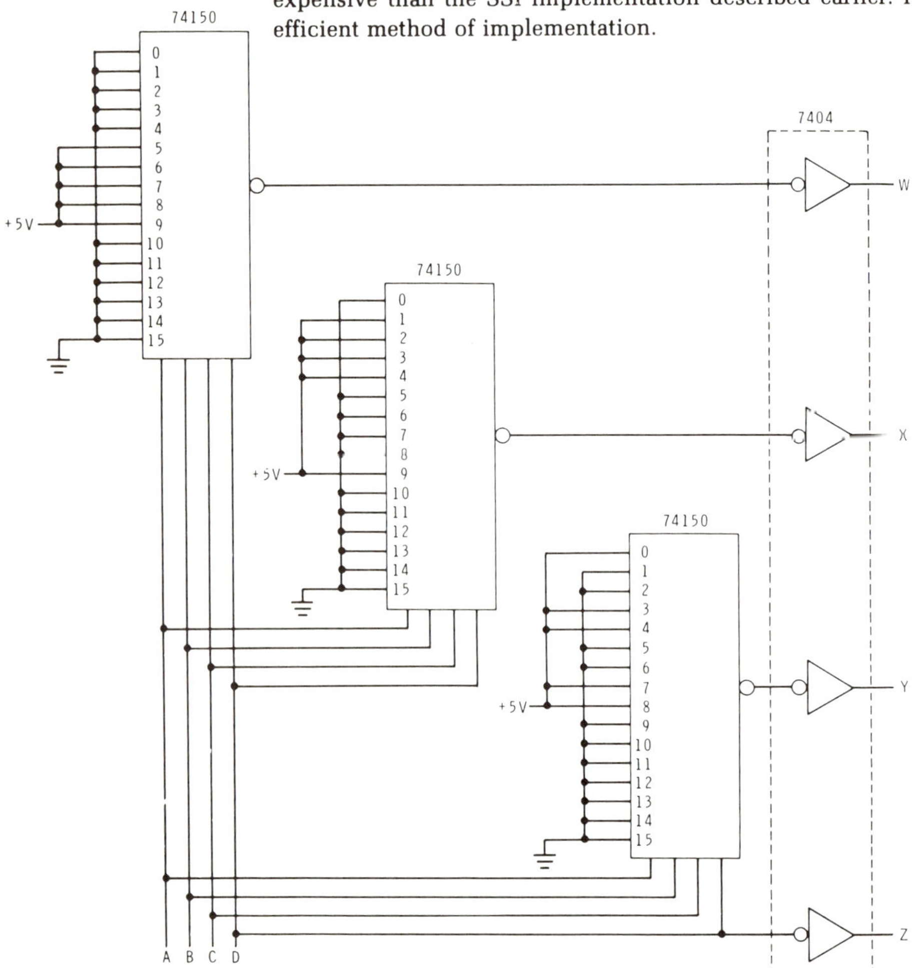

Figure 9-36
BCD to XS3 code converter implemented with MSI data selectors.

Perhaps the easiest way to implement this code converter circuit is to use a ROM. The 8421 BCD input code can be applied to the address lines of the ROM. The corresponding XS3 output code can be stored in the memory location specified by the input address. Since there are ten input states and ten output states, ten memory locations are required. The four-bit output code means that a total of forty bits are required in the ROM to implement this function.

Our guidelines for determing the applicability of a ROM to a combinational logic design is that the circuit have four or more inputs and outputs. Such a criterion specifies a minimum 64-bit ROM. Four input lines can specify a total of sixteen memory locations. Four output lines specifies four bits per memory word or $4 \times 16 = 64$ bits. Commercial ROMs this small are not available. The smallest available commercial ROM is a 256-bit unit organized as 32 eight-bit words. Such a ROM could be used for implementing the BCD to XS3 code converter.

Figure 9-37 shows the block diagram of a 32×8 ROM used to implement this function. The 32 word memory is addressed by five address input lines. The fifth or E input line is not required so it is simply connected to ground. The BCD input code is applied to the A, B, C, and D input lines. Each memory location can hold up to eight bits. Therefore, there are eight output lines. Only four of these are required for this application. These are labeled W, X, Y and Z to correspond to the desired output code signals. When the ROM is manufactured, the XS3 code will be stored in the memory location specified by the 8421 BCD input code.

With this arrangement only forty of the total possible 256 bits are used. This means that a significant amount of the memory is wasted. However, if this circuit is to be used in high volume, the cost of this device can be very low. Since it requires only a single 16-pin dual-in-line package, it may be the most desirable means of implementing this function.

The other method of implementing this logic equation is with a PLA. This is not a practical means in this application since the requirement is not large enough or complex enough to warrant the use of a PLA. Therefore it should not be considered.

In considering the various means of implementing the circuit we have outlined, the two most desirable means appear to be the SSI implementation of the minimized equations or a ROM. The SSI implementation is the lowest cost approach but does require four integrated circuit packages and the associated printed circuit board design. The ROM method is more expensive but occupies less space. Depending upon the quantities used and the size and space limitations of the project, the ROM method should be carefully considered.

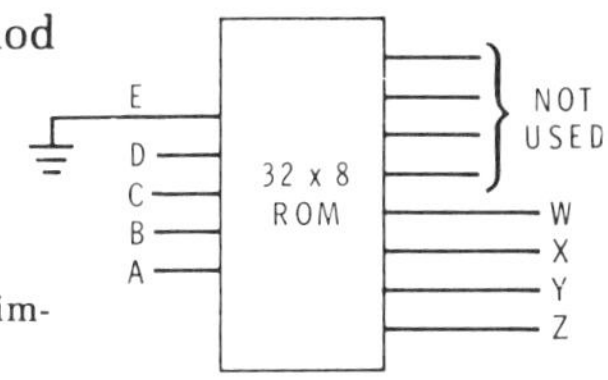

Figure 9-37
BCD to XS3 code converter implemented with a ROM.

SEQUENTIAL LOGIC CIRCUIT DESIGN

A sequential logic circuit is one designed to deal with a sequence of logic operations occuring over a period of time. The circuit may generate a sequence of timing pulses for use in controlling the operation of other circuits. In such an application the sequential logic circuit is used to **automate** a particular function. It will carry out a certain programmed sequence of events in the proper order and in the proper time sequence. Alternately, a sequential logic circuit may **process** logic signals occuring in a particular sequence. A sequential circuit for example, may be designed to detect a certain sequence of events and respond by generating output signals to other circuits.

In order to carry out these functions, a sequential logic circuit must in some cases be capable of making logical decisions. Logical decision-making of course is carried out by combinational logic circuits. This can be a simple logic gate or a more complex functional logic circuit. Most sequential circuits will contain some form of combinational logic circuit for decision making purposes.

The key feature of a sequential logic circuit is its ability to store data. In order to generate a desired sequence of output pulses, the sequential circuit must have some type of memory so that it can keep track of its sequence. The major element in any sequential logic circuit is the flip-flop. Flip-flops are interconnected with the combinational logic circuit to perform the desired function.

A sequential logic circuit responds to the various inputs applied to it. In return it generates specific output pulses depending upon its function. The output signals are a function not only of the input states, but also the current state of the sequential circuit as stored in the flip-flop memory.

Figure 9-38 shows a general block diagram of a sequential logic circuit. The heart of the circuit is the flip-flop memory where the state of the circuit is determined. The flip-flops are generally controlled by clock pulses. The flip-flop outputs drive the combinational logic circuits. The combinational logic circuits are also driven by various input signals. The combinational logic outputs in turn drive the flip-flops and can also drive external circuits as required.

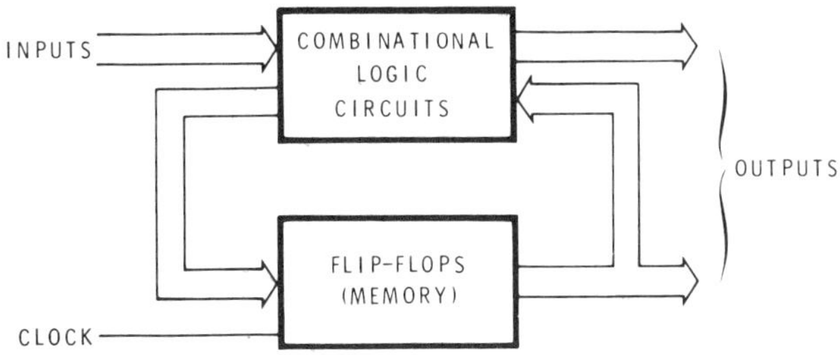

Figure 9-38
General block diagram of a sequential logic circuit.

The number of flip-flops in the sequential logic circuit determines the total number of different states in which the circuit can exist. These states are defined by a particular binary code which is stored in the flip-flops. As the inputs are applied and clock pulses occur, the state of the sequential logic circuit will change. Flip-flops will become set or reset according to a particular desired pattern. The flip-flop states may be interpreted as a special form of binary code. These states, in turn, control the combinational logic circuits to cause the correct sequence of events to occur and to generate the sequence of output pulses.

You have already studied the most common forms of sequential logic circuits. These are counters and shift registers. Standard binary and BCD counters are made up of flip-flops and in some cases combinational logic circuits. Shift registers are another form of sequential logic circuit. In most design situations a standard counter or shift register can generally be used to meet the need for a sequential logic circuit. However, there are many applications where special sequential circuits can result in benefits. Unusual applications and special functions are readily implemented with special forms of sequential logic circuits. When unique codes and sequences are required, special sequential logic circuits can often result in a more efficient design using fewer parts which operate at higher speeds.

The most common way of classifying sequential logic circuits is the process by which the circuit changes from one state to the next. The state transition can occur synchronously with the application of an external clock pulse or asynchronously. In a synchronous sequential circuit, the speed of operation is a function of the clock frequency. The states of the circuit change in step with the clock pulses. In an asynchronous circuit, the state changes occur as a result of previous changes. The speed of operation is strictly a function of the propagation delay times in the circuit and the rate of occurence of any external signals. Because most IC logic circuitry has such short propagation delays, very high speed operations can be obtained with asynchronous circuits. However, because of the unequal propagation delays of the various gates and circuit paths, unreliable operation can occur. False triggering and invalid states can prevent circuits from operating correctly. The design of an asynchronous circuit is also more difficult since it requires an analysis of the possible fault conditions that can occur and some means of correcting these faults. Generally, synchronous circuits are easier to design, implement and control, and therefore are recommended over asynchronous designs. This section emphasizes the design and application of synchronous sequential circuits using JK flip-flops.

Design Procedure

Most of the sequential logic circuits in use today are some form of special counter. Like the standard binary and BCD counters you have studied before, these special counters will sequence through a number of states in response to the input signals applied. The flip-flops in the counters will set and reset according to a specific count sequence and generate the output pulses called for by the application. Such sequential circuits can be counters that implement a specific code or frequency dividers that have a required number of states. When sequential circuits like this are used for control purposes, the counters are normally referred to as sequencers or controllers. A wide variety of different code types can be implemented in this application, but in controllers and sequencers the codes most often used are cyclical codes where only one flip-flop in the counter changes state at a time.

The design procedure to be described here shows one method of designing special counters and sequencers. This procedure will provide you with a means of designing such circuits for most of the applications that you will encounter. Keep in mind that it is only one of many different techniques. Alternate design methods are available but most applications can be manipulated such that the type of circuit described here and the procedure to design it can be used.

In describing and illustrating this design procedure we will assume the use of synchronous circuits using JK flip-flops. This is the most commonly used approach and the one that will result in the most versatile and reliable design.

The design procedure for sequential logic circuits is as follows:

1. State the problem and completely define the design objectives.
2. Develop a state table from the problem definition.
3. Develop Karnaugh maps for the flip-flop inputs from the state table.
4. Write the input equations for the flip-flops from the Karnaugh maps.
5. Draw the complete logic diagram from the logic equations.
6. Implement the circuit using standard integrated circuits.

Defining the Problem. The first step in designing a sequential logic circuit is to define the required objective. This is best done by writing down a complete but concise description of the function to be performed. State explicitly what operations are to occur. There are many different ways of expressing the logic function to be performed. There is no standard method for doing this, but the most important part of it is to include all possible conditions. As part of the problem definition you should specify the characteristics of the input and output signals and specify or determine the number of states that the circuit must assume.

The circuit inputs and outputs can be expressed in several different ways. These may take the form of logical waveforms that define the sequence of functions to be performed. Alternately, these may be logic levels that occur at specific times which cause certain operations to take place. The inputs and outputs can also be expressed in the form of a truth table or state table. A state table is similar in format to a truth table, in that the input and output signals are expressed in terms of 1's and 0's in table form that shows the sequence of change at each of the desired steps or states defined by the problem.

Specifying the input and output signals and writing out a description of the function to be performed will generally decide or specify the number of states in which the circuit can exist. The number of steps in the sequence of operations to take place will generally determine the number of circuit states. When the number of states have been determined, this will tell you the number of flip-flops that the memory section will contain. These flip-flops will define a binary word. The sequence of the steps designates how the flip-flops change state. This in turn defines a specific code sequence. The code sequence may be the standard binary code or it could be any one of a number of special codes such as XS3 or Gray Code. Of course, any binary sequence can be selected and implemented if it is required by the application.

Developing a state or flow table. Having determined the number of states required by the application, you should now be able to develop a state or flow table that completely defines all of the states in the circuit. A state or flow table is simply a truth table that expresses the outputs of the flip-flops in the circuit for each of the states required by the application. Knowing the number of states required, the number of flip-flops can be calculated using the techniques described earlier. For example, an application requiring seven states would require a 3-bit counter. A 3-bit counter will produce $2^3 = 8$ states. Two flip-flops are insufficient since they will produce a total of only $2^2 = 4$ states. Three flip-flops will handle the application properly. One of the eight states will not be used.

Figure 9-39 shows a state table or flow table for an application requiring seven states. The states have been specified by the application, and are numbered 0 through 6. The eighth state (7) whose binary output is 100 is not used but is generally included in the table and labeled as not being used. Note that the circuit contains three flip-flops labeled A, B, and C. The binary code for each state bears no relationship to the decimal state assigned to it. The flip-flop outputs in such a case are simply treated as a bit pattern rather than a binary number.

A

	FLIP FLOP OUTPUTS	
STATE	A B C	
0	0 0 0	←
1	1 1 0	
2	0 1 0	
3	1 0 1	RECYCLE
4	0 0 1	
5	0 1 1	
6	1 1 1	→
7	1 0 0	NOT USED

Figure 9-39

B

	t	t + 1
STATE	A B C	A B C
0	0 0 0	1 1 0
1	1 1 0	0 1 0
2	0 1 0	1 0 1
3	1 0 1	0 0 1
4	0 0 1	0 1 1
5	0 1 1	1 1 1
6	1 1 1	0 0 0
7	1 0 0	NOT USED

Figure 9-40

To interpret the flow table in Figure 9-39, you simply observe the flip-flop states as the circuit changes from one state to the next sequentially moving from top to bottom. The state column indicates that the 0 state is the initial condition state and the other states occur sequentially as shown. When state six occurs (binary output 111) the circuit will then recycle back to the initial zero state upon the application of the next clock pulse.

Figure 9-40 shows another method of constructing a flow or state table. The state column and the section labelled t are similar to the flow table in Figure 9-39. An additional section labelled t + 1 is also included. The t indicates the state of the flip-flop outputs **before** the application of a clock pulse. The section t + 1 indicates the state of the flip-flops **after** the occurrence of one clock pulse. The information in this table is exactly the same as that in the flow table of Figure 9-48. Only the format is different.

The flow tables shown illustrate one of many possible special codes that can occur as the result of developing a sequential circuit for a specific application. It is possible that the code illustrated was derived strictly from the desired flip-flop output states at each state in the circuit. This table may have been developed by observing output waveforms that were originally specified as required by the application to produce the specific timing and sequencing required. Studying the table we see that no recognizable standard code exists. It is simply a random or special code that meets the particular application.

There will be other applications where a standard code may be used or specified. The problem may call for the pure binary code, a BCD code, excess 3 code, or the Gray code. In still other applications no standard code will be specified or required. In addition, a particular application may not be code sensitive, that is any code can be used. In such situations it is generally desirable to use some form of standard code. The conventional binary code is desirable since a counter can be easily constructed. This results in hardware simplifications.

For most sequential circuit applications where a special counter or sequencer is required, the most desirable approach is to implement a circuit in which only one flip-flop changes state at a time when the circuit is stepped from one state to the next. The Gray code is an example of such a code. Such a circuit can be made to operate faster than other types of counters. Unequal propagation delays in the various circuit components will not cause false triggering or spurious undesirable output pulses known as **glitches**. If more than one flip-flop changes state at a time, the unequal propagation delays of the flip-flops can cause momentary false states. Gates used to decode the various counter states can then produce very short duration pulses equal in length to the difference between the propagation delay time changes in the various flip-flops. These glitches can cause false circuit triggering. By changing only one flip-flop at a time such pulses are eliminated or very greatly minimized. In addition, it is also desirable to use a synchronous circuit so that all flip-flops in the counter are clocked at the same time. This too helps to minimize glitches.

The Gray code is only one of many cyclical codes in which only one bit of the code word changes at a time from one state to the next. There are many different combinations possible and you may use any combination to achieve the desired end result.

Another useful guideline in developing a special code for a counter or controller is that the initial state be assigned 0. All flip-flops should be reset for the 0 or initial state of the circuit. Most circuits will have a reset, resting or wait state from which all operations begin. By making this state equal to 0, it becomes easy to identify. The direct clear inputs on most JK flip-flops can also be tied together to generate this initial or reset state by simply bringing this line low. This line can then be controlled from a pushbutton or by a special circuit that will automatically place the counter in the 0 state when power is applied.

Developing a Karnaugh Map For The Counter. Another way of showing the states of a special counter or sequencer is to use a Karnaugh map. Each cell or square in the map indicates a specific state. The decimal number corresponding to each state can be written into the cell corresponding to the binary code produced by the circuit. Arrows can then be drawn on the table to indicate the sequence of flow as defined by the state table. The Karnaugh map produces a visual means of indicating the states of the sequential circuit. Figure 9-41 shows a Karnaugh map that plots the seven-state controller defined by the tables in Figure 9-40.

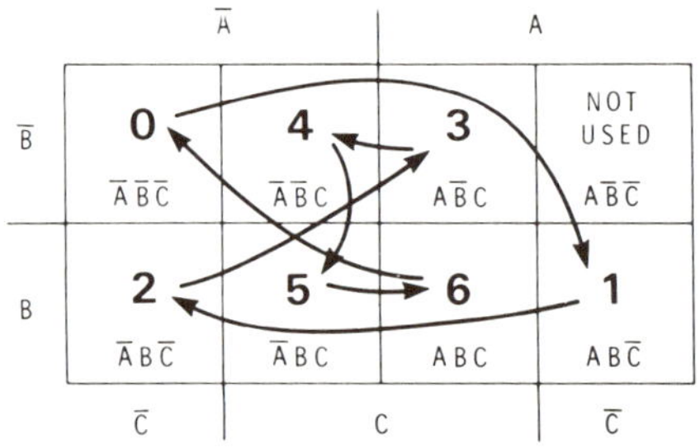

Figure 9-41
Karnaugh map for the seven-state counter defined by the table in Figure 9-50.

Besides helping you visualize what your counter or sequencer is doing, it can also help you in selecting a suitable cyclical code where only one bit changes from one state to the next. Recall from our earlier discussion of Karnaugh maps that adjacent cells in the map represent a change in only one of the variables. In moving from one cell to the next only one of the variables will change. Therefore to create a special code all that is necessary is to choose an initial starting point and then move from one cell to the next as many times as required to generate the special code.

Figure 9-42 shows several examples of how the Karnaugh map can be used to develop a special code. In all cases the initial starting point is 000 or $\overline{A}\,\overline{B}\,\overline{C}$. In each example, the counter or sequencer has 6 states. Six of the cells in the map, therefore are involved. Note that in moving from one cell to the next only one variable changes at a time. Note particularly the state change involved in recycling from the sixth state to the initial state. Remember that the Karnaugh map can be treated as a cylinder where the left and right edges of the map are considered to be adjacent.

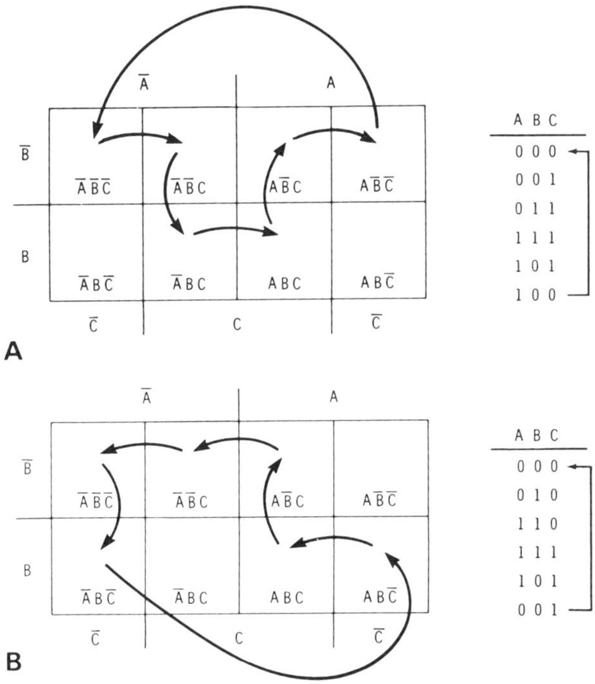

Figure 9-42
Examples of Karnaugh maps defining six-state cyclical counters.

While we have indicated six states in these examples, controllers of any number of states can be developed using this same technique. It is not always possible to generate a code where only one bit changes from one state to the next. This is particularly true of counters or sequencers with an odd number of states. It may be possible to generate a code that changes from only one state to the next, but in recycling more than one bit may have to change in order to have the code return to its initial state.

Figure 9-43 shows a five-state code where only one bit changes from the initial state through the five code states. But in recycling from the last state 110 to the initial state 000, two bits change. Generally, such conditions are not detrimental. Where they are, an even number of states could be introduced. The extra or unneeded state, called a **dummy** state, would not be used by the application but would serve only as a means of recycling the counter by having only one flip-flop change state at a time.

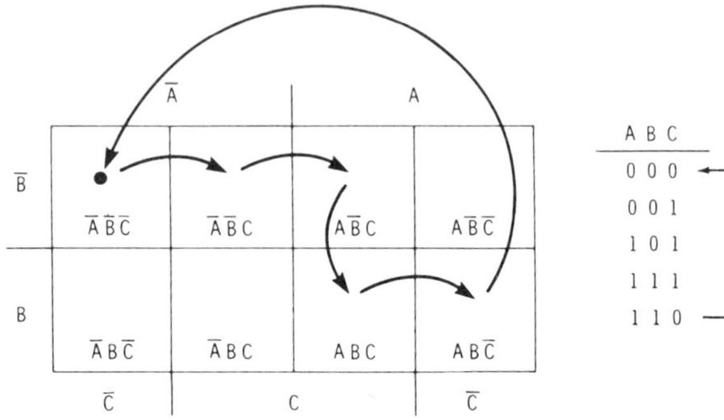

Figure 9-43
Examples of a Karnaugh defining a five-state code.

The most important application of a Karnaugh map in designing a counter or sequencer is in determining the input states to the flip flops required to produce this special code specified. In using JK flip-flops, certain states must be applied to the J and K inputs to cause the flip-flops to set and reset in the desired code sequence. To determine these inputs, we observe the state table and indicate which flip-flop must be set or reset in changing from one state to the next. The set or reset conditions are then plotted on the Karnaugh map. Then, by properly grouping these plotted input states, the input logic equations for the various flip-flops can be found.

To plot the state change Karnaugh maps for each flip-flop in the counter, you will mark each cell in the map with a symbol that designates the state change that is to take place. There are five possible conditions that can occur. These are:

1. Flip-flop changes from reset to set.
2. Flip-flop changes from set to reset.
3. Flip-flop is set and remains set.
4. Flip-flop is reset and remains reset.
5. Don't care.

To indicate these five state changes, or conditions, the symbols below are used.

1 Flip-flop changes from reset to set.
/ Flip-flop is in the set state and remains in the set state.
ϕ Flip-flop changes from set to reset.
0 Flip-flop is initially reset and remains reset.
X Input conditions do not occur or a "don't care" state exists.

These state changes are summarized by the table in Figure 9-44. The left-hand column shows the symbol used to indicate the state change. The t column represents the state of the flip-flop prior to the application of a clock pulse. The t + 1 column indicates the state of the flip-flop after the clock pulse.

SYMBOL TO REPRESENT STATE CHANGE	t	t + 1
1	0	1
/	1	1
Ø	1	0
0	0	0

t = BEFORE CLOCK PULSE
t + 1 = AFTER CLOCK PULSE

Figure 9-44
Symbols representing flip-flop state changes.

The state table is then analyzed to determine how each flip-flop change changes from one state to the next. These state changes are then plotted on the Karnaugh maps, one map for each of the flip-flops in the counter.

The symbol to be plotted in each cell of the Karnaugh map designates the state change that must take place in the output variable associated with the map in moving from one state to the next. For example, if flip-flop A is presently reset as indicated by its condition in the cell of interest and must set in order to transfer to the next state, a 1 will be marked in that cell. **The symbol in a cell represents the state change that must occur to move to the next state.**

This is best illustrated by developing the Karnaugh maps for the flip-flops in the special 7-state counter discussed earlier. The maps for this circuit are shown in Figure 9-45.

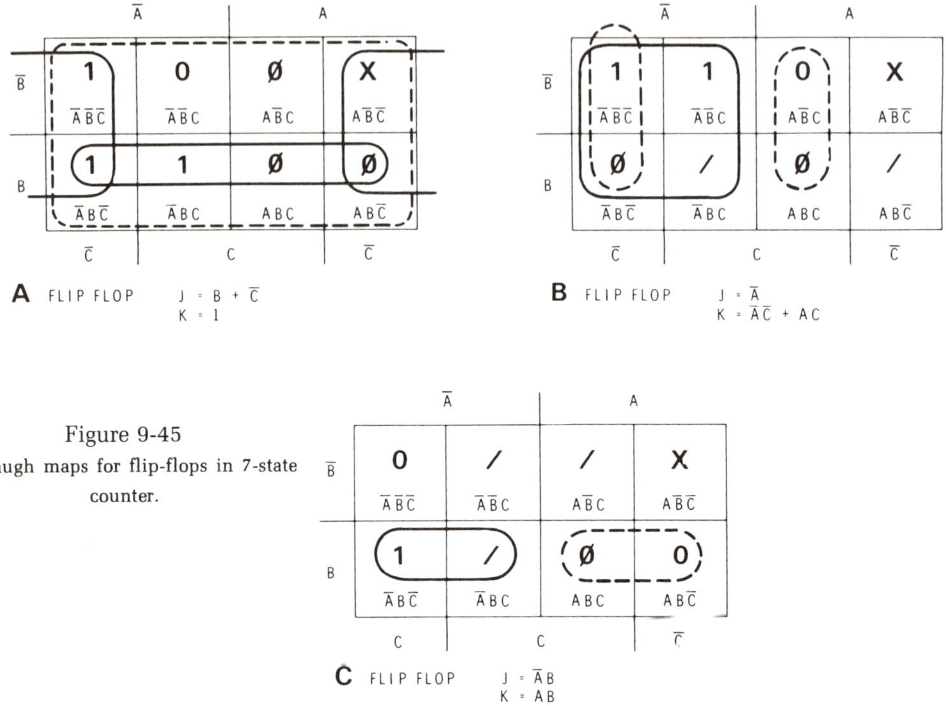

Figure 9-45
Karnaugh maps for flip-flops in 7-state counter.

In each cell is recorded the symbol that designates the state change that will take place in moving from the current state to the next state. Using the state tables in Figure 9-40 as a guide, verify the use of the correct symbol in each case. As an example, consider state 2 for the B flip-flop. State 2 is defined by the code 010 or $\bar{A}$ B $\bar{C}$. Locate this cell on the Karnaugh maps for the B flip-flop. To move from this state to the next state, state 3 (Code 101 or A $\bar{B}$ C) the B flip-flop will change from its current set state to the reset state. Therefore, in cell $\bar{A}$ B $\bar{C}$ you will record the symbol indicating that a reset condition must occur. This is the Ø symbol.

As another example consider the condition of flip-flop C in going from state 4 to state 5. Here you see that the C flip-flop is set in state 4 (code 001 or $\bar{A}$ $\bar{B}$ C). Locate state 4 on the Karnaugh maps for the C flip-flop. To move to the next state, state 5 (011 or $\bar{A}$ B C), the C flip-flop does not change state, it remains set. Therefore, a / is recorded in the fourth state cell. Be sure to verify that the proper symbol is used in each cell of each Karnaugh map to obtain practice in reading and developing these maps.

Once you have completely plotted the maps for each flip-flop, you will use them to develop the J and K input logic equations for each. To do this you will group the various terms in the maps together in groups of 2, 4, 8, or higher powers of 2 as you did in minimizing combinational equations. There are some special rules that you must follow with regard to grouping the variables in the maps. For a JK flip-flop these rules are as indicated below.

J input equation

1 Each 1 square must be accounted for in the J equation.
/ Optional
Ø Optional
0 Must not be used
X Optional

K input equation

1 Optional
/ Must not be used
Ø Each Ø square must be accounted for in the K equation.
0 Optional
X Optional

In developing the J input logic equation you must consider each 1 term marked in the cells of the map. Each 1 must be used in some way to account for all of the necessary input states. Cells marked with a 0 must not be used. All other input symbols such as the /, Ø, and X can be used in the same way as "don't care" states in any other Karnaugh map.

In determining the K input logic equations, all Ø terms must be accounted for and used in one grouping or another. The / cells must not be used. All other cells marked with symbols 1, 0 and X can be used as "don't care" states.

The Karnaugh maps in Figure 9-45 show the proper groupings of the terms of both the J and K input equations. The J input equations are identified by groups marked with solid lines. The K input equations are designated by the dashed lines. The Karnaugh map is read in the same way as you read Karnaugh maps for combinational logic circuits. Simply look at each marked group and determine which variable or variables does not change when moving from one cell to the next within that group. Make a product term with these variables. Then OR together each of these terms, one term for each group. The equations corresponding to the J and K input states are designated adjacent to each map. Using the rules given earlier, verify the correct grouping of the variables. Then derive the input equations for the J and K inputs of each flip-flop yourself from the map to check the equations.

A special case in the map of flip-flop A is that all 8 cells are valid for the K input. This means that we can form one single large group of eight representing the input term for the K input of the A flip-flop. When all terms in a map can be looped together, it simply indicates a binary 1 condition. In other words, input A,B, or C could be in either the binary 1 or binary 0 state and the output would still be binary 1.

Drawing the Logic Diagram. Knowing the number of flip-flops in the counter and having the equations of the J and K inputs for each you can draw a logic diagram for the counter or sequencer. Keep in mind that we are dealing strictly with the JK flip-flops and a synchronous circuit. Figure 9-46 shows the logic diagram of the seven-state counter. The T inputs of each flip-flop are connected together to a common clock input line. This indicates a synchronous circuit. Note that all direct clear inputs of the flip-flops are connected together so that all of these flip-flops can be cleared, reset or initialized from a common input line if desired. The remaining logic circuitry represents logic gates that implement the J and K input equations for each flip-flop. Gate 1 through 4 are positive NAND negative NOR gates such as the type 7400 quad two-input NAND gate. Gates 5 and 6 are positive NOR negative NAND circuits such as type 7402 IC. JK flip-flops like the dual flip-flop 7476 could be used. The resulting circuit is a counter that will sequence itself in the given code with the minimum amount of hardware.

Figure 9-46

Logic diagram of the seven-state counter.

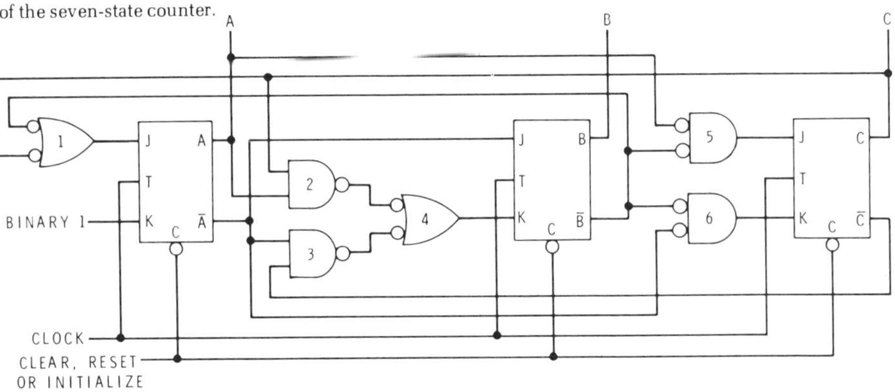

Design Examples

To further illustrate the design procedure for sequential logic circuits, the example designs given here are presented. They show that a counter can be designed with any number of states and to sequence or step from one arbitrary binary bit pattern to another. Special codes or arbitrary sequences are equally easy to design and implement.

Two-Bit Gray Code Counter. Assume that we wish to design a four-state Gray Code Counter. In attempting to achieve high speed we choose the Gray Code since only one bit changes from one state to the next. The fact that we need a Gray Code counter is determined by the application itself. Perhaps the application simply calls for a four-state counter and based on your knowledge of the application you determined that a Gray Code sequence was the most desirable for high speed operation. The Gray Code sequence could also have been determined by the desired output waveforms.

STATE	A	B
0	0	0
1	0	1
2	1	1
3	1	0

RECYCLE

Figure 9-47
State table for 2-bit Gray Code counter.

Once the problem has been completely specified and stated, a state table is developed. Figure 9-47 shows a state table for a two-bit Gray Code counter. The states are labeled 0 through 3 and as usually desirable the initial or first state, state 0, is made equal to 00. It takes two bits to define four states. Note that in changing from one state to the next only one of the two bits changes at a time. This includes the state change from the fourth state, state 3, back to the initial state, state 0.

The next step of the design procedure is to plot the state changes for each flip-flop on a Karnaugh map. To do this you examine the state changes that must occur in each flip-flop in moving from one state to the next. These state changes are then plotted in the appropriate cells on the Karnaugh map. In each cell you plot a symbol that designates the transition that must take place to move to the next state. Use the symbols given earlier for designating the state changes.

Figure 9-48 shows the Karnaugh maps for the A and B flip-flops with the appropriate state changes plotted. The various symbols in this table are then grouped according to the directions given earlier to determine the input expressions for the J and K inputs on each flip-flop. The minimized input expressions are given adjacent to the Karnaugh maps.

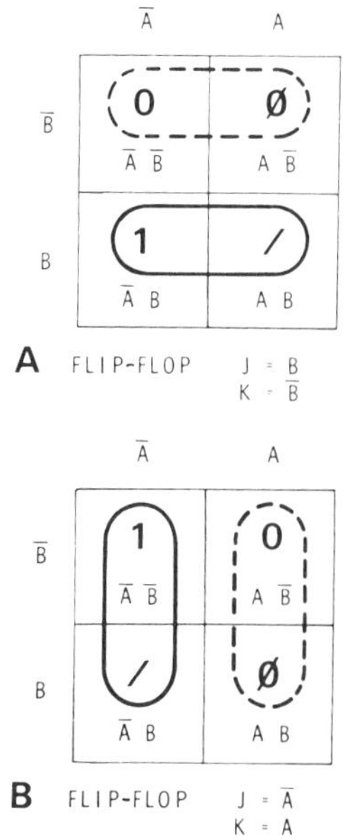

Figure 9-48
Karnaugh maps for flip-flops in 2-bit Gray Code counter.

From the information derived from the Karnaugh maps, a logic diagram can be drawn. This is illustrated in Figure 9-49. The two JK flip-flops are interconnected as specified by the J and K input expression given in Figure 9-48. Since this is a synchronous circuit, the toggle inputs to the flip-flips are tied together to the clock circuit.

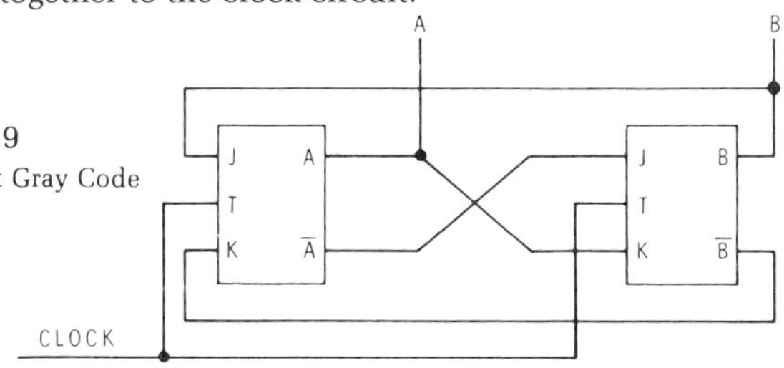

Figure 9-49
Logic diagram of 2-bit Gray Code counter.

Be sure to work through this example problem yourself. This includes developing the initial state table, plotting the state changes on the Karnaugh maps, developing the input equations and sketching the logic circuit. Verify each step of the example to be sure that you understand what was done.

XS3 Code BCD Counter. Design a counter circuit that will count in the standard XS3 BCD Code. We know from the statement of the design problem that a ten-state counter is required. A BCD counter is a decade counter with ten states. The code is also defined for us. The standard XS3 code is specified. From this information we can immediately develop a state table. This is shown in Figure 9-50. The initial or zero state is 0011. From there the counter steps in a standard binary code sequence until the tenth state (1100) is reached. The counter then recycles on the tenth input pulse.

The next step in the design process is to translate the flip-flop transitions in the state table into the symbols that can be plotted on the Karnaugh map. Use Figure 9-44 as a guide.

Figure 9-51 shows four sixteen-cell Karnaugh maps used to plot the state changes for each of the four flip-flops. Since the counter has only ten states, six of the four states will not be used and therefore can be treated as "don't care" conditions. These are states 0000, 0001, 0010, 1101, 1110, 1111. X's are placed in the appropriate squares in all four Karnaugh maps so that these "don't care" conditions can be used in minimizing the input equations for each flip-flop.

Next, the transitions of each flip-flop from one state to the next are analyzed and plotted on the appropriate Karnaugh map. Go through each of these yourself to see how the various transition changes were determined.

Using the rules given earlier about grouping the symbols in the Karnaugh maps, develop the input expressions for each flip-flop. The Karnaugh maps indicate the appropriate groupings and the resulting logic equations for the JK inputs. Solid line groups are for the J inputs while dashed line groups are for the K inputs.

	OUTPUTS	
STATE	A B C D	
0	0 0 1 1	
1	0 1 0 0	
2	0 1 0 1	
3	0 1 1 0	
4	0 1 1 1	RECYCLE
5	1 0 0 0	
6	1 0 0 1	
7	1 0 1 0	
8	1 0 1 1	
9	1 1 0 0	

Figure 9-50
State Table for Excess 3 counter.

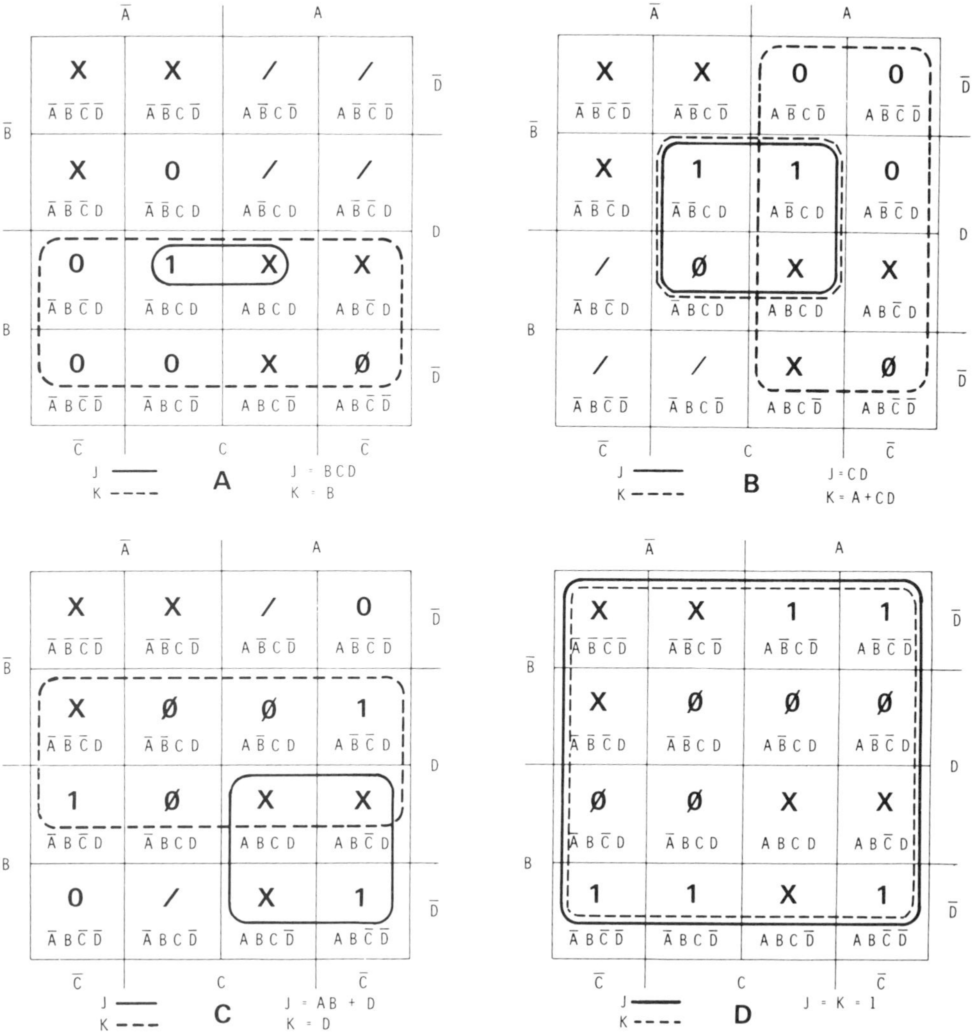

Figure 9-51
Karnaugh maps for the XS3 counter.

From the input equations developed from the Karnaugh maps, the logic diagram can be sketched. Figure 9-52 shows one method of implementing this circuit. All of the T inputs to the flip-flops are connected together to form a synchronous circuit. The J and K inputs to each flip-flop are specified by the input equations. This gating assumes the use of standard SSI logic packages.

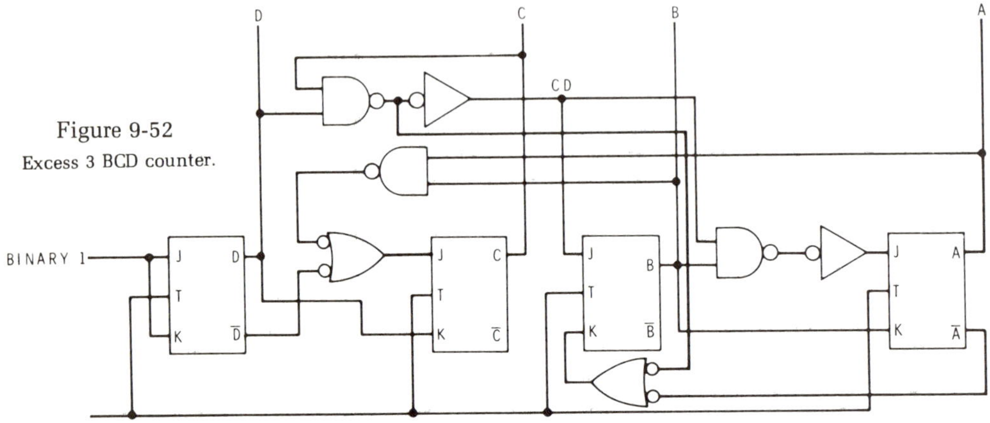

Figure 9-52
Excess 3 BCD counter.

Design Variations

The sequential circuits that we have considered are special counters that can have any number of states and any special code sequence. All of these circuits have a single input, the clock. However, there are other sequential circuits where external control signals are used to control the counter. These external inputs essentially determine when a counter or sequencer steps from one state to the next. The only modification needed in our design procedure to handle external inputs is to include these signals as variables in the JK input expressions. In order to cause a circuit to change from one state to the next, the JK inputs of the various flip-flops must have the appropriate input signals as determined by the count sequence of the counter. If an external input signal is to have control over the change from one specific state to the next, then that external input signal becomes one of the product terms in the expressions for the JK inputs on each of the relevant flip-flops.

This concept can be simply illustrated with the two-bit Gray Code counter discussed earlier. Assume that we wish to have an external start signal to control the counter. In other words we wish the counter to remain in its initial 00 state until it receives a binary 1 signal on the START input line. Once the START signal goes high, the counter will be incremented by the clock pulses from one state to the next. This sequence will continue until the START line is brought low. Then the counter will stop counting.

Figure 9-53 shows one way that this circuit could be implemented. An AND gate is connected to the J input of the B flip-flop. Normally this J input is connected directly to the $\overline{A}$ output. The $\overline{A}$ output is used as one input to the control AND gate. A START signal is also applied to the AND gate. Now in order for the B flip-flop to set, the $\overline{A}$ output must be high and the START input must be high. With the counter initially in its 00 or reset state, the JK inputs on each flip-flop are 1 and 0 respectively. As clock pulses are applied both flip-flops will continue to be reset. When the START input line goes high, the AND gate output goes high making the J input to the B flip-flop high. The K input to the B flip-flop is low as it is connected directly to the normal output of the A flip-flop. The conditions are now correct for the B flip-flop set on the occurrence of the next clock pulse. When this happens, the normal count sequence of the Gray code counter will begin. The counter will continue to count as indicated by the waveforms in Figure 9-63.

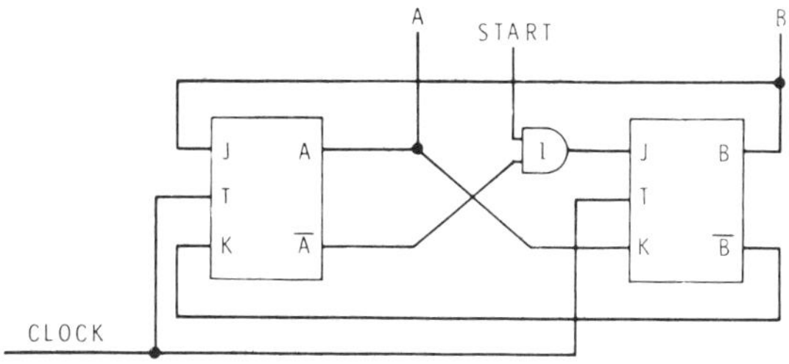

Figure 9-53

Two-bit Gray Code counter with external control input.

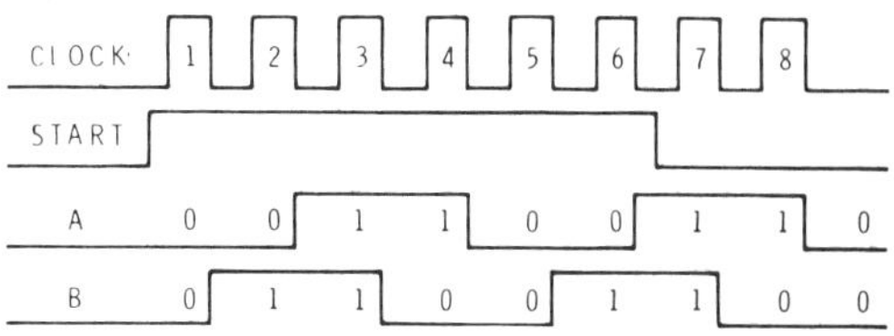

Figure 9-54

Waveforms for two-bit Gray Code counter with external control input.

If the START line should go low during the count sequence as indicated in the waveforms of Figure 9-54, the counter will continue to run until the 00 state is reached at which time the counting sequence will stop. The counter will remain reset until the next start pulse is applied. While this is a simple example, it does illustrate the concept of using external signals to control the occurrence of the state changes in a special counter or sequencer.

In using these special counter or sequencer circuits, we will often be able to use the flip-flop outputs directly to control external circuits. In such a case no additional circuitry is required. It is possible to define the control waveforms required and then design a counter to produce the desired count sequence thereby minimizing the circuitry. Another approach to obtain a sequence of timing pulses is to decode the state of the special counter. AND gates can be connected to the flip-flop outputs to recognize each unique state produced by the counter. The outputs of these decode gates can then be used to control the sequence of operations in external circuits.

Figure 9-55 shows how all four states of the two-bit Gray Code counter are decoded. The output signals produced by the decoder gates are illustrated in Figure 9-56. Note that as the counter steps from one state to the next, a sequence of timing pulses is generated. These pulses are then used to control the external circuits. In some applications all states of the counter must be decoded to create the necessary timing pulses. In other applications only specific states may be required thereby minimizing the number of decode gates required. For three- and four-bit counters, MSI decoder circuits can be used to reduce the amount of circuitry required to decode the desired states.

Figure 9-55

Two-bit Gray Code counter with all states decoded.

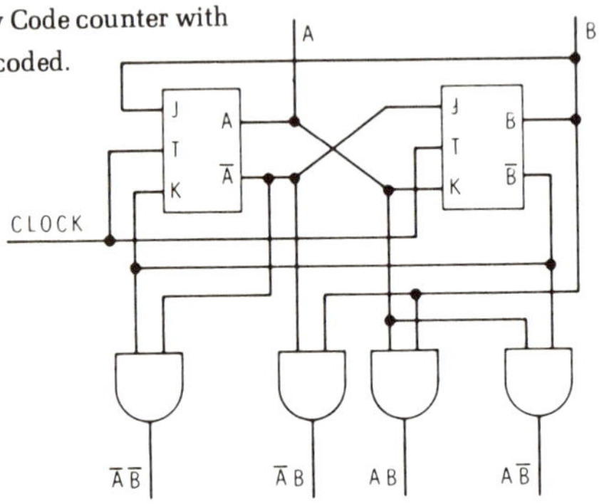

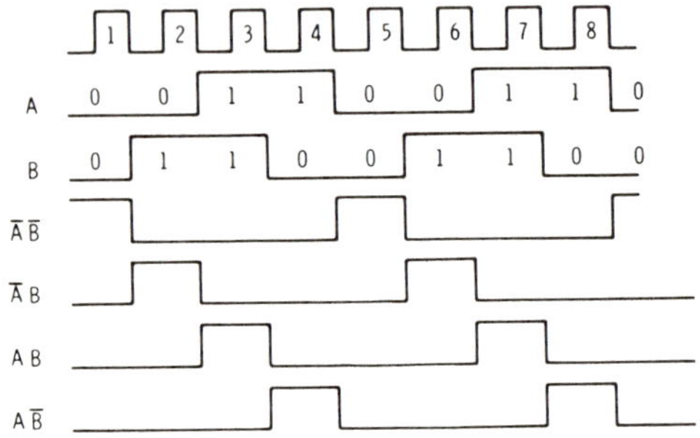

Figure 9-56

Waveforms for two-bit Gray Code counter and decoder.

Unit 10

DIGITAL APPLICATIONS

INTRODUCTION

In this final unit of "Digital Techniques," you are going to study some typical applications. Throughout the program we have emphasized component operation and circuit details, but in this unit we will put all of this information together and show you examples of how digital circuitry performs useful functions.

Perhaps the greatest application for digital techniques is in computers. The development of computers and digital circuits has been a parallel effort, each area being a benefit to the other. Another area of electronics affected by digital technques has been test/measurement equipment. Most of the high quality, precision test equipment available today is based on digital techniques.

It is impossible to cover the complete spectrum of digital applications in a unit this size. For that reason we are selecting what we feel are two of the most important applications for digital techniques. These are test and measuring equipment and computers. In this unit we will give you typical examples of each. These examples clearly illustrate the power and benefits of digital techniques.

DIGITAL TEST EQUIPMENT: The Frequency Counter

Digital test equipment has greatly improved the resolution and accuracy of electronic measurements. At the same time, digital techniques have made them faster and more convenient. The quantity being measured is displayed directly with numerical digits, thereby eliminating the interpolation associated with meter-type analog instruments.

While digital instruments have been designed to measure almost every electronic quantity, there are several quantities which are more commonly measured than others. These are voltage, current, resistance and frequency. Two basic types of digital test instruments have been developed to measure these quantities. These are the digital multimeter (DMM) and the frequency counter.

A digital multimeter is an electronic test instrument that is used for measuring voltage, current, and resistance. The most commonly measured quantity is voltage. A digital instrument used to measure voltage is referred to as digital voltmeter (DVM). A digital multimeter is used like any analog multimeter in that the test leads are connected to the circuit under test. The quantity being measured is displayed on a decimal readout instead of being indicated by the position of a pointer on a meter scale. Both the resolution and accuracy of the digital display is better than that associated with even the highest quality analog test instruments.

Figure 10-1 shows a low cost digital multimeter for measuring current, voltage and resistance. Special single function digital meters are also available to replace standard analog panel meters for either current or voltage measurements. These are referred to as digital panel meters (DPM). Any digital multimeter is essentially an analog-to-digital converter. The meter converts analog quantities of voltage, current or resistance into an equivalent BCD word so that the quantity can be displayed in decimal form. The analog-to-digital conversion technique employed in digital multimeters can also be used in measuring other electronic quantities such as capacitance, inductance, reactance, impedance, power and others.

Figure 10-1
Typical digital multimeter for measuring voltage, current and resistance. (Heathkit IM-1212)

The other widely used digital test instrument is the frequency counter. This instrument is designed primarily to measure the frequency of a periodic input signal. It displays the frequency in Hz, KHz, or MHz in decimal form. A frequency counter literally counts the number of pulses or cycles of an input signal that occurs in a known period of time and displays the frequency directly. Figure 10-2 shows a typical frequency counter.

Figure 10-2
The new Heathkit Portable
5 Hz-512 MHz Frequency
counter.

While frequency measurement is the basic function of a digital counter, many counters can make other time related measurements. Some general purpose counters can also measure the period of an input signal. Frequency counters can also be used for totalizing operations where the counter functions to keep tally of input events that occur. Some counters also permit frequency ratio measurements where two input frequencies can be compared and their ratio displayed. Time interval measurements can also be made with some counters.

The digital counter is one of the most versatile electronic test instruments available. Time and frequency measurements are vital to the proper testing and evaluation of electronic circuits and equipment. The digital counter has greatly improved the resolution and accuracy of frequency and time measurements, and at the same time has made them faster and more convenient.

The digital counter is an excellent example of the use of digital circuitry to perform a useful measurement function. For that reason, we're going to analyze the operation of a typical digital counter. Many of the digital circuits described in this program are used to implement the various counting functions.

A general block diagram of a digital counter is shown in Figure 10-3. It consists of four main sections; the input circuit, the gate and control circuits, the time base, and the decimal counter and display. These circuits work together in various ways to provide measurement of frequency or time interval as determined by the application. Let's take a detailed look at each of these major sections.

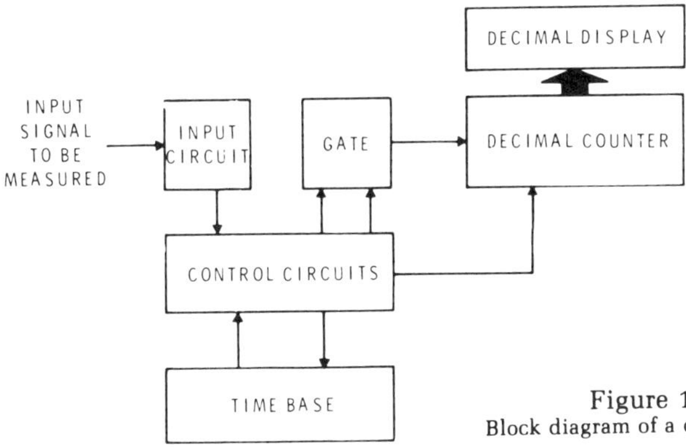

Figure 10-3
Block diagram of a digital counter.

Input Circuit

The input circuit is a signal conditioner designed to make the signal to be measured compatable with the digital circuitry in the counter. The input circuit must be flexible enough to permit the counter to accept signals whose amplitude and wave shape vary considerably.

The input circuit incorporates some form of amplification and isolation. The isolation is obtained with a high input impedance follower stage to minimize the loading on the circuit generating the input signal. Amplification is used to improve the sensitivity of the counter so that very low amplitude signals can be measured.

The input circuitry also incorporates an attenuator and protection circuitry for handling large amplitude signals. A resistive attenuator is used to reduce high amplitude signals to a level compatable with the input circuitry. Diodes are also incorporated to clamp or limit the input voltage. This protects the circuitry from excessively high amplitude signals.

Perhaps the most important part of the input circuitry is the wave shaper. This is a circuit that converts the input signal into a rectangular pulse whose output logic levels are compatable with those of the rest of the counter circuitry. This circuit is used to shape and square sine wave and other non-rectangular input signals. A typical circuit used for such squaring operations is the Schmitt trigger. The Schmitt trigger is a special form of bistable multivibrator circuit. The binary output state is a function of the input amplitude. When the input is below a certain threshold level the output is binary 0. When the input exceeds this threshold level the output switches to binary 1. If the input voltage then drops below another threshold level, the output switches back to binary 0. Schmitt trigger circuits are available in TTL, ECL and CMOS form. The output of the Schmitt trigger is a signal whose frequency is identical to that of the input signal but whose shape and amplitude are compatable with the remaining counter circuitry. The output of the input circuit is applied to the control circuitry which then determines how that input signal is used in the measurement process.

Gate and Control Circuits

The gate is nothing more than a standard AND circuit that controls the application of the input pulses or a precision timing signal to the decimal counter. The control circuit determines how the input circuitry and precision timing signals from the time base are used to control the gate. The control circuits are also used to operate the decimal counter. The control circuits generate all of the pulses required to permit the counter to operate in a variety of modes.

Time Base

The time base consists of a precision crystal oscillator and a series of frequency dividers that generate highly accurate and stable timing signals. These signals are used as the reference or standard in making the frequency and time measurements. The crystal oscillator which usually runs at 1 MHz or 10 MHz provides this standard. BCD counters used as frequency dividers generate decade sub-multiples of the crystal oscillator frequency. The precision time signals generated by the time base are then used to control the gate for frequency measurements or are counted by the counter in period and time interval measurements.

The quality and accuracy of the time and frequency measurements made by the counter is a direct function of the accuracy and stability of the crystal oscillator. Most good counters use a temperature compensated crystal oscillator (TCXO) to insure that the frequency remains stable over a wide range of ambient temperature variations. In higher quality counters the crystal oscillator itself is contained within a small oven or temperature controlled chamber where the temperature is maintained at a constant level.

In lower quality, less accurate counters, other signal sources can be used as a time base. For example, the ac power line voltage can be used as a reasonably accurate time and frequency standard. The frequency of most ac power line signals is controlled to better than .1 percent. The normal 60 Hz power line signal can be divided down to produce timing signals of 100 ms, 1 second, and 10 seconds.

Decimal Counter and Display

The heart of the digital counter is the decimal counter circuit and its display. The decimal counter consists of a number of cascaded BCD counter stages. The counter counts the pulses received from the gate. These pulses can be the shaped input signal or a reference signal from the time base. The decimal counter accumulates these pulses and stores them as a multiple digit BCD word. The output of each BCD counter is fed to a storage register where the BCD word can be stored. The output of these registers are then used to drive BCD-to-decimal decoder-driver circuits. These operate the display readout elements. Seven-segment light emitting diodes and gas discharge readouts are the most commonly used displays in digital counters.

The operation of the decimal counter is determined by the control circuits. The control circuit generates signals to reset the counter and to transfer the contents of the BCD counters into the storage registers.

Modes of Operation

Digital counters are capable of operating in a variety of modes, each permitting a different type of time and frequency measurement. The most commonly used mode of operation is frequency measurement. Many counters are wired to perform only this function. This is true of the counter shown in Figure 10-2. Other more flexible counters permit a variety of time and frequency measurements to be made. Such counters are widely used in laboratory testing and development applications. A front panel switch selects the desired mode of operation. In this section, we discuss each of the most commonly used digital counter modes.

Totalized Mode. The simplest form of operation for a digital counter is the totalized mode. This is an events counting mode where pulses appearing at the input are counted with the sum being accumulated in the counter and displayed. Figure 10-4 shows the interconnections in the counter for the totalize mode. The electrical events to be counted are applied to the input circuit. The output of the input circuit is connected to the count input of the gate. A front panel pushbutton is used to reset the counter and clear the display to zero. As the input signals occur they will be counted and the total displayed.

In some counters the count control input to the gate can be enabled or inhibited by an external gating pulse. The totalized mode provides a simple means of counting or keeping tally of the input events that occur. In this mode the time base section of the counter is not used.

An application example of the counter used in the totalized mode will more clearly illustrate its operation. Consider the situation in a manufacturing company where it is desired to count the number of items transported along a conveyor belt. As the products come off a production line they are placed on to a conveyor belt approximately one foot apart. At one point in their trip down the conveyor belt, they will pass between a light source and a photo cell. The photo cell is connected to the input circuit of the counter. As each item passes between the light source and the photo cell, it breaks the light beam. The photo cell generates an input pulse that is used to increment the counter. The counter will be incremented as each unit passes the photo cell. By using the counter in this way, an accurate tally of the number of items coming off the production line is maintained in the counter. The counter automates the counting function thereby eliminating the time, effort and error of a human operator.

Frequency Measurement. The most commonly used counter function is frequency measurement. The counter circuit configuration for frequency measurement is given in Figure 10-5. The signal whose frequency

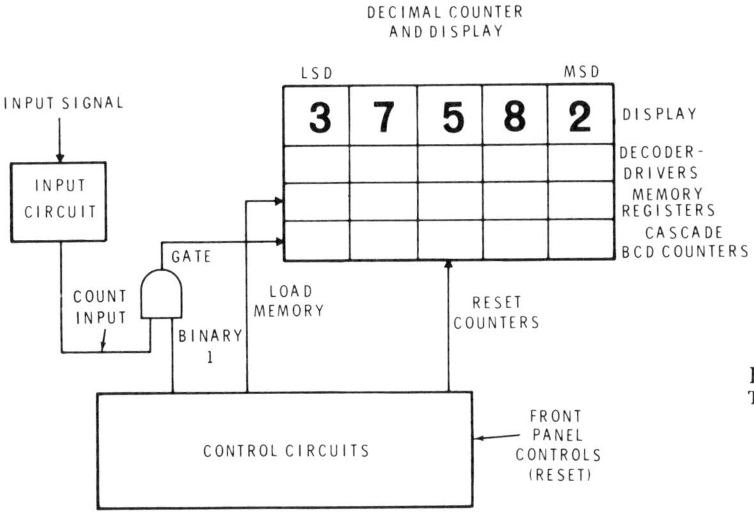

Figure 10-4
Totalize Mode

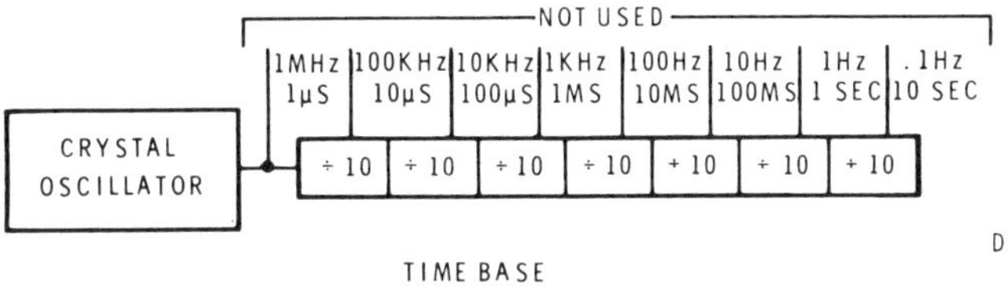

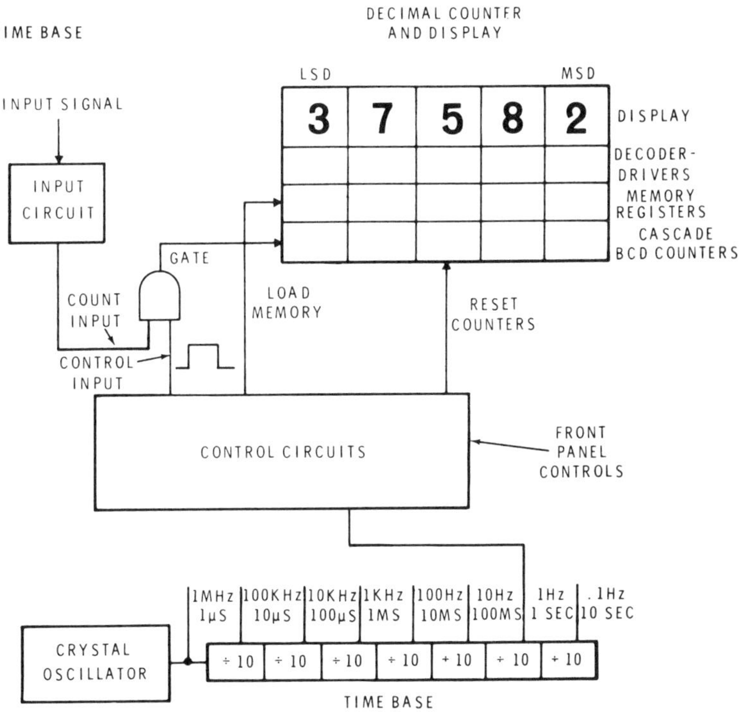

Figure 10-5
Frequency Measurement

is to be measured is applied to the input circuit. The output of the input circuit is then applied to the count input of the gate. The control input to the gate is a pulse derived from the time base. In order to perform frequency measurement, the control input to the gate is an accurate time interval. For example if the control input is a pulse that is exactly one second in duration, the counter will count the number of input pulses or cycles that occur during the one-second time interval that the gate is enabled. This will cause the display to read the frequency in cycles or pulses per second. Assuming a one second gate control pulse on the counter in Figure 10-5, the frequency shown on the display is 28573 Hz. The one-second pulse for the control input on the gate is derived from the 1 Hz output of the time base.

Higher frequencies can be measured and read out on the same display by using a gate control input of shorter duration. For example, by using a gate control pulse with a duration of 1 ms, the display will read frequency in kilohertz (kHz). The counter counts the number of input pulses that occur in one millisecond of time. Assuming a one-millisecond gate pulse for the circuit in Figure 10-5, the frequency being measured by the counter as indicated by the display is 28,573 kHz or 28.573 MHz. If the counter counts 28573 pulses in one millisecond (1/1000th second) then the number of pulses that occur in one second is $28573 \times 1000 = 28573000$ Hz or 28.573 MHz. The one millisecond pulse is derived from the 1 kHz output of the time base. Other time base frequencies can be used to generate control pulse intervals that are some multiple of a power of ten.

The selection of the time base output frequency determines the measurement resolution and how the frequency is displayed. In some counters the time base frequency is selected with a rotary switch or a series of pushbuttons. By making all of the time base frequencies available, much flexibility is obtained. However, optimum resolution as well as wide range can be obtained by using only two time base frequencies: 1 Hz and 1 kHz. Several counters provide a two position switch to select these two time bases. When the time base frequency is switched, the decimal point in the display is switched to the appropriate position to provide the corrected readout frequency.

Some counters such as the one in Figure 10-2 have an auto-ranging feature. Here the time base selection is made automatically, based on the input signal frequency. The auto-ranging circuitry automatically determines the correct time base frequency for maximum measurement resolution without over-ranging. Over-ranging refers to exceeding the count capability of the counter. The number of digits in a counter determines the over-ranging point for a given time base. Five and six digit counters are the most common.

Period Measurement. Another commonly used counter function is period measurement. In this mode of operation, the counter measures and displays the amount of time that it takes for the input signal to complete one cycle. Figure 10-6 shows the counter arrangement for this mode of operation. The input signal after being shaped is applied to the control circuits. The control circuits permit the input signal to operate the control input of the gate. One of the outputs of the time base is then connected to the count input of the gate through the control circuitry.

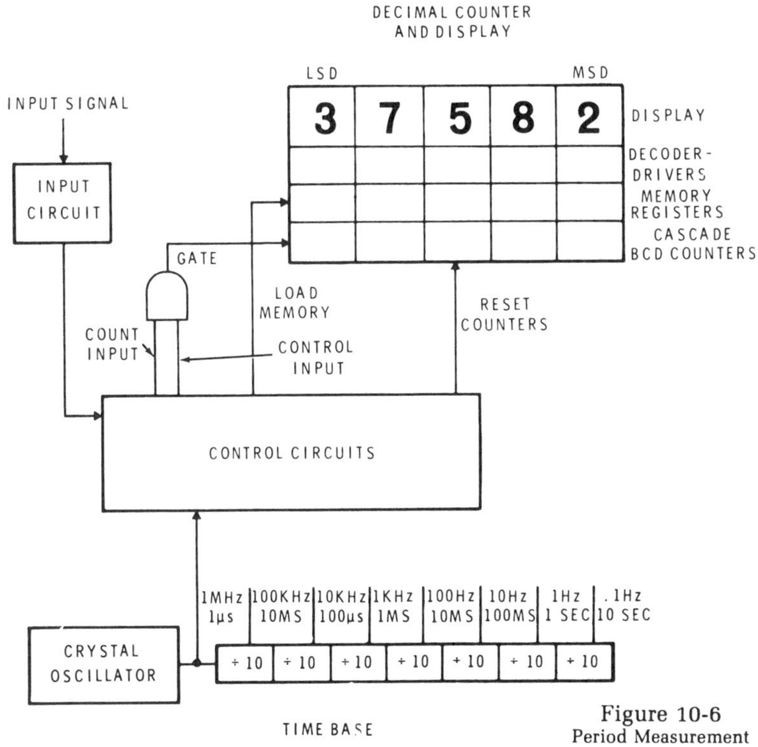

Figure 10-6
Period Measurement

The control circuitry causes the counter to be initially reset. Then the gate is opened or enabled by one period of the input signal. During this time interval, the counter is incremented by the time base signal. As indicated in Figure 10-6, the counter counts the 1 MHz pulses from the time base. According to the display in Figure 10-6, the period (t) of the input signal for this example is 28,573 microseconds. This corresponds to a frequency of $f = 1/t = 1/_{28573} \times 10^{-6} = 34.99$ Hz. The control circuit generally causes the period measurement to be made repeatedly during the time the input signal is applied.

The purpose of the period mode is to provide a means of measuring and displaying the time that it takes for an input signal to complete one cycle. However the primary benefit of the period mode is to provide improved resolution and accuracy of measurement of low frequency signals. It is difficult to obtain a highly accurate measurement of low frequency signals with most counters. For example in measuring the 60 Hz power line frequency with a counter whose gate time is 1 second, the count displayed on the counter would be 60. Since the least significant digital is 1Hz, then the resolution of the measurement is no better than 1 out of 60. This is not a very accurate measurement of the power line frequency and some means must be provided to improve the resolution.

There are several ways that the accuracy of low frequency measurements can be improved. The most direct way is to increase the measurement interval, that is increase the gate pulse duration in the frequency mode. For example, the gate pulse could be made equal to ten seconds. The counters would then count the number of cycles of the input signal that occur over a ten second interval. The display on the counter would then read 600 for a 60 Hz input. In other words, the least significant digit would then indicate tenths of a cycle rather than one cycle. By increasing the measurement interval, we provide greater resolution of measurement. The trade off, of course, is in the length of time involved in making the measurement. While ten seconds doesn't sound like a long time, for an electronic measurement it is extremely long. It is a distinct disadvantage to have to wait this length of time for a measurement to take place. For example, an increase in gate time to 100 seconds would provide an improvement in the measurement resolution and thereby the accuracy. However the measurement time is further increased.

To improve the resolution and accuracy of frequency measurements at low frequencies, the period mode can be used. The period of the input signal is measured quickly and provides much greater resolution. For example the period of the 60 Hz power line frequency is 16.667 ms. By counting the number of 1 MHz pulses of the time base that occur during one period, the counter display would read 16666. Then by taking the reciprocal of the period, the frequency is obtained. This method of measurement is much more accurate and faster than measuring the frequency directly. However, measuring the period does necessitate the computation of the frequency once the measurement has been made. This can be quickly accomplished with an electronic calculator.

There are many situations where it is desirable to obtain a highly accurate measurement of some low frequency ac signal. Audio frequency applications involving filters and musical instruments often require accurate low frequency measurements. In the field of geophysics and the vibration testing of equipment, accurate low frequency measurements are often needed. The period mode of a standard counter can produce these low frequency measurements. However to improve the readout convenience, special low frequency counters have been developed. A typical unit is shown in Figure 10-7. This is a computing counter that is designed primarily for making accurate low frequency measurements. This counter operates in the period mode. However, once the period is measured, it automatically computes the frequency and displays it directly. The counter circuitry contains a MOS LSI calculator integrated circuit that is set up to perform the period-to-frequency computation automatically. To the operator, the computing counter appears to be measuring frequency directly. This counter provides a maximum resolution of .00001 Hz. To obtain the same resolution with a standard frequency counter would require a gate period greater than 27 hours.

Figure 10-7
A computer frequency counter measures period then computes frequency to provide high measurement resolution at low frequencies.

A further improvement in the resolution and accuracy of low frequency measurements can be made with a multiple period measurement. In this mode of operation, the time base pulses are counted for a duration equal to some multiple of the period of the input signal. This is done by feeding the signal whose period is to be measured into a frequency divider that produces division by 10, 100, 1000 or higher power of 10. This causes the control circuitry to generate a gate interval equal to 10, 100, 1000 or more times the period of the input signal. The counter counts the accurate pulses from the time base during this interval. The effect is to produce a period averaging measurement. For example if the input signal were applied to a divide by 100 circuit, the counter would accumulate pulses over a period of time equal to 100 times the period of the signal. The actual time for one cycle then would be equal to the display number divided by 100. In some counters a special period averaging mode is included. The counter display is made to read the period directly by simply shifting the decimal point the correct number of positions to compensate for the selected period interval.

Time Interval Measurements. The time interval measurement mode of a digital counter is a variation of the period mode. In this application, the count input to the gate is derived from the time base. The resolution of measurement of the time interval is determined by the time base frequency. For example, with a 1 MHz time base signal, the accuracy or resolution of measurement is 1 μs. The time resolution is the reciprocal of the time base frequency ($t = 1/f$). To the control input of the gate is applied a signal that will determine the length of time that the gate is open. This can come from a variety of sources depending upon the application. One application may be the measurement of pulse width. The signal whose pulse width is to be measured is applied to the input circuit. It causes the gate to be enabled for a period of time equal to the pulse duration. During this time interval, the counter counts the time base pulses. The display reads the pulse duration directly.

The counter can also be used to measure the time interval occurring between two independent events. A typical application is drag racing where the measurement of a car's performance is its elapsed time over a quarter mile distance. Light beams and photo cells are used at the start and finish lines to control the counter that will measure the elapsed time. The car is initially at rest at the starting line where its front wheel breaks the light beam. When a start signal is given to the driver, the car moves forward and the light strikes the photo cell. This generates an electrical impulse that is used to start the counter. The counter is initially reset and then begins to count the pulses from the time base. A 1 kHz signal is normally used to provide a measurement resolution of one millisecond. The car then traverses the quarter mile distance and breaks the light beam as it crosses the finish line. This generates another pulse that is used to close the gate and stop the time interval measurement. In this application there are two independent input pulses, one from the starting line photo cell and the other from the finish line photo cell. These start-stop pulses can be used as the inputs to a set-reset flip-flop. The starting line photo cell will set the flip-flop. The flip-flop then enables the gate. When the car crosses the finish line, the pulse from the finish line photo cell will reset the flip-flop and inhibit the gate. The count in the counter is the elapsed time.

There are many applications for the time interval mode of measurement. Another typical electronic application involves the measurement of the pull-in or release time of a relay. The counter could also be used to measure shutter opening intervals in a camera. In this mode of operation the counter in effect acts as a high resolution stop watch.

Frequency Ratio Measurement. Another mode of operation for the digital counter is frequency ratio. In this mode, the counter is used as a means of comparing two external frequencies. One of the frequencies is used to control the duration of time that the gate is enabled while the other signal is fed to the gate to be counted by the counter. The number displayed on the readout is the ratio of the two frequencies. In this mode of operation, the internal counter time base is not used. Instead, one of the external signals acts as the time base. To provide high resolution of the ratio of two frequencies that are close to one another, one of the frequencies is fed to the time base frequency divider. This allows the gate to remain open for a period of time equal to 10, 100, 1000 or more times longer than the period of the reference input frequency.

Counter Specifications

When selecting a digital counter for a specific application or in comparing counters there are certain important specifications which must be considered. In this section we define and explain these important specifications.

Input Sensitivity. Input sensitivity is a specification that refers to the amount of input voltage required to cause the counter to operate properly. This specification usually calls out the minimum value of the input voltage required for the counter to trigger reliably. Most good counters have an input sensitivity of less than 100 mV. Higher quality counters have input sensitivity as low as 1 mV. This means that the frequency or period of a signal whose amplitude is as low as a few millivolts can be determined. The input circuit in the counter generally incorporates some amplification that permits such low level signals to be used.

Generally a counter with high sensitivity is most desirable. However, the greater the sensitivity the more susceptible the counter is to noise problems. Noise pulses riding on top of a signal to be measured can cause false triggering of the counter and an inaccurate reading. For this reason it is generally desirable to use a counter with a sensitivity level to match the application. If ultra-high sensitivity is not required the cost of the counter will be less and there will be less susceptibility to the effects of noise. Many counters have an adjustable sensitivity level that permits the input triggering level to be varied over a wide range to match a given application.

Input Impedance. The input impedance of a counter is the impedance looking into the input circuit of the counter. This is the impedance that is connected across the signal source whose frequency or period is to be measured. Most commercial counters have an input impedance of one megohm of resistance in parallel with a small value of capacitance in the 10 to 100 picofarad range. This high input impedance is chosen to minimize loading effects on the circuit under test. At high frequencies, the most important part of the input impedance becomes the shunt capacity. At frequencies above 1 MHz it is also desirable to take into consideration the effect of any cable impedance. Normally, a coaxial cable is used to connect the signal being measured and the counter input. Such cables generally have a high capacity (above 30 pF per foot) which can significantly affect the amplitude and characteristics of the signal being measured. A high input impedance, low capacity attenuator probe normally used with oscilloscopes can also be used with most general purpose counters to minimize the effect of capacitive loading at the expense of input signal attenuation.

Frequency Range. The upper and lower frequency limits of a given counter define the counters frequency range. Because of circuitry limitations, the upper and lower frequency limits are restricted. For example a typical medium priced counter has a frequency range of 5 Hz to 30 MHz. The lower frequency limit is generally determined by the size of the input coupling capacitor and input resistance. In most counters, the input signal is ac coupled to the input circuit through a series capacitor. At the lower frequencies, the reactance of this capacitor increases and, with the input impedance, forms a voltage divider. The higher the input impedance and the larger this capacitor, the better the lower frequency response. Most typical counters have a lower frequency limit in the 1 to 5 Hz range.

The upper frequency limit of a counter is determined by the high frequency response of the input amplifier circuit, the propagation delay of the gate, and the upper frequency counting limit of the input decade counter. Some simple low cost counters have an upper frequency limit of only several megahertz. Standard direct counters are available for measuring frequencies as high as one gigahertz (1 GHz). Special counting techniques permit counters to measure frequencies up to several hundred GHz. (1 GHz = 10^9 Hz.)

Display Digits. The number of read-out display digits is an important counter characteristic. The greater the number of digits, the higher the resolution. Keep in mind that the resolution is also a function of the gate time and time base frequencies available in the counter. Most low and medium priced counters have a minimum of five digits in the readout. Counters for measuring very high frequencies in the GHz range have as many as nine digits.

Time Base. An important characteristic of any counter is the time base characteristics. This includes the time base frequencies and intervals as well as information on the stability of the time base oscillator. As a general rule, the greater the number of time base signals available, the greater the flexibility in making time and frequency measurements. The higher the frequency, the greater the resolution that can be obtained in period and time interval measurements.

The accuracy and stability of the crystal oscillator in the unit is an important factor in the quality of the counter. The crystal oscillator frequency is generally made adjustable over a narrow range to permit setting the counter to the exact frequency against an accurate known standard. From this point the stability of the counter will determine how much this frequency changes due to temperature changes and aging. Both long term and short term stability is considered.

Short term stability is affected mainly by the inherent losses in the oscillator circuit itself. Temperature variations greatly affect the short term stability of the oscillator. Temperature compensating techniques are used to help stabilize the frequency changes with temperature.

The long term stability of a crystal oscillator is due to aging of the crystal. Most crystal aging takes place in the first several months of operation causing frequency drift. After this, it stabilizes to a very low level. A wide range of stability characteristics are available in commercial counters. The degree of stability desired depends upon the application.

Modes. An important characteristic in selection of a counter is the availability of the various measurement modes. The large percentage of available commercial counters have only the frequency measurement mode. Since this is the most common mode of operation it is no disadvantage. If your application requires only the measurement of frequency such a counter will suffice. However if the counter is to be used as a general purpose laboratory or test instrument, it is desirable to incorporate other modes of measurement such as period, time interval, totalizing and frequency ratio.

A Typical Digital Counter

The best way to learn the operation of a digital counter is to analyze a typical unit. In this section we describe in detail the operation of the Heathkit Model IM-4100 Digital Counter. This is a low cost digital counter with a five digit display and a frequency range of 5 Hz to 30 MHz. It incorporates the frequency, period and totalize modes of operation. It has a sensitivity of 15 mV rms and an input impedance of 1 megohm shunted by less than 35 pF.

General Circuit Description. Refer to the block diagram in Figure 10-8 and the schematic diagram in Figure 10-9 as you read the following sections.

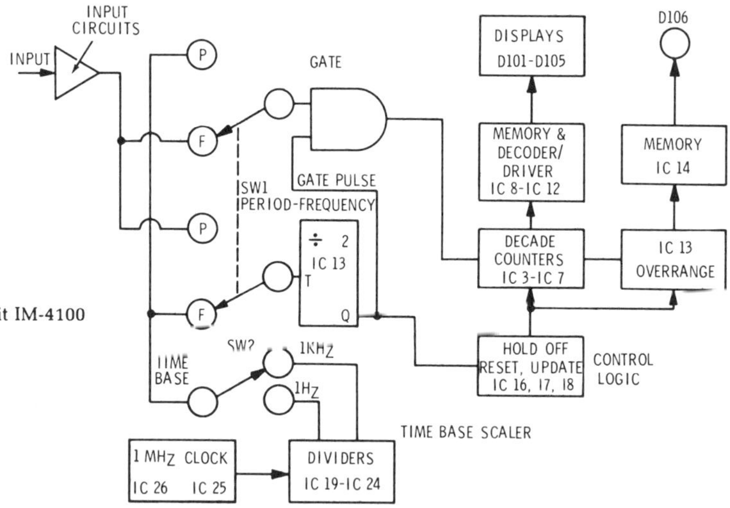

Figure 10-8
Block diagram of Heathkit IM-4100 digital counter

The signal to be counted is applied to the input circuits which consist of the input attenuator, impedance converter, and the Schmitt trigger. There the signal is squared and applied to the GATE. During the time the gate pulse from IC 13 is also present, the GATE is open and the frequency is counted by the decade counters. At the end of the GATE pulse, the count in the decade counters is transferred to the memories by a transfer pulse. At this time, proper segments of the display units turn on and the frequency is displayed. A reset pulse then clears the decade counter so they are ready for the next time the GATE is open. The duration of the GATE pulse is determined by the position of the POWER/TIME BASE switch (SW2). The pulse is of one second duration in the kHz position and of one millisecond duration in the MHz position. A one megahertz signal is also produced for the period measurements.

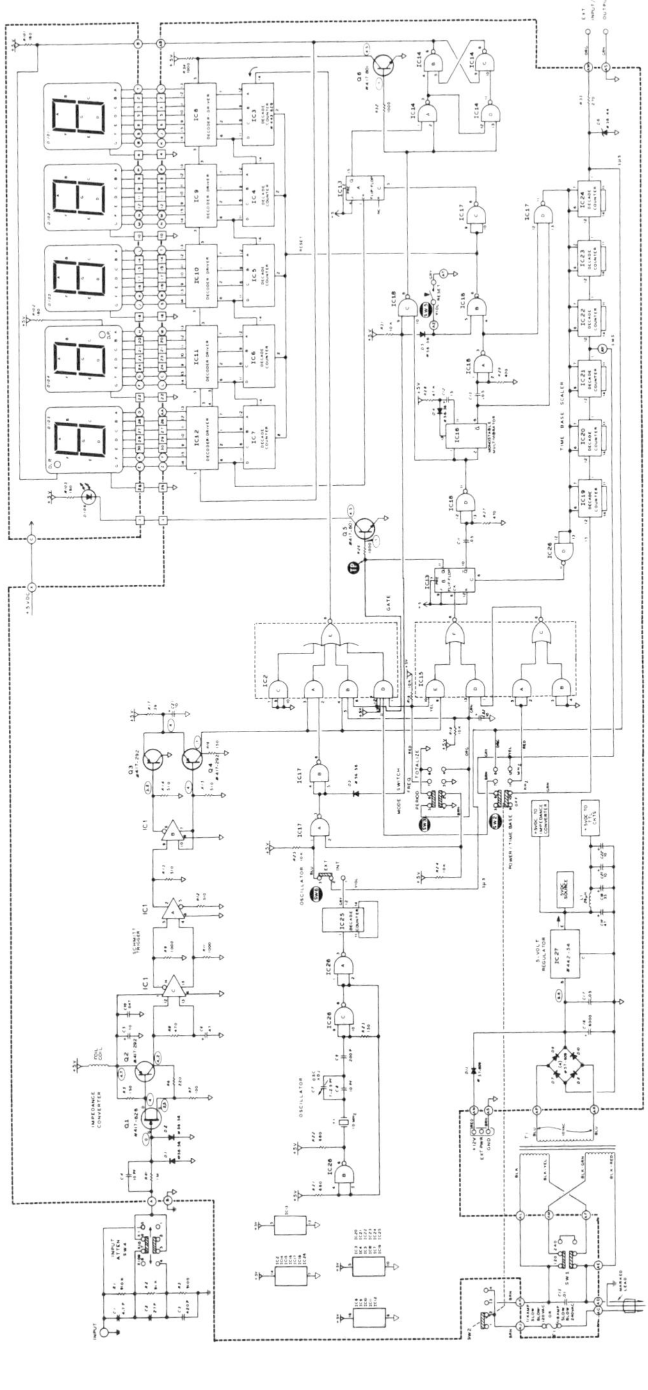

Figure 10-9
Schematic of the Heath Model IM-4100/SM-4100 frequency counter.

IC13 converts either the time base or the input period into a gate pulse depending on the position of the mode switch (SW3). By counting the input signal gated by the time base signal, frequency is displayed. By counting the time base signal gated by the input signal, the period is displayed. In the totalize mode, the counter simply counts the input signal without any gating.

Input Circuit and Schmitt Trigger. Refer to Figure 10-9. The input signal is applied to an input circuit that consists of a switchable voltage divider (R1, R2, R3) that is frequency compensated by C1, C2, and C3. This attenuates the input signal by a factor of 1, 10, or 100 depending upon the position of SW4. The signal is then coupled through C4 and R4 to D1 and D2 which provide over-voltage protection for Q1.

Transistors Q1 and Q2 are direct coupled with 100% negative feedback to form a unity gain follower circuit. The transistors provide wide bandwidth, high input impedance, low output impedance, and a gain of one. IC1C and IC1A then amplify the signal up to the input trigger threshold limits of IC1B which is wired as a Schmitt trigger. IC1 is an integrated circuit containing three differential amplifier circuits. Each amplifier has high gain and wide bandwidth. IC1B drives Q3 and Q4 which translate the signal into TTL levels making it compatible with the remaining logic circuitry. The conditioned input signal appears at the collector of Q4 and is applied to the gating circuitry.

10 MHz Clock and Scaler. A 10 MHz crystal and gates B and C of a type 7400 TTL gate (IC26) form a TTL-compatible clock oscillator. Capacitors C7, C8 and C9 provide the proper capacitive load for the crystal. C7 is variable to permit precise calibration of the oscillator. Resistors R21, R22, and R23 bias the IC and ensure efficient starting of the oscillator. Gate A of IC26 provides buffering action between the oscillator and the first decade divider of the time base scaler, a type 7490 (IC25) BCD counter. The 10 MHz clock signal is then further divided by other 7490 decade dividers, IC19 through 24, to provide appropriate gate times for the frequency mode. This divider chain is referred to as the time base scaler.

The 1 MHz time base output signal from decade counter IC25 is applied to the control circuitry and to the decade dividers in the time base scaler to provide the timing pulses to operate the counter in the various modes. This 1 MHz signal passes through switch SW6 when it is in the internal (INT) position to the input of IC24, the first decade divider in the time base scaler. The scaler then develops the 1 kHz and 1 Hz signals that will produce the 1 ms and 1 s gate pulses for frequency mode operation. The 1 kHz and 1 Hz pulses appear on the terminals of switch SW2. When the kHz position is selected, the 1 Hz signal output from IC19 is connected to gate IC15A. In turn it passes through gates IC15C and D. In the frequency mode, gate

IC15D is enabled and the 1 Hz signal passes through IC15F to the gate flip-flop IC13B. Setting SW2 to the MHz position causes the 1 kHz signal from IC22 to be connected to gate IC15A and therefore transmitted to the gate flip-flop IC13B.

In the period mode gate IC15D is inhibited by the binary 0 applied to pin 13 from SW3. This prevents the time base signal from IC15C from reaching the gate flip-flop. Instead, the input signal whose period is to be measured will operate gate flip-flop IC13B. The input signal from Q4 is applied to gate IC15E which is enabled during the period mode.

In the totalized mode the internal time base circuitry is disabled entirely. This is done by inhibiting gates IC2B and D and IC15D and E. In this mode the input signal passes through gates IC2A and E directly to the counter.

Decade Counter and Display. Figure 10-9 indicates that the main counting unit of the IM-4100 counter is made up of five integrated circuit decade counters, IC3 through IC7. IC3 is the least significant digit (LSD). The pulses to be counted are applied to pin 14 of this counter. Each of the IC's in the counter is a type 7490 TTL BCD counter similar to the one you have worked with earlier in this program. The counters are cascaded and can achieve a maximum count of 99999. Note that all of the reset pins (pin 2) are connected together so that the counter can be reset by one of the pulses from the control circuitry.

The four-line BCD output of each decade counter is applied to a memory-decoder/driver integrated circuit. These are IC8 through IC12. Each of these devices is a type 9368 decoder-driver circuit. As you can recall, this device contains a 4-bit storage register and a BCD to 7-segment display device. At some time during the operation of the counter, the BCD data stored in the decade counter is transferred to the storage registers in the decoder-driver IC's. It is the BCD data stored in the registers of IC8 through IC12 that is displayed on the LED readouts. The loading of the registers in the decoder-driver IC's is accomplished by a control pulse applied to pin 3 of these circuits.

Control Circuitry. Now let's take a look at the operation of the control circuitry for the counter in each of its three operating modes. We will discuss the gating, memory and reset functions for each mode.

Assume that the mode switch SW3 is in the frequency mode and the power-time base switch SW2 is in the MHz position. The input signal from the collector of Q4 is applied to gate B of IC2. This gate acts as the main gate for the counter in the frequency mode. The output of gate B is coupled through gate E in IC2 and is applied to the input of the LSD decade counter IC3. Gate IC2B is enabled by a one second or one millisecond pulse from pin 11 of IC13B. IC13B is a J-K flip-flop that is set and reset by the control circuitry. With IC13B set, transistor Q5 is turned on. This enables D106, the gate LED indicator. Flip-flop IC13B remains set for one millisecond or one second depending upon the setting of the time base switch. During this time the counter counts or accumulates the pulses applied to the input.

When IC13B resets, its $\overline{Q}$ output at pin 10 will go high. This causes the output of IC18D to go low momentarily. IC18D is a TTL gate connected as a pulse generator. The 470 ohm resistor R27 holds both inputs of the gate at a low enough potential so that to the gate it appears that a binary 0 is applied. This keeps the output of IC18D normally high. When the input to the capacitor C11 goes high, the gate inputs also go high. This forces the output low. However capacitor C11 charges quickly through R27 thereby leaving little voltage across R27. As a result the input again appears to be a binary 0 and the output switches back to its original high state. As you can see IC18D generates an output pulse that switches from high to low and back again. The duration of the pulse is a function of the time constant of C11 and R27. This pulse is produced when IC13B resets.

The output from IC18D is a negative going pulse that is applied to IC18C which drives Q6. This momentary pulse is applied to pin 3 of the decoder/driver ICs, IC8 through IC12. This causes the count stored in the decade counters to be transferred to the storage registers. The LED's then display the count.

The trailing edge of the pulse from pin 11 of IC18D triggers IC16. This monostable multivibrator produces a 200 ms output pulse.

The termination of the 200 ms output pulse from IC16 triggers IC18A. This circuit generates a negative going pulse which is used to reset the decade counters IC3 through IC7. This pulse is applied to the counters through IC18B. The reset pulse is also used to reset flip flop IC13A through gate IC17C. Flip-flop IC13A is used in the over-range circuit which will be discussed later.

The complement output of the monostable (IC16) and the pulse from IC18A are also used to gate IC17D. This gate causes all of the decade dividers in the time base scaler to be set to 1001. The next input pulse from the time base oscillator will then cause all of these time base decade dividers to be cycled to zero (0000). The signal from IC17D that sets the time base dividers to 1100 is also used to reset flip-flop IC13B through gate IC26D. The timing pulses for the frequency mode are shown in Figure 10-10.

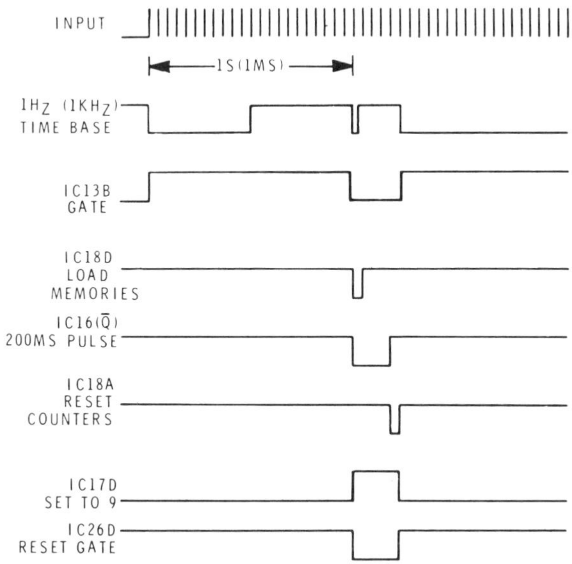

Figure 10-10
Timing diagram of control pulses of IM-4100 counter in the frequency mode.

In the period mode it is the input signal that controls the GATE flip-flop IC13B and therefore determines the period of time during which the GATE is open. During this time the counter is incremented by the 1 kHz or 1MHz time base signals. This permits period measurements with a resolution of 1 millisecond or 1 microsecond to be made.

Assume that the mode switch SW3 is in the period position as shown in Figure 10-9. This inhibits gates IC2B and IC15D which are used in the frequency mode. Gate IC17A is also inhibited. At this time gate IC15E is enabled. The input signal from the collector of Q4 is coupled through this gate and IC15F to the gate flip-flop IC13B. The 1 kHz or 1MHz signal from the time base is applied to gate IC2D pin 11. Here gate IC2D is enabled by the gate flip flop IC13B. The time base signal is allowed to pass through gates IC2D and E to the counter for a period of time equal to one cycle of the incoming signal.

In the period mode, the control circuitry for the memory transfer and reset operation is similar to that for the frequency mode. At the end of the gate interval, a pulse is generated by IC18D as before to cause the data in the decade counter to be transferred to the storage latches and a display. The termination of the 200 millisecond pulse generated by IC16, in turn, causes the reset pulse to be generated by IC18B. See Figure 10-11 for the control pulses in the period mode.

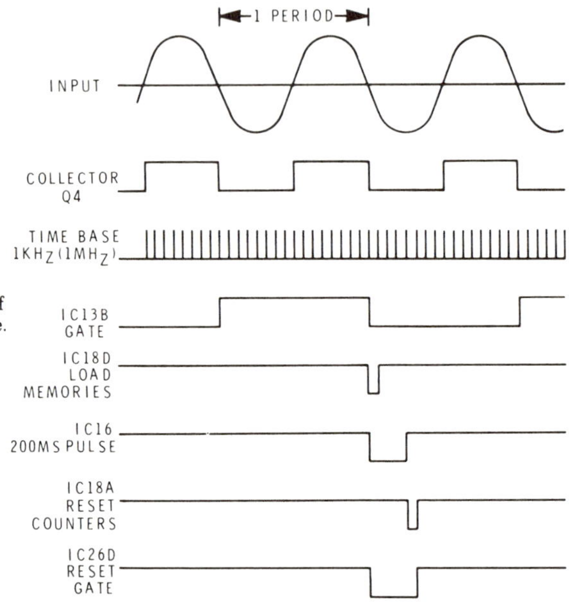

Figure 10-11
Timing diagram of control pulses of IM-4100 counter in the period mode.

In the totalized mode, the input pulses to be counted are passed through the input circuitry and appear at the collector of Q4 as in the other modes. The input signal is then applied to gate IC2A. With the mode switch SW3 in the totalized mode, gates IC2D and IC15E are inhibited. This prevents the gate function of the counter from operating. Gate IC2A is enabled at this time so that the signals to be counted are applied directly to the decade counter without any control.

When the totalized mode is used, both inputs to IC17A are high thereby holding the output of this gate low. This low level is coupled through diode D3 to the input to gate IC18C. This holds the output of IC18C high thereby causing the output of Q6 to go low. What this does is to enable the memory storage latches in IC8 through IC12 so that the pulses being counted in the decade counters will be transferred directly to the display. As the input pulses occur you can see the count change on the LED readout.

The only control available in the totalize mode is the front panel reset switch SW5. When this manual pushbutton is depressed, it forces the cathode end of D5 low. This generates a reset pulse through IC18B which clears the decade counters. In operating the counter in the totalized mode, the counter should be initially reset to zero prior to the counting of any input events.

In the totalize mode, it is possible to gate the input signal to the counter by using an external input signal connected to the input normally associated with the time base. The gate pulse is coupled through resistor R33 and switch SW6 to gate IC17A. If the external control signal is low, the output of IC17B will be low and will therefore inhibit gate A and prevent the input signals from appearing at the counter input. Figure 10-12 shows the control pulses in the totalize mode with an external gating signal.

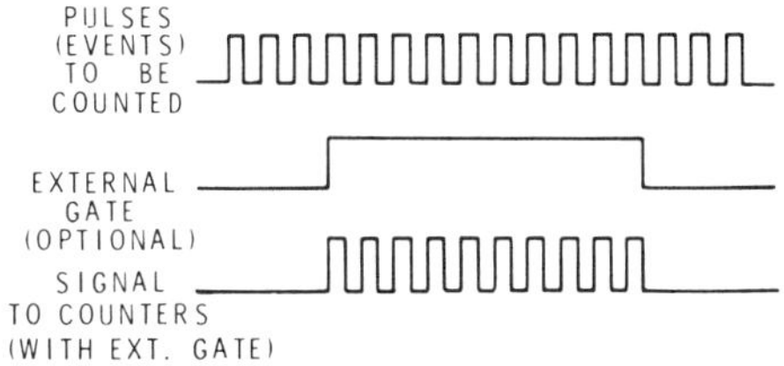

Figure 10-12
Timing diagram of control pulses of IM-4100 counter in the totalize mode.

Over-Range Detection. The IM-4100 counter has an over-range circuit that will give the user an indication of when the count input exceeds the count capability of the counter. If the gate duration and the time base signals are such that the count passes from 99999 to 00000, a carry output pulse occurs at the D output of the most significant digit decade counter IC7. This will toggle flip-flop IC13A. When this flip-flop is set, it indicates that an over-range condition has occurred. This flip-flop will remain in the binary 1 state indicating an over-range condition until the normal reset pulse is generated by the control circuitry. At this time IC13A is reset via gate IC17C.

The over-range condition is transferred to a d-type flip-flop (data latch) made up of the gates in IC14. When the control circuitry generates the pulse to transfer the data from the decade counter to the storage registers, data latch input gates IC14A and D will be enabled. If an over-range condition exists at this time as determined by the condition of IC13A, the data latch will be set. The output of the data latch at IC14 pin 6 normally keeps the over-range indicator LED shorted. This LED indicator is the decimal point element in the seven-segment LED display D105. It is normally lighted by the supply voltage through the 180 ohm resistor R101. If the data latch is reset, its output at IC14 pin 6 is low and therefore the over-range indicator is off. However when the data latch is set, the decimal point element is lighted thereby indicating an over-range condition.

DIGITAL COMPUTERS

The greatest application for digital circuits today is in digital computers. Digital circuits were originally developed to provide a means of implementing digital computers. As new circuits and techniques have been developed, computer performance has improved. The greatest impact on the digital computer has been the development of integrated circuits. IC's have greatly reduced the size and cost of digital computers and has made them more powerful. Over the years digital computers have continued to decrease in price. Their size and power consumption have also decreased significantly. At the same time, their performance and sophistication have increased, making them practical for a wider range of applications.

Recent technological advances in semiconductor techniques have created a unique digital product. Large scale integration of digital circuits has permitted the semiconductor manufacturers to put an entire digital computer on a single chip of silicon. These computers are known as microprocessors. We normally think of digital integrated circuits as being the gates and flip-flops used to implement a computer. Now, the computer itself is a single low cost integrated circuit. But the power of this device is significant, and for many applications it can replace hundreds of SSI and MSI circuits. This significant development will further broaden the applications for digital computers. Best of all, it will increase the sophistication and capabilities of the electronic equipment that uses them.

Today digital computers (microprocessors) are as much a part of digital techniques and digital design as any of the smaller and simpler circuits. While it is impossible to cover all aspects of this exciting field in this unit, we will introduce you to the digital computer and related techniques as they apply to implementing digital systems. Our primary emphasis will be on the microprocessor and its ability to replace standard hard-wired control logic systems.

What is a Digital Computer?

A digital computer is an electronic machine that automatically processes data by the use of digital techniques. Data refers to any information such as numbers, letters, words, or even complete sentences and paragraphs. Processing is a general term referring to a variety of ways in which the data can be manipulated. The computer processes the data by performing arithmetic operations on it, editing and sorting it, or evaluating its characteristics and making decisions based upon it. In addition to being able to manipulate data in a variety of ways, the computer contains an extensive memory where data is stored. The key characteristic of a digital computer is its ability to process data automatically without operator intervention.

The manner in which the data is manipulated is determined by a set of instructions contained within the machine. These instructions form a program that tells the computer exactly how to handle the data. The instructions are executed sequentially to carry out the desired manipulations. Most computers are general purpose in that the instructions can be assembled into an almost infinite variety of application programs. Each computer has a specific instruction set. These instructions are then put into the proper sequence to perform the required calculation or operation. The process of writing the desired sequence of instructions is called programming.

How Computers are Classified

There are many different types of digital computers and a variety of ways in which they can be classified. One method of classifying computers is by size and computing power. There is a broad spectrum covering all types of computers. At one end of the spectrum are large scale computers with extensive memory and high-speed calculating capabilities. These machines can process huge volumes of data in a short period of time and in any desired manner. At the other end of the spectrum are the small scale, low cost digital computers such as the microprocessor whose application and computing power is more limited.

Computers are also classified by function or application. The most commonly known digital computer is the electronic data processor that is used by most business, industry and government organizations for maintaining records, performing accounting functions, maintaining an inventory, and providing a wide variety of other data processing functions. Then there are the scientific and engineering computers that are used primarily as mathematical problem solvers. They greatly speed up and simplify the calculations of complex and difficult scientific and engineering problems.

Another way to classify digital computers is general purpose or special purpose. General purpose machines are designed to be as flexible as possible. This means that they can be programmed for virtually any application. Special purpose computers, on the other hand, are generally dedicated to a specific application. They are designed to carry out only a single function. General purpose computers with a fixed program become special purpose computers.

Most digital computers are of the general purpose type. Most have versatile instruction sets which permit the computer to be programmed to perform almost any operation. With the proper program, a general purpose computer can perform business data processing functions, scientific and mathematical calculations, or industrial control functions.

The most widely used computers are the small scale machines. These include the minicomputer, the microcomputer, the programmable calculator, and the microprocessor. While all of these small scale machines together account for less than 10% of the total computer dollar investment, they represent more than 95% of the unit volume of computers. Small scale computer systems are very low priced and are within the reach of almost everyone. Today a complete computer system can be purchased for less than the price of a new automobile. Microprocessors and programmable calculators are even less expensive. There are many thousands of small computers in use today. Your own personal contact with a digital computer will no doubt be through some type of small scale computer.

Minicomputers. The largest of the four types of small computers is the minicomputer. A minicomputer is a general purpose digital computer usually constructed of TTL or ECL bipolar logic circuits. It is supported with software and peripheral units. Software refers to the programs supplied with the computer that make it easy to use. Peripheral units are the input-output devices that allow an operator to communicate with the computer. Typical peripheral units are typewriters, card readers and printers. Minicomputers are similar to the larger digital computers, but their memory capacity, speed and applications are more limited.

A complete but minimum minicomputer can be purchased for less than $1,000. This does not include peripheral equipment. However, such machines are often purchased to be built into a larger piece of equipment or a system for use as a controller. The users of such computers are referred to as original equipment manufacturers (OEM). A complete stand-alone minicomputer with sufficient memory, peripheral devices and software to be used for general purpose computing can be purchased for less than $5,000.

Microcomputers. A microcomputer is similar in many respects to a minicomputer in that it is a general purpose machine that can be programmed to perform a wide variety of functions. However, the microcomputer is normally smaller and more restricted in its application. Its speed and memory capacity is less than a minicomputer. As a result, microcomputers are substantially less expensive than minicomputers. Microcomputers are more often used in dedicated, single function applications. Software and peripheral support is at a minimum. Most microcomputers are implemented with MOS LSI circuitry.

Programmable Calculator. A programmable calculator can be classified as a special purpose microcomputer. These machines are similar in many respects to the hand-held and desk top electronic calculators widely available. The programmable calculator has an input keyboard for entering data and a decimal display for reading out the results of calculations. In a standard calculator, an operator enters the numbers to be manipulated and the functions to be performed by depressing keys on the keyboard in the proper sequence. The solution to the problems then appears on the display. A programmable calculator can be used this way also, but it contains a memory and control unit that is used to automate the problem solving process. The data to be operated upon and the functions to be performed are entered via the keyboard and stored in the memory in the proper sequence. When enabled, the programmable calculator will then automatically solve the problem stored in its memory without operator control. Programmable calculators offer the advantage of improved speed and convenience over standard calculators when the same problem must be computed several times with different data. Long problems requiring complex data and many mathematical operations are also best solved by a programmable calculator as they relieve the operator from the tedious work and greatly minimize errors. Another advantage of the programmable calculator over other types of digital computers is its ability to communicate directly with the operator through the keyboard and decimal readout display.

Microprocessors. A microprocessor is the smallest and least expensive type of digital computer that still retains all of the basic features and characteristics of a computer. It can be implemented with standard digital integrated circuits or it is available as a single large scale integrated (LSI) circuit. While the capabilities of a microprocessor are limited when compared with a microcomputer or minicomputer, this device is still a very powerful unit. It extends the applications of computer techniques to many areas where minicomputers and microcomputers are not economically feasible.

Microprocessors are generally designed to perform a dedicated function. These devices are built into electronic equipment that will be used for some specific application. Some typical dedicated applications include traffic light controllers, electronic scales and cash registers, and electronic games. In addition, engineers are finding that low cost microprocessors can be used to replace standard hard-wired digital logic. Design time and cost can be significantly reduced in the design of a digital system when microprocessors are used. A microprocessor can be used economically if the design is equivalent to thirty or more standard TTL integrated circuit packages. Such hard-wired logic designs are replaced by a microprocessor with a stored program. The program stored in a read only memory permits the microprocessor to carry out the same functions as a hard-wired logic controller. Microprocessors can also be used as the main component of a minicomputer or microcomputer.

Digital Computer Organization

All digital computers are made up of four basic units. These are the memory, the control unit, the arithmetic-logic unit (ALU), and the input-output (I/O) unit. These major sections and their relationship to one another are illustrated in Figure 10-13. An understanding of digital computer operation starts with a knowledge of how these sections operate and how they affect one another.

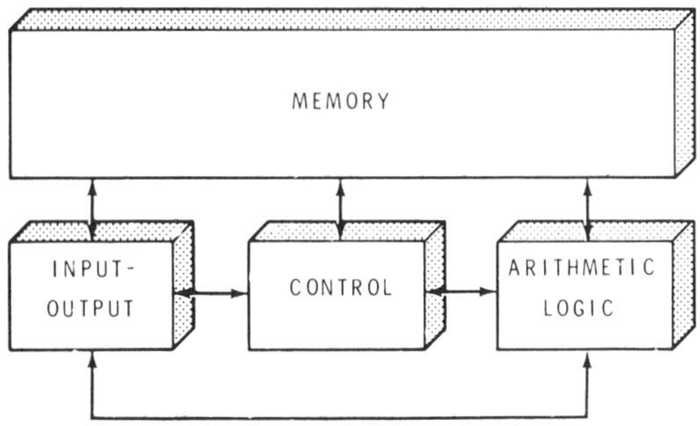

Figure 10-13
General block diagram of a digital computer.

Memory. The heart of any digital computer is its memory. It is in the memory where the program and data are stored. As indicated earlier, the program is a series of instructions that are stored and executed in sequence to carry out some specific function. The instructions cause the computer to manipulate the data in some way.

Computer memories are organized as a large group of storage locations for fixed length binary words. A computer instruction is nothing more than a binary word whose bit pattern defines a specific function to be performed. The data to be processed by the computer is also a binary word. A computer memory is an accumulation of storage registers for these instruction and data words. Most computers have memories capable of storing many thousands of words.

Digital computers typically have a fixed word size. A 32-bit word is common for many large computers. Minicomputers usually have a 16-bit word. Microprocessors widely use an 8-bit word. Memory sizes range from approximately several hundred words to several hundred thousand words of storage. A typical minicomputer may provide 4096 16-bit words. A microprocessor may use 1024 words of 8-bit memory. The number of words in memory is generally some power of two.

Each memory location appears to be like a storage register. Data can be loaded into the register and retained. The word can also be read out of memory for use in performing some operation. Each memory word is given a numbered location called an address. The address is a binary word used to locate a particular word in memory. The normal procedure is to store the instruction words in sequential memory locations. The instruction word generally contains an address which refers to the location of some data word to be used in carrying out the operation specified. The instructions stored in the sequential memory locations are executed one at a time until the desired function is performed.

Most modern digital computers use semiconductor memory. These are MOS LSI circuits where data is stored in latch flip-flops or as the charge on a capacitor. Semiconductor memories are small, fast, and inexpensive. Many computers, however, still use magnetic core memories. In these memories, binary data is stored in tiny donut-shaped magnetic cores. By magnetizing the core in one direction a binary zero is stored. Magnetizing the core in the opposite direction causes it to store a binary one. Electronic circuitry associated with the cores is used to store data into the memory and read it out. The advantage of core memories over semiconductor memories is their non-volatility. When power is removed from a semiconductor memory, all of the data is lost. Removing the power from a magnetic core memory has no effect on the data contents. Because the cores are permanently magnetized in one direction or the other, all data is retained.

The typical organization of a computer memory is shown in Figure 10-14. It consists of the semiconductor or magnetic core elements that retain the binary data. The memory used in a digital computer is generally referred to as a random access read/write memory. Random access refers to the ability of the computer to directly seek out and access any specific word stored in the computer memory. Read/write refers to the ability of the memory to store data (write) or to retrieve data for use elsewhere (read).

As you can see from Figure 10-14, the access to a specific word in memory is achieved through the memory address register (MAR) and memory address decoder. The memory address register is a flip-flop register into which is placed a multi-bit binary word that designates the location of a desired word in memory. If the address 00010011 is stored in the MAR, the content of memory location 19 is referenced. The address word may refer to the location of an instruction or a data word. The size of the address word determines the maximum memory size. For example, if the memory address word is 12-bits in length, the maximum number of words that the computer memory can contain is $2^{12} = 4096$ words. (called a 4K memory)

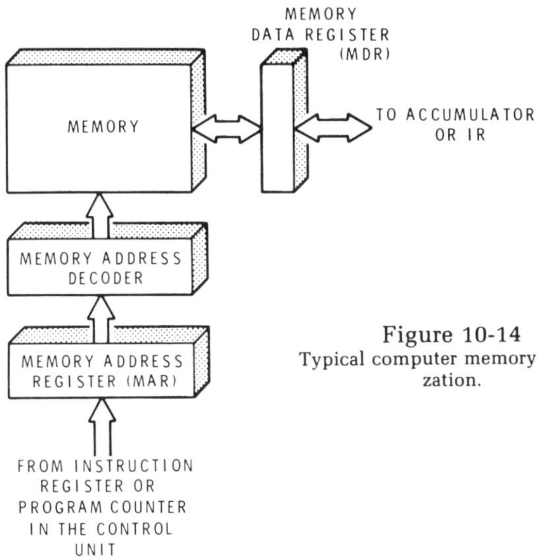

Figure 10-14
Typical computer memory organization.

The output of the memory address register drives the memory address decoder which recognizes one unique memory address word at a time and enables the appropriate location. In semiconductor memories, the memory address decoder is generally a fixed part of the integrated circuit memory itself. When an address word is loaded into the MAR, the specific location in memory designated by that address is enabled. Data can then be written into or read out of that memory location.

The access to the addressed memory location is made through a memory data register (MDR) or memory buffer register (MBR). This is a flip-flop register into which the data or instruction word is stored on its way into or out of the memory. A word to be stored in memory is first loaded into the MDR and then stored in the addressed memory location. If a read operation is being carried out by the memory, the data stored in the addressed location is first loaded into the MDR. From there it is sent to other portions of the computer as needed. Many computers do not use an MDR. Instead the data or instruction goes to or comes from another register in the computer.

Control Unit. The control unit in a digital computer is a sequential logic circuit. Its purpose is to examine each of the instruction words in memory, one at a time, and generate the control pulses necessary to carry out the function specified by that instruction. The instruction, for example, may call for the addition of two numbers. In this case, the control unit would send pulses to the arithmetic-logic unit to carry out the addition of the two numbers. If the instruction calls for the storage data in memory, the control unit would generate the necessary control pulses to carry out that storage operation. As you can see, it is the control unit that is responsible for the automatic operation of the digital computer.

Almost any type of sequential logic circuit can be used to implement the control unit. However, most modern digital computers incorporate a microprogrammed control unit using a ROM. Here special binary words known as microinstructions are stored in the read only memory. When an instruction is analyzed by the control unit, that instruction will cause a certain sequence of microinstruction words in the ROM to be executed. The result is the generation of logic signals that will carry out the operation designated by the instruction. The instructions set for any digital computer is defined by the operation of the control unit.

The exact logic circuitry used in the control unit varies widely from one machine to another. However, the basic elements are shown in Figure 10-15. The control unit consists of an instruction register, a program counter, an instruction decoder, a clock oscillator, and some type of sequential logic circuit used for generating the control pulses.

Figure 10-15
Typical control unit organization.

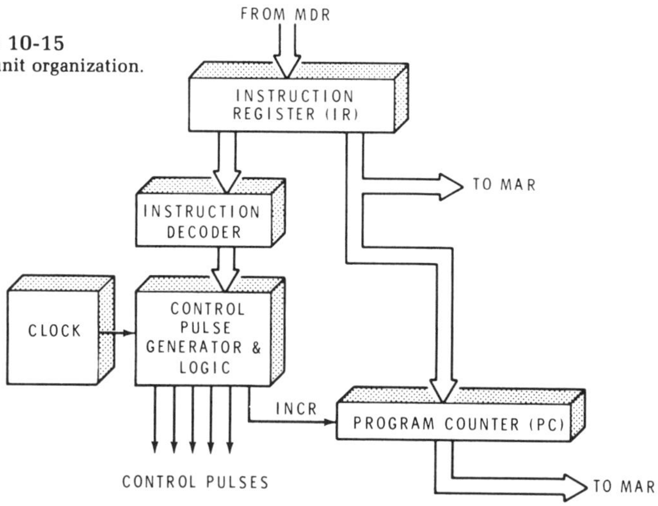

The instruction register is a multi-bit flip-flop register used for storing the instruction word. When an instruction is taken from memory, it passes through the MDR and then into the instruction register. From here the instruction is decoded by the instruction decoder. This logic circuitry recognizes which instruction is to be performed. It then sends the appropriate logic signals to the control pulse generator. Under the control of the clock oscillator, the control pulse generator then produces the logic signals that will enable the other circuitry in the machine to carry out the specified instruction.

The program counter is simply a binary up counter that keeps track of the sequence of instructions to be executed. The program consists of instructions that are stored in sequential memory locations. To begin a program, the program counter is loaded with the starting address. The starting address is the location of the first instruction in the program to be executed. The first instruction is then read out of memory, interpreted and carried out. The control circuitry then increments the program counter. The contents of the program counter is then fed to the memory address register that then permits the next instruction in sequence to be addressed. Each time an instruction is executed, the program counter is incremented so that the next instruction in sequence is fetched and executed. This process continues until the program is complete.

In Figure 10-15 you will notice a connection between the instruction register and the program counter. There are times when the instruction itself will modify the contents of the program counter. Some instructions specify a jump or branch operation that causes the program to deviate from its normal sequential execution of instructions. The instruction register will contain an address that will be loaded into the program counter to determine the location to which the program jumps.

Arithmetic-Logic Unit. The arithmetic-logic unit (ALU) is that portion of the digital computer that carries out most of the operations specified by the instructions. The arithmetic-logic unit performs mathematical operations, logical operations and decision making functions. Most arithmetic-logic units can perform addition and subtraction. Multiplication and division operations are generally programmed. The ALU can also perform logic operations such as inversion, AND, OR, and exclusive OR. In addition, the ALU can make decisions. It can compare numbers or test for specific quantities such as zero or negative numbers.

The arithmetic-logic unit and control unit are very closely related, so much so that it is sometimes difficult to separate them. Because of this, the ALU and control unit together are often referred to as the central processing unit (CPU). Most microprocessors are single chip LSI CPUs.

The arithmetic-logic unit in a digital computer varies widely from one type of machine to another. Figure 10-16 shows the ALU circuitry associated with a very simple, minimum digital computer. The heart of the arithmetic-logic unit is the accumulator register. It is in this register where most of the computer operations take place. Here the data is manipulated, computations are carried out, and decisions are made.

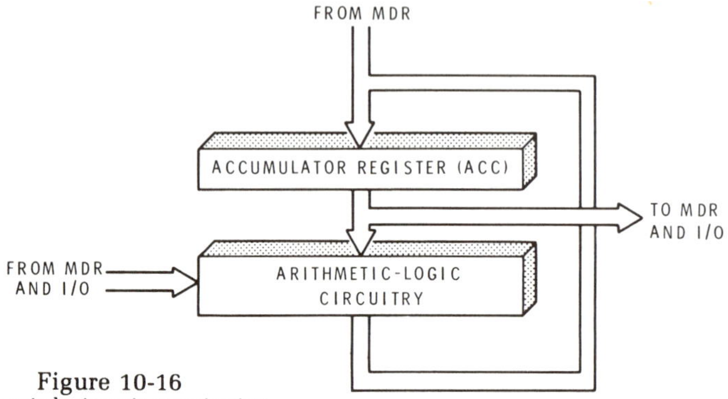

Figure 10-16
Arithmetic logic unit organization.

The accumulator register is a flexible unit that can usually be incremented or decremented. It can also be shifted right or shifted left. Many of the instructions define operations that will be carried out on the data stored in the accumulator register. The size of the accumulator register is generally determined by the basic computer word size, which is the same as the memory word size.

Associated with the accumulator register is the arithmetic-logic circuitry. For the most part, this circuitry is a binary adder. With the binary adder, both binary addition and subtraction can be accomplished. The arithmetic-logic circuitry is also usually capable of carrying out logic operations such as AND, OR, and exclusive OR on the data stored in the accumulator register.

The arithmetic-logic circuitry is capable of adding two binary words. One of the binary words is stored in the accumulator. The other binary word is stored in the memory data register. The sum of these two numbers appears at the output of the arithmetic-logic circuitry and is stored in the accumulator register replacing the number originally contained there. Most other operations with the arithmetic-logic circuitry is carried out in this manner. The two words to be manipulated are initially stored in the accumulator and the MDR with the results of the operation appearing back in the accumulator replacing the original contents.

Input-Output Unit. The input-output (I/O) unit of a computer is that section that interfaces the computer circuitry with the outside world. In order for the computer to communicate with an operator or with peripheral equipment, some means must be provided for entering data into the computer and reading it out. Data and programs to be stored in the memory are usually entered through the input-output unit. The solutions to calculations and control output signals are usually passed to the external equipment through the I/O unit.

The I/O unit is generally under the control of the CPU. Special I/O instructions are used to transfer data into and out of the computer. More sophisticated I/O units can recognize signals from extra peripheral devices called interrupts that can change the operating sequence of the program. Some I/O units permit direct communications between the computer memory and an external peripheral device without interference from the CPU. Such a function is called direct memory access (DMA).

The input/output section of a digital computer is the least clearly defined of all digital computer sections in that it can vary from practically no circuitry at all to very complex logic circuitry approaching the magnitude of the remainder of the computer itself. For our explanation of digital computer operation here, we will assume the simplest form of input/output circuitry. Data transfers between the computer and external peripheral devices take place via the accumulator register. Data to be inputted and stored in memory will be transferred a word at a time into the accumulator and then into the memory through the MDR. Data to be outputted is first transferred from the memory into the MDR, then into the accumulator, and finally to the external peripheral device. These data transfers into and out of the accumulator register take place under the control of the CPU and are referred to as programmed I/O operations. Special input/output instructions cause the proper sequence of operations to take place.

Most digital computers can also perform I/O operations at the request of an interrupt. An interrupt is a signal from an external device requesting service. The external device may have data to transmit to the computer or may require the computer to send it data. When an interrupt occurs, the computer completes the execution of its current instruction, then jumps to another program in memory that services the interrupt. Once the interrupt request has been handled, the computer resumes execution of the main program. Data transfers occurring in the interrupt mode can also take place through the accumulator.

Digital Computer Operation

Now that you are familiar with the basic architecture of a digital computer, you are ready to see how the various sections operate together to execute a program. The units we described in the previous section, together, actually form a simple, hypothetical digital computer that we will use here to demonstrate how a computer operates. We will assume that a program for solving a specific problem is already stored in the computer memory. The computer will execute each instruction in sequence until a solution is reached. We will describe the operation of the computer and show the contents of each of the registers as the program is carried out.

Assume that the problem to be solved is a simple mathematical operation that tells us to add two numbers, subtract a third number, store the result, print the answer, then stop. The numbers that we will work with are 36, 19 and 22. The program calls for adding 36 and 19, subtracting 22, then storing and printing the answer.

The solution to this simple problem as it is solved step-by-step by the computer is illustrated in Figure 10-17. Here we show a simplified block diagram of the digital computer showing the memory and the major registers. The program is stored in memory. The contents of each memory location, either instruction or data, is shown adjacent to the memory address. To solve this problem, the computer sequentially executes the instructions. This is done in a two-step operation. First, the instruction is fetched or read out of memory. Second, the instruction is executed. This fetch-execute cycle is repeated until all of the instructions in the program have been executed. The figures in Figure 10-17 illustrate the contents of the various registers and memory locations for each fetch and execute operation.

In Figure 10-17A, the first instruction of the program is fetched. The instruction word is read out of memory and appears in the memory data register (MDR). It is then transferred to the instruction register (IR) where it is interpreted. Note that the memory address register (MAR) contains 0, which is the address of the first instruction. The accumulator register (ACC) is set to 0 prior to the execution of the program.

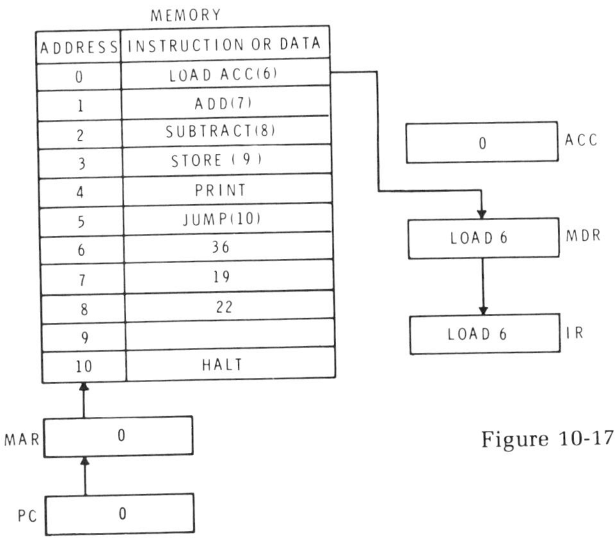

(A) Fetch first instruction (LOAD)

Figure 10-17

Figure 10-17B shows the execution of the first instruction. The first instruction LOAD ACC (6) tells us to load the accumulator with the data stored in memory location 6. In executing this instruction, the number 36 is transferred to the accumulator. Note how this is done. The address specified by the instruction word (6) is transferred from the instruction register to the memory address register (MAR). This causes the number 36 stored in that location to be transferred to the MDR and then to the accumulator. During this step, the program counter (PC) is incremented by one so the next instruction in sequence will be fetched.

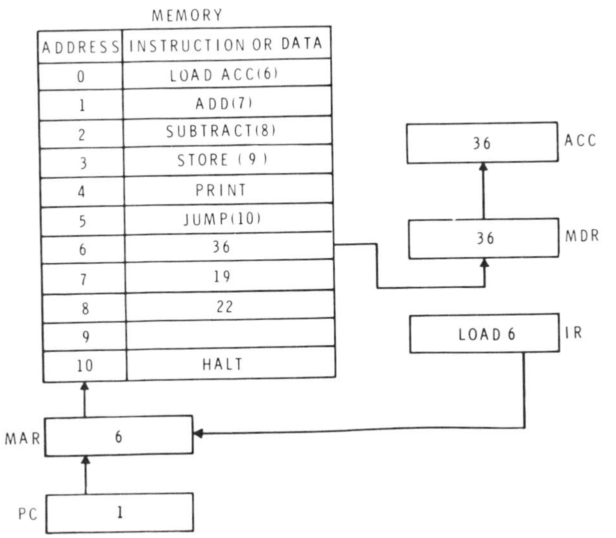

(B) Execute first instruction (LOAD)

Figure 10-17C shows the fetch operation for the second instruction. The contents of the program counter is transferred to the MAR so that the ADD(7) instruction is fetched. This instruction passes through the MDR into the instruction register.

(C) Fetch second instruction (ADD)

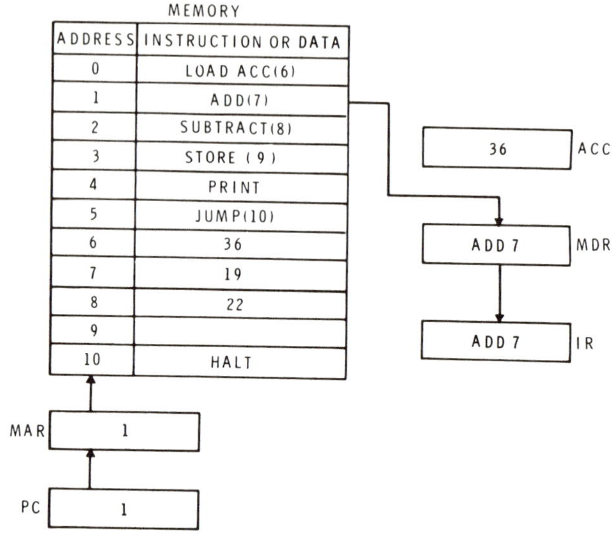

The execution of the add instruction is shown in Figure 10-17D. This instruction tells us to add the contents of memory location 7 to the contents of the accumulator. The address of the add instruction is trans-

(D) Execute second instruction (ADD)

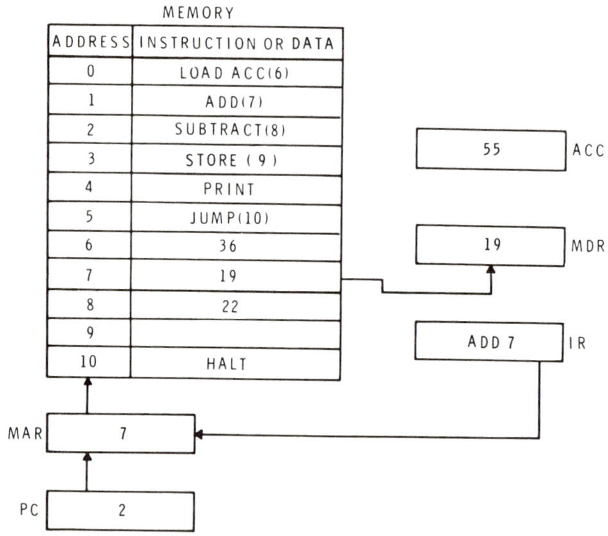

ferred to the MAR. This causes the contents of memory location 7, the number 19, to be transferred to the MDR. The contents of the MDR are added to the contents of the accumulator with the sum appearing back in the accumulator. As you can see, the sum of 36 and 19 is 55. Note that the program counter is again incremented so that the next instruction in sequence will be fetched.

The remaining instructions in the program are fetched and executed in a similar manner. The third instruction, a subtract, causes the memory contents of location 8 to be subtracted from the contents of the accumulator with the resulting remainder appearing in the accumulator. This produces an answer of 33. This fetch-execute sequence is shown in Figure 10-17E and F.

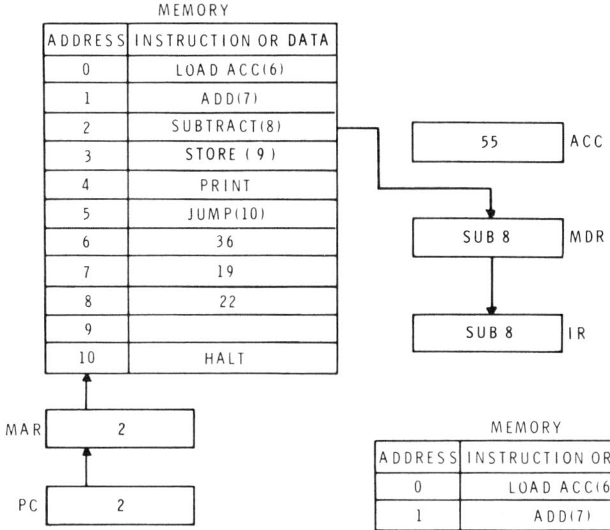

(E) Fetch third instruction (SUB)

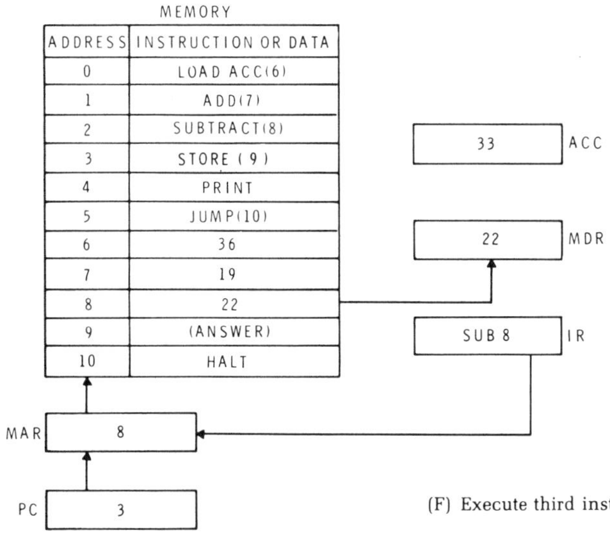

(F) Execute third instruction (SU

The next instruction in sequence, STORE(9), tells us to store the contents of the accumulator in memory location 9. The number 33 in the accumulator is transferred to the MDR and stored in location 9, as indicated in Figure 10-17G and H.

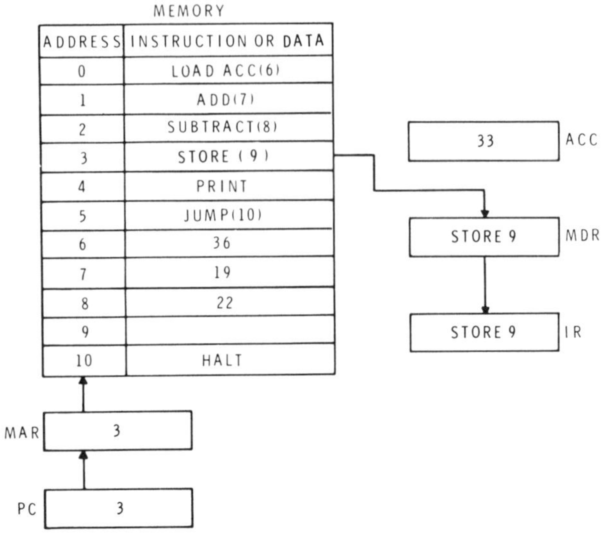

(G) Fetch fourth instruction (STORE)

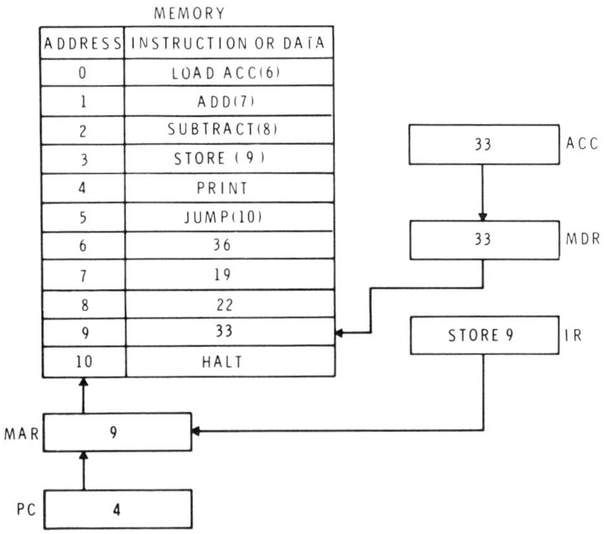

(H) Execute fourth instruction (STORE)

The fifth instruction in the program, PRINT, tells us to print the contents of the accumulator on the external printer. The number stored in the accumulator will then be transferred to a printer where it is printed. The fetch-execute cycle for this operation is shown in Figures 10-17I and J.

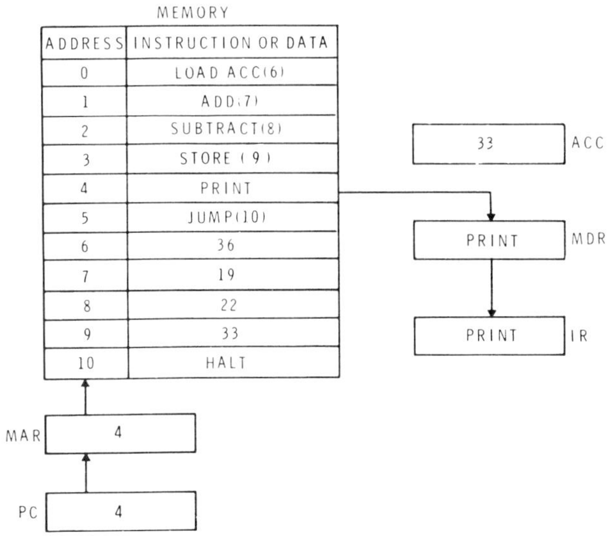

(I) Fetch fifth instruction (PRINT)

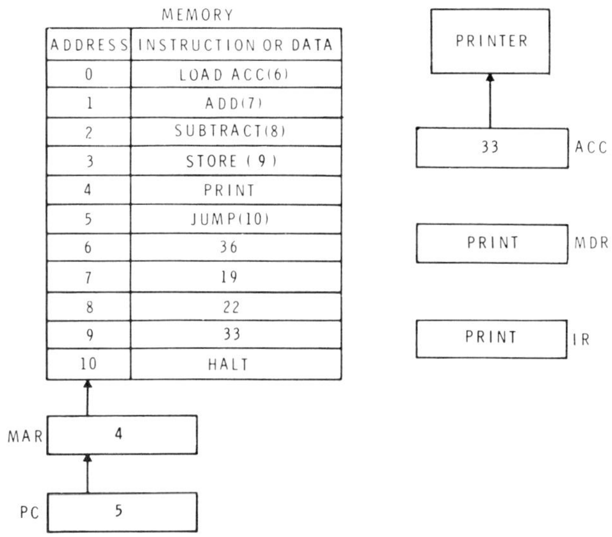

(J) Execute fifth instruction (PRINT)

The sixth instruction in sequence is a JUMP(10) instruction that causes the normal sequence of program executions to change. The jump instruction tells us not to execute the contents of the next memory location in sequence. Instead it tells us to take the next instruction from memory location 10. You can see by referring to Figure 10-17K that the contents of the next memory location in sequence (address 6) contains a data word. The computer, being a dumb machine, would simply interpret a data word as an instruction and attempt to execute it. If this ever happens, the result of the computation will be erroneous. The purpose of the jump instruction in our program is to jump over the data words in the program stored in locations 6, 7, 8, and 9. The program is continued in location 10, where a

(K) Fetch sixth instruction (JUMP)

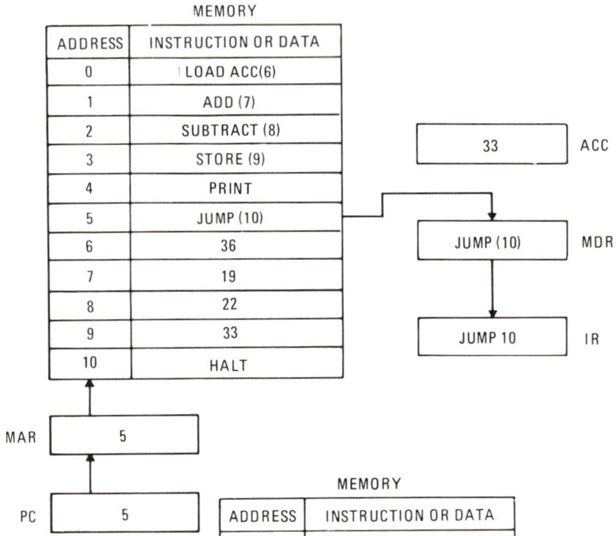

(L) Execute sixth instruction (JUMP)

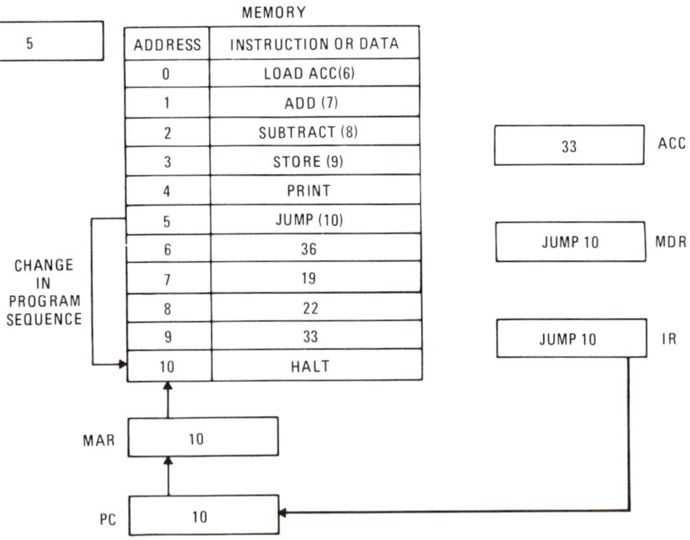

HALT instruction is stored. By executing the jump instruction, the program counter is loaded with the address portion of the jump instruction (10) instead of being incremented as it normally is. See Figure 10-17L. This causes the computer to fetch and execute the instruction stored in location 10. This is illustrated in Figures 10-17M and N.

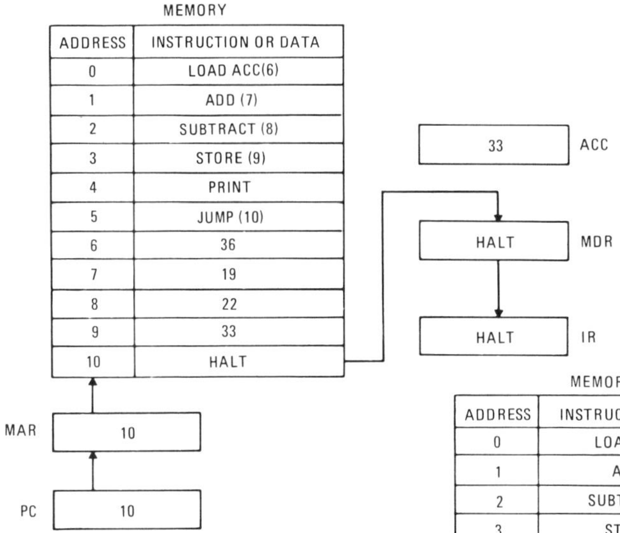

(M) Fetch seventh instruction (HALT)

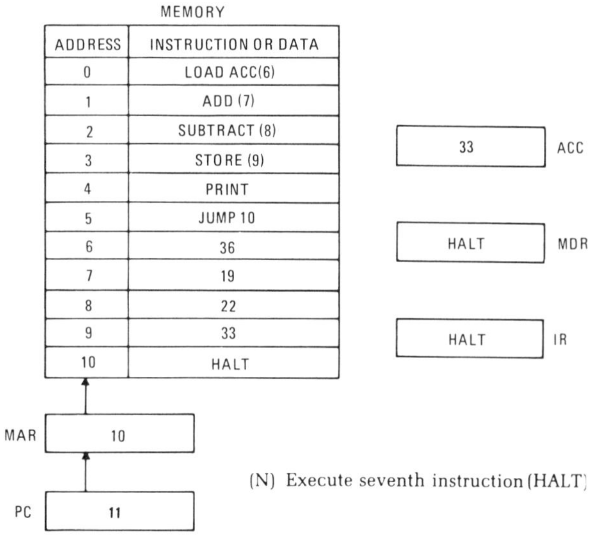

(N) Execute seventh instruction (HALT)

The last instruction in the program is a HALT. This instruction has no effect other than to stop the operation of the machine. Note in Figure 10-17N that the program counter was incremented so that it contains the memory location (11) of the next instruction in sequence to be fetched.

Study the program shown in Figure 10-17. Trace through each of the fetch and execute cycles for each instruction to be sure that you fully understand the operation. All digital computers operate in this same way with minor variations.

Programming

A digital computer without a program is a useless piece of electronic hardware. The logic circuitry making up the computer is incapable of performing any useful end function without a program. It is this characteristic of a digital computer that sets it apart from other types of digital circuitry. And it is this characteristic that makes the digital computer the versatile machine that it is. For this reason, a discussion of digital computers is not complete without information on programming.

The process of using a digital computer is mainly that of programming it. Whether the computer is a simple microprocessor or a large scale system, it must be programmed in order for it to perform some useful service. The application of the computer will define the program. Programming is the process of telling the computer specifically what it must do to satisfy our application.

Programming is a complex and sophisticated art. In many ways it is almost a field apart from the digital circuitry and the computer hardware itself. There are many different levels of programming and many unique methods that are employed. For that reason it is impossible to cover them all here. The purpose of this section is to give you an overview of the process of programming a computer. As indicated earlier, our emphasis will be on the programming of small scale digital computers such as the microprocessor.

Programming Procedure. There are many different ways to program a digital computer. The simplest and most basic form of programming is machine language programming. This is the process of writing programs by using the instruction set of the computer and entering the programs in binary form, one instruction at a time. Programming at this level is difficult, time consuming, and error prone. It also requires an in-depth understanding of the computer organization and operation. Despite these disadvantages, however, this method of programming is often used for short simple programs. While most computer applications do not use machine language programming, it is desirable to learn programming at this level. It helps to develop a thorough knowledge of machine operation and generally results in the shortest, most efficient programs. Many microprocessors and minicomputers are programmed in machine language.

To illustrate the concepts of programming in this unit, we will use machine language programming. Other more sophisticated methods of programming will be discussed later.

Programming a digital computer is basically a seven step process. These seven steps are: (1) define the problem; (2) develop a workable solution; (3) flow chart the problem; (4) code the program; (5) enter the program into the computer; (6) debug the program; (7) run the program. Let's take a look at each of these steps in detail.

The first and perhaps the most important step in programming a digital computer is defining the problem to be solved. The success of the program is directly related to how well you define the operation to be performed. There is no set standard for the problem defining procedure, and any suitable method can be employed. The definition can be a written statement of the function to be carried out, or it may take the form of a mathematical equation. In some cases, the problem may be more easily defined by graphical means. For control applications, the problem may be expressed in the form of a truth table. The form in which you place the definition is strictly a function of the application.

Once the problem is analyzed and defined, you can begin thinking in terms of how the computer may solve the problem. Remember that a computer program is a step-by-step sequence of instructions that will lead to the correct results. You should think in terms of solving your problem in some step-by-step sequential manner. What you will be doing in this phase of the programming procedure is developing an algorithm. An algorithm is a method of procedure of solving a problem.

An important point to remember is that there is usually more than one way to solve a given problem. In other words, there is more than one algorithm suitable for achieving the goal that you have set. Much of the job of programming is in determining the alternatives and weighing them to select the best suitable approach. The simplest and most direct algorithms are usually the best.

The next step is to flowchart the problem. A flow chart is a graphical description of the problem solution. Various symbols are used to designate key steps in the solution of the problem. Figure 10-18 shows the basic flowcharting symbols. An oval defines the starting and finishing points. A rectangular box defines each individual computational step leading to the solution. Each rectangle contains some basic operation or calculation that is to take place. The diamond shaped symbol represents a decision point. It is often necessary to observe the intermediate results in a problem solution and make a decision regarding the next step to be taken. There are generally two exits to the diamond shaped decision making symbol. These represent a yes or no type of decision.

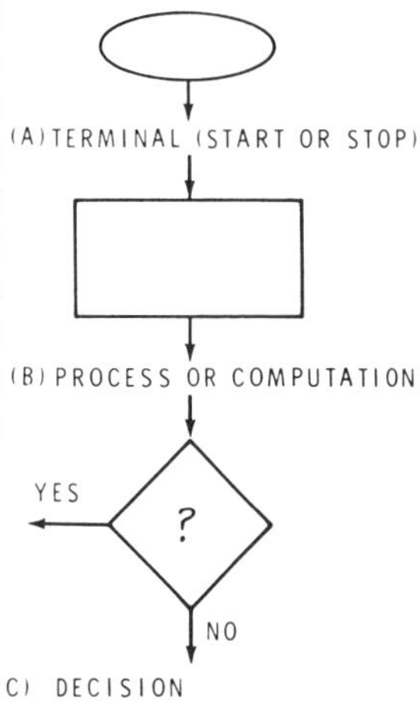

Figure 10-18
Basic flowcharting symbols.

Figure 10-19 shows a simple flow chart for the problem solved in Figure 10-17. No decision was made in this program.

As you can see, the flow chart is a graphical representation of the basic method used to solve the problem. The flow chart permits you to visualize the algorithm you developed. In many cases, the flowcharting of a problem helps to determine the best approach to solving a problem since it forces you to think in a logical sequence and express the solution in a step-by-step form.

At this point in the programming procedure, your problem is quite well defined and a basic method of solving the problem has been determined. You are now ready to convert your flow chart and algorithm into a machine language program. This process is called coding. Coding is the procedure of listing the specific computer instruction sequentially to carry out the algorithm defined by the flow chart. This requires a familiarity with the instruction set of the computer you plan to use.

The next step in the programming procedure is to load the program into the computer memory. Once you have written the program with the computer instructions, you have all of the information necessary to enter that program into the computer memory. If you are dealing with machine language programming, you will convert the instruction words into their binary equivalents and then load them into the computer. If the program is a simple one, it can be loaded by using the binary switches on the front panel of the computer. However, for long, complex programs, this manual loading procedure is difficult and time consuming.

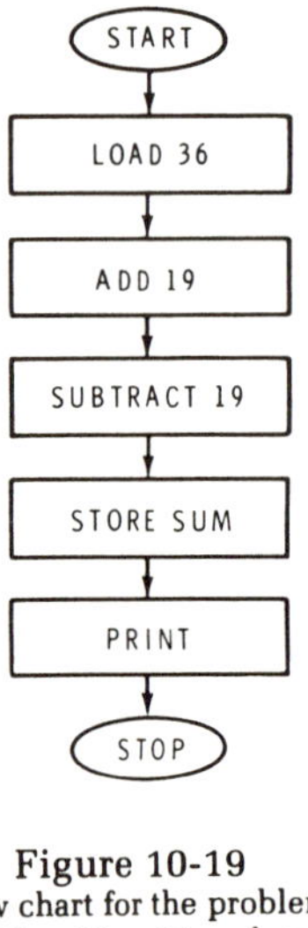

Figure 10-19
Flow chart for the problem
36 + 19 − 22 = 33 and print.

Most computers make it easy for the programmer to enter his program. Because of the availability and use of support programs residing within the machine, the program can usually be entered automatically. One of the most common ways of entering data into a computer is by the use of a keypunch machine. This is a typewriter-like machine that punches a standard computer card with the instructions to be entered. Teletype input/output machines using perforated paper tape are also commonly used for program entry. The instruction designations are typed on the machine and as they are typed, a paper tape is punched.

Once the cards or paper tape are punched, they are then fed into a tape reader or card reader and loaded into the computer memory. A special program residing within the computer memory called a loader causes the program to be loaded automatically.

With the program now in the computer memory, you can begin to run it. However, before you use it to obtain your final answer, it is often necessary to run through the program slowly a step at a time to look for programming errors and other problems. This process is called debugging. You test the program to see that it produces the desired results. If the program produces the correct result, it is ready to use. Often, programming mistakes are encountered and it is necessary to modify the program by changing the instruction steps. Often the entire program may be discarded and a new one written, using a different algorithm.

Once the program has been debugged, it is ready for use. With the program stored in memory, your problem can be solved. The computer is started and the desired results are produced.

Writing Programs

Before you can begin coding programs, you must be familiar with the instruction set of the computer you are using. Most digital computer instruction sets are basically alike in that they all perform certain basic functions such as addition, branching, input/output and the like. But each instruction set is different because the logic circuits unique to each computer carry out these operations in different ways. To code the program properly, you must know exactly what each instruction does. You can get this information by studying the instruction set as it is listed and explained in the computer's operation and programming manuals. By studying the instruction set you will learn how the computer is organized and how it operates. The insight you gain from this will be valuable to you not only in coding the program but also in developing the best solution to a problem with a given machine.

Computer Instructions. A computer instruction is a binary word that is stored in the computer memory and defines a specific operation that the computer is to perform. The instruction word bits indicate the function to be performed and the data which is to be used in that operation.

There are two basic types of computer instructions: memory reference and non-memory reference. A memory reference instruction specifies the location in memory of the data word to be used in the operation indicated by the instruction. A non-memory reference instruction simply designates an operation to be performed. Non-memory reference instructions generally refer to internal housekeeping operations to be performed by the computer and manipulations on data stored in the various registers in the computer.

Figure 10-20 shows typical instruction word formats for an 8-bit microprocessor. The format shown in Figure 10-20A is a memory reference instruction. The instruction is defined by three 8-bit words which are stored in sequenctial memory locations. The first 8-bit word is the op code or operations code which is simply a binary bit pattern specifying some operation. The second and third 8-bit words specify the memory address of the data or operand to be used. The 8-bit op code defines 256 possible operations or functions. It is the op code that designates the operation that is to be performed. The 16-bit address specifies the memory location of the data to be operated upon. The size of the address generally indicates the maximum memory size of the computer. With 16 bits of address information, $2^{16} = 65{,}536$ words can be directly addressed. We usually say that the maximum memory size is 65K.

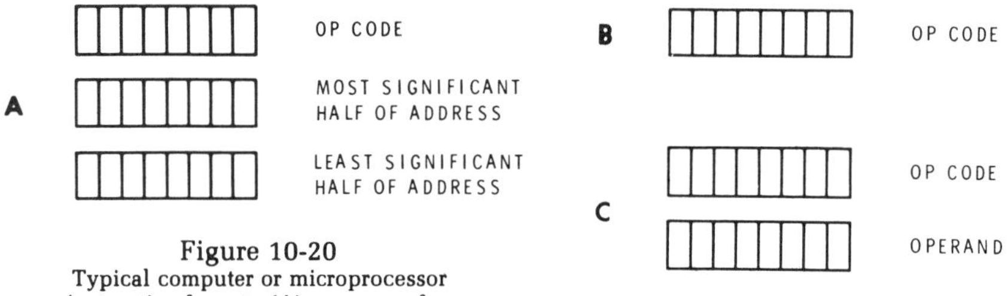

Figure 10-20
Typical computer or microprocessor instruction formats. (A) memory reference, (B) non-memory reference, (C) immediate.

The word format in Figure 10-20B is the typical format for non-memory reference instructions. Only an 8-bit op code is used. In this type of instruction, an address is not needed since we do not reference a location in memory where data is stored. Instead, the bits in this field are used to specify various operations that are to take place within the CPU. For example, such an instruction might call for the resetting (clearing) of a register or the transfer of data from one register to another. Certain types of input-output instructions have this format.

Another instruction type is the immediate instruction which is widely used in microprocessors. The format for this instruction is shown in Figure 10-20C. It consists of an 8-bit op code that specifies the operation. The second 8-bit part of this instruction is the data or operand to be used in the operation called for. The immediate instruction is like the memory reference instruction in that it specifies the use of some data word. The data to be used is in the instruction word itself rather than being referenced by an address in the instruction word. Immediate instructions save memory space and shorten instruction fetch and execution times.

Another method of classifying computer instructions is to group them according to the type of functions that they perform. These include arithmetic and logic, decision-making, data moving, and control. Let's consider each of these in more detail.

An arithmetic instruction defines a specific mathematical operation that is to take place. The most commonly used arithmetic instructions are add and subtract. In larger computers the multiply and divide functions are also included. Multiply and divide operations in smaller computers such as minicomputers and microprocessors are carried out by special subroutines. As an example, multiplication can be performed by repeated addition. Division can be programmed by the use of repeated subtractions. Arithmetic instructions are generally of the memory reference type.

Logical instructions specify digital logic operations that are to be performed on computer data. These include the standard logic functions of AND, OR and invert (complement). Many computers include the exclusive OR function. Other logic instructions include shift right and shift left operations. The AND, OR and XOR logical instructions are usually memory reference type. The shifting and inversion instructions are of the non-memory reference type as they generally refer to operations carried out on data stored in one of the computer's registers.

A decision making instruction is one that permits the computer to test for a variety of results and based upon these tests make a decision regarding the next operation to be performed. It is the decision-making instructions that set the computer apart from the standard calculator. Decision-making instructions allow the computer to automate its operations. A decision-making instruction generally follows a sequence of other instructions that perform some arithmetic or logical operation. Once the operation is performed, the decision-making instruction tests for specific results. For example, decision-making instructions test for positive or negative numbers, zero, odd or even numbers or equality. These tests are generally made on the data stored in various registers in the machine. If the test for a specific condition exists, the computer is usually instructed to deviate from its normal sequential execution of instructions. Jump or branch instructions are memory reference instructions that test for certain conditions and then specify a memory location where the next instruction to be executed is located. Skip instructions also change the computing sequence. These instructions test for a specific condition and then, if that condition exists, direct the computer to skip the next instruction in sequence. Skip instructions are non-memory reference types.

A data moving instruction is one that causes data words to be transferred from one location to another in the computer. It is these instructions that are used to take data from memory and load it into one of the operating registers in the computer. Other data moving instructions cause data stored in a register to be stored in a specific memory location. These are memory reference instructions. Other data moving instructions specify the transfer to data words between registers in the machine. These are non-memory reference instructions. The data moving instructions provide a flexible means of transferring data within the machine to prepare it to be processed as required by the application. A special class of data moving instructions are the input/output instructions. I/O instructions cause data to be transferred into and out of the computer. These non-memory reference instructions often specify one of several input/output channels or a specific peripheral device. Input/output operations can be programmed to take place through the operating registers of the machine, or in some computers, directly between the memory and the peripheral unit.

A control instruction is a non-memory reference instruction that does not involve the use of data. Instead, it designates some operation that is to take place on the circuitry in the computer. Clearing a register, setting or resetting a flip-flop or halting the computer are examples of control instructions.

A Hypothetical Instruction Set. A typical but hypothetical instruction set for a minicomputer or microprocessor is shown in Table I. Only a few of the most commonly used instructions are listed so that you can become acquainted with them quickly. Real instruction sets are far more extensive. Nevertheless, the instruction set in Table I is representative. We will use it to demonstrate the writing and coding of programs.

The instruction set in Table I can apply to the hypothetical computer described earlier or a typical microprocessor. For the instructions listed here, we assume that the computer has an 8-bit word length and 65K of memory. The accumulator and memory data registers are 8 bits in length. The program counter and MAR are 16 bits in length. I/O transfers take place through the accumulator. A single instruction may occupy one, two or three consecutive memory locations depending upon its format as shown in Figure 10-20. Study the instructions in Table I so that you will be familiar with the operation each performs. Note that each instruction is designated by a three letter mnemonic. The type of instruction is designated by the letters R (memory reference), N (non-memory reference), A (arithmetic-logic), T (data moving or transfer), D (decision) and C (control). Despite the simplicity of this instruction set, it can be used to program virtually any function.

Hypothetical Computer Instruction Set

Table I

MNEMONIC	TYPE OF INSTRUCTION	OPERATION PERFORMED
LDA	R, T	Load the data stored in the specified memory location (M) into the accumulator register.
STA	R, T	Store the data in the accumulator register in the specified memory location (M).
ADD	R, A	Add the contents of the specified memory location (M) to the contents of the accumulator and store the sum in the accumulator.
SUB	R, A	Subtract the contents of the specified memory location (M) from the contents of the accumulator and store the remainder in the accumulator.
AND	R, A	Perform a logical AND on the data in the specified memory location (M) and the contents of the accumulator and store the results in the accumulator.
OR	R, A	Perform a logical OR on the data in the specified memory location (M) and the contents of the accumulator and store the results in the accumulator.
JMP	R, D	Unconditionally jump or branch to the specified memory location (M) and execute the instruction stored in that location.
JMZ	R, D	Jump to the specified memory location if the content of the accumulator is zero (reset). Execute the instruction stored in that location. If the accumulator is not zero, continue with the next instruction in normal sequence.
CLA	N, C	Clear or reset the accumulator to zero.

CMP	N, A	Complement the contents of the accumulator.
SHL	N, A	Shift the contents of the accumulator one bit position to the left.
SHR	N, A	Shift the contents of the accumulator one bit position to the right.
INP	N, T	Transfer an 8-bit parallel input word into the accumulator.
OUT	N, T	Transfer the contents of the accumulator to an external device.
HLT	N, C	Halt. Stop computing.
INC	N, C	Increment the contents of the accumulator.
DCR	N, C	Decrement the contents of the accumulator.
SKO	N, D	If the number in the accumulator is odd (LSB = 1), skip the next instruction and execute the following instruction. If the accumulator is even (LSB = 0), simply execute the next instruction in sequence.

Example Programs. The following examples illustrate the use of the instruction set in writing programs. The program description, flowchart and instruction code are given in each example. Study each program, mentally executing the instructions and imagining the outcome. The format of the instruction coding is shown below.

3 ADD (7)

The number on the left is the memory address. The mnemonic specifies the instruction. The number in parenthesis is the address of the operand called for by a memory reference instruction. This line of instruction coding tells us that in memory location 3 is an add instruction that tells us to add the content of location 7 to the content of the accumulator.

The program shown below is a repeat of the program given in Figure 10-17. The only differences are the memory location numbers of the instructions, the use of mnemonics, and the substitution of the OUT instruction for the PRINT instruction.

```
 0  LDA  (16)
 3  ADD  (17)
 6  SUB  (18)
 9  STA  (19)
12  OUT
13  JMP  (20)
16  36
17  19
18  22
19  ANSWER
20  HLT
```

The difference in memory addresses is the result of the assumption that our computer uses an 8-bit word and that memory reference instructions occupy three sequential memory locations. In the program of Figure 10-17 we assumed that one memory address contained one instruction. In the program above, the LDA (16) occupies memory locations 0, 1 and 2. The op code is in 0, the most significant part of the address (0000 0000) is in location 1, and the least significant part of the address (0001 0000) is in location 2. The ADD, SUB, STA and JMP memory reference instructions each occupy three sequential locations. The OUT and HLT instructions do not reference memory so they occupy only a single location.

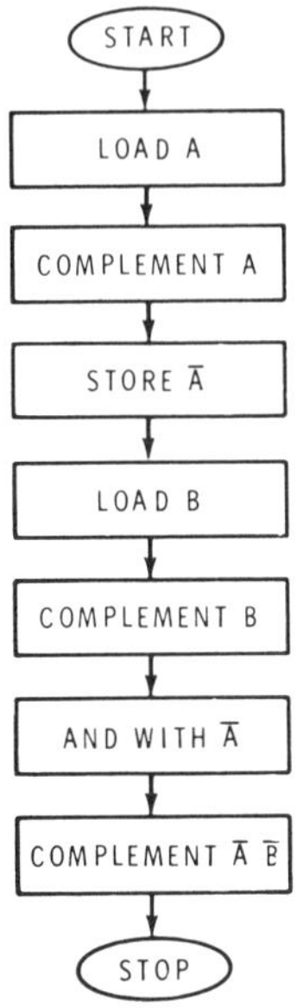

Figure 10-21

Flow chart illustrating a method of performing th OR function with AND and invert instructions.

The program below illustrates the use of the logical instructions.

Assume that the only logical instructions that your computer has are AND and CMP (complement). We need to perform the OR function on the two words A and B stored in locations 16 and 17. The flow chart in Figure 10-21 and the program below illustrates how this is done. By DeMorgan's theorem we know that $\overline{A+B}=\bar{A}\bar{B}$. Complementing both sides gives us the OR function.

$$A + B = \overline{\bar{A}\ \bar{B}}$$

It is the right-hand part of this equation that is our algorithm.

```
 0  LDA (16)
 3  CMP
 4  STA (18)
 7  LDA (17)
10  CMP
11  AND (18)
14  CMP
15  HLT
16  A
17  B
18  Intermediate storage of Ā
```

Note the use of memory location 18 as temporary storage for an intermediate result ($\overline{A}$). This frees the accumulator to process other data.

The next program illustrates several important concepts. First, it shows how the computer makes decisions. Second, it demonstrates the use of a program loop. A loop is a sequence of instructions that is automatically repeated. The sequence is executed once and a jump instruction causes the program to branch back (loop) to the beginning of the sequence and repeat it again.

The program below is designed to enter two 4-bit BCD numbers and store them in a single 8-bit memory location. The desired memory format is shown in Figure 10-22. The BCD digits are entered, one at a time, into the four least significant bit positions of the accumulator as shown in Figure 10-23. The first digit entered must be moved to the four most significant bit positions of the accumulator before the second digit can be entered. This is accomplished with a series of shift left instructions.

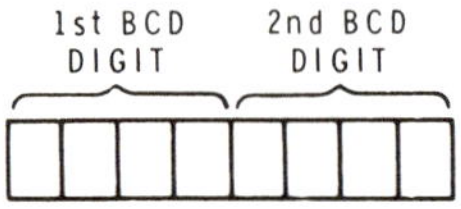

Figure 10-22
Memory format for two BCD digits.

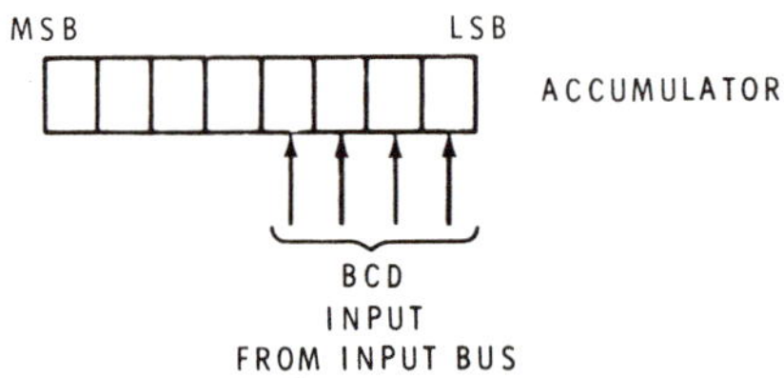

Figure 10-23
Loading the accumulator from the data bus.

The flow chart in Figure 10-24 shows how this is accomplished. The program is given below.

```
 0  CLA
 1  INP
 2  SHL
 3  STA (25)
 6  LDA (24)
 7  DCR
 8  STA (24)
11  JMZ (20)
14  LDA (25)
17  JMP (2)
20  LDA (25)
21  INP
22  STA (26)
23  HLT
24  4
25  Temporary storage
26  Storage for two BCD digits
```

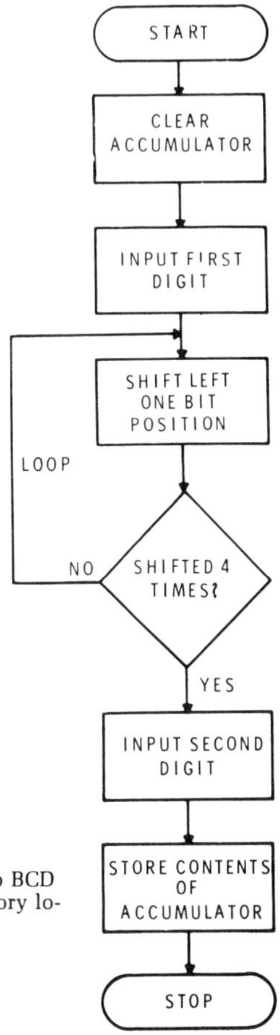

Figure 10-24
Flow chart showing how two BCD words are stored in one memory location.

The first instruction (CLA) clears the accumulator. The second instruction (INP) loads the first BCD digit. This digit is then shifted left one bit position. We need to shift it four positions to the left. The sequence of instructions in locations 2, 3, 6, 7, 8, 11, 14 and 17 form a loop and a decision-making test to accomplish this. Stored in memory location 24 is a number that tells us how many times to shift. That number is loaded into the accumulator decremented by one and restored each time a shift occurs. We test that number with a jump on zero instruction (JMZ). When the BCD digit has been shifted four times, the number in location 24 has been reduced to zero. The JMZ instruction detects this condition and branches the program to location 20 where the data word is retrieved from temporary storage and the second BCD digit is inputted.

Let's consider the loop and decision-making process in more detail. After the first input digit is loaded, it is shifted left once. We then store it temporarily in location 25. This is to prevent loss of the data while we are in our decision-making loop. Next, the number of desired shifts is loaded into the accumulator. We decrement it by one, indicating that we have shifted left once. Next, we restore this number (now 3) in location 24. This number, which is still in the accumulator, is now tested with the JMZ instruction. At this time the accumulator is not zero so the program does not branch. Instead, the next instruction in sequence is executed. This is a load accumulator instruction that retrieves the data words which we temporarily stored in location 25. Then the JMP instruction is executed. This instruction returns us to location 2 to produce another shift. It is the jump instruction that creates the loop.

The loop is then repeated three more times. On the fourth pass through the loop, the number in location 24 is decremented to zero. The JMZ instruction detects this condition and causes the program to branch to location 20. We have now escaped from the loop. Next we reload the shifted data from location 25. Finally, we load the second BCD digit. Both digits are now in the accumulator so we can store them in location 26 with the STA (26) instruction. The program is now complete and the HLT instruction terminates it.

These examples show how a computer performs its work. It does it laboriously, one step at a time. The only thing that makes it practical is its high speed operation. With each instruction taking only microseconds, even long complex programs are executed quickly. To an operator, the execution appears almost instantaneous.

Software

The steps that we have just described make up the procedure typically used in developing application programs for the digital computer. The program may be solving a mathematical equation, sorting and editing a large volume of data, or providing some type of automatic control to an external machine. These application programs fall into a larger category of computer programs called software. Software is a broad general term used to describe all of the programs used in a digital computer. Besides the specific applications programs, there are many special programs supplied by the computer manufacturer which are used to simplify and speed up the use of the computer. These support programs eliminate much of the drudgery from programming and using a computer. It was determined early in the development of the digital computer that the computer itself with special internal control programs could assume much of the responsibility for the detailed translation of a problem into the binary language of the computer.

Subroutines. Many digital computer manufacturers supply what are called software libraries of subroutines and utility programs. A subroutine is a short machine language program that solves a specific problem or carries out some often used operation. For example, typical subroutines in many minicomputers and microprocessors are the multiply and divide programs. Instead of using the multiply subroutine each time it is required in a problem, the subroutine is stored in the computer memory only once. This saves a substantial amount of memory space. Each time the multiply subroutine is required, a jump instruction in the program causes the program to branch to the multiply subroutine. Once the multiplication has been performed, the computer jumps back to the normal program sequence.

There are many different types of commonly used subroutines. Multiplication and division are two of the most commonly used. Other subroutines include binary to BCD and BCD to binary code conversions. To communicate with external peripheral devices which use the decimal number system and the alphabet, a code such as ASCII is used. Data is entered into the computer in the ASCII code. In order for the computer to process this data, it must first be converted into pure binary numbers. Solutions to computer programs are in the pure binary form. A subroutine is used to convert the binary numbers into the ASCII format and then they are sent to an external peripheral device such as the printer.

Utility Programs. Utility programs refer to the short routines used to run the computer. Input/output programs for specific types of peripheral devices fall into this category. A loader is another utility program. This is a short sequence of instructions that allows data to be loaded into the computer. In order to operate, a computer must be programmed. But to load a program into the computer automatically requires that the loader program exist in the memory to begin with. Such loader programs are often entered manually from the computer front panel. The short loader program in memory then permits longer, more complicated programs to be loaded automatically.

There are many different types of utility programs used in digital computers. These also include editor programs for manipulating data and moving it from one part of the memory to another. These programs also permit you to conveniently modify a program through a peripheral device such as a teletypewriter. Utility programs often include diagnostic programs which provide a means of testing all of the functions of the computer. Diagnostic routines test each computer instruction and all memory locations. Other diagnostic programs permit peripheral devices to be exercised and their operation verified.

Assembler. The most sophisticated software supplied with most computers are large complex conversion programs called assemblers and compilers. These programs allow the computer to be programmed in a simpler language. Machine language programming is completely impractical for many modern applications. In order to simplify computer programming and remove the necessity for a dependence upon a knowledge of binary numbers and the computer architecture, computer manufacturers have developed easier methods of programming the computer. These methods involve higher level languages, which are special systems for speeding up the programming process. The higher level language permits someone with no computer expertise whatsoever to use the computer. The higher level language permits the programmer to express his problem as a mathematical equation or in some cases as an English language statement. These equations and statements are then fed to the computer, which then automatically converts them into the binary instructions used to solve the problem.

The simplest form of higher level programming language is called assembly language. This is a method of programming the computer an instruction at a time as you do in machine language programming. However, instead of binary designations for each instruction, short multiletter names called mnemonics are given to each of the computer instructions. These are then written sequentially to form the program. Mnemonics are also given to memory addresses to avoid the use of specific memory locations.

Once the computer program is written in the assembly language, it is then entered into the machine along with an assembler program. The assembler program resides in the computer memory and is used to convert the mnemonics into the binary instruction words that the computer can interpret. As you can see, the assembler is a program that eliminates the necessity of dealing with binary numbers in the digital computer. However, since the machine is still programmed an instruction at a time, it provides wide flexibility in solving a given problem.

Compiler. The compiler, like the assembler, is a complex conversion program that resides in the computer memory. Its purpose is to convert a simplified statement of the program into the binary machine code that the computer can understand. The difference between the compiler and the assembler is that the compiler is capable of recognizing even simpler problem statements.

In one type of compiler programming language known as Fortran, the program can be written as an algebraic equation. This algebraic equation is then entered into the computer through a teletypewriter or via punched cards. The compiler program then analyzes the formula and proceeds to construct a binary program to solve this equation at some location in memory.

Another higher level programming language known as Cobol uses English language statements to describe the problem. These English language statements are punched into cards and then read into the computer memory. The compiler interprets them and converts them into the binary program. Unlike an assembly language program which has a one-to-one correspondence of instruction steps with machine language, a single compiler language program statement often causes many binary instructions to be generated.

There are many different types of higher level programming language used with computers. All of them have the prime function of simplifying the programming procedure. They greatly speed up and expedite the communications with the computer. They allow anyone who is capable of defining his problem to use the computer.

Cross Assemblers and Compilers. There are several special types of higher level programming language that have been developed to aid in programming microprocessors. For simple applications, microprocessors are programmed at the machine language level. However, when longer or more complex programs are required, it is desirable to use an assembler or compiler if it is available. For minicomputers and larger scale computers, compilers and assemblers that reside within the computer memory itself are available to aid in the programming process. However, most microprocessors do not have sufficient memory to accommodate such large complex programs. In addition, the microprocessor is generally to be dedicated to a specific application and therefore its memory will only be large enough to hold the application program required. In order to simplify the development of programs for use in a microprocessor, special programs called cross-assemblers and cross-compilers have been developed. These are special programs that reside in the memory of a larger general purpose digital computer. The application programs are written in these higher level languages and the larger machine then converts the application program into the binary machine language required by the microprocessor. The output of the larger scale computer is generally a paper tape containing the binary program, which is later loaded into the microprocessor memory.

Some of the larger more sophisticated microprocessors have been used as the primary component in a microcomputer that can be used as a software development system for that microprocessor. A large random access memory is added to the microprocessor along with appropriate peripheral devices. Resident assembler programs have been developed for these machines. In this way, the microcomputer based on the microprocessor can be used to develop application programs that will be used later in another system employing the same microprocessor.

Microprocessors

As we indicated earlier, a microprocessor is the simplest and least expensive form of digital computer available. However, now we want to be more specific. In this section, we are going to discuss exactly what a microprocessor is, the types that are available, and how they are used.

Types of Microprocessors. Most microprocessors are the central processing unit (CPU) of a digital computer. That is, the microprocessors usually contain the arithmetic-logic and control sections of a small scale digital computer. Most of these microprocessors also contain a limited form of input-output circuitry which permits them to communicate with external equipment. To make the microprocessor a complete computer, external memory and input-output devices must be added. An external read only memory is normally used to store the program to be executed. Some read/write, random access memory may also be used. The external input-output circuitry generally consists of registers and control gating that buffer the flow of data into and out of the CPU.

Microprocessors come in a wide variety of forms. However, the most popular and widely used microprocessor is a MOS LSI circuit. These circuits are made with both P-channel or N-channel enhancement mode MOS devices. The entire CPU is contained on a single chip of silicon and mounted in either a 16, 24 or 40-pin dual in-line package. Such microprocessors are available with standard word lengths of 4, 8 and 16 bits. Other more sophisticated types of microprocessors are contained within two or more integrated circuit packages. When combined, they form a complete, small scale digital computer.

While most microprocessors are of the single chip MOS variety, there are numerous bipolar microprocessors available. These are inherently faster than the MOS devices but occupy more chip space and consume more power. Where high speed is required, these bipolar devices can be used. A recently developed integrated circuit technology, referred to as integrated injection logic (I^2L), combines both the speed of bipolar devices and the high density characteristics of MOS devices. These new I^2L LSI circuits offer many benefits, and their potential for microprocessor applications is great.

Microprocessors can also be constructed with the standard TTL or ECL integrated circuits. Standard MSI packages can be combined to construct a small CPU. Figure 10-25 shows a computer of this type. While this kind of microprocessor takes more circuitry and consumes more power, it generally offers several advantages. First, the microprocessor can be constructed to execute a special instruction set designed specifically for the application. With a standard off-the-shelf CPU, the instruction set is fixed. Special instruction sets are often necessary for some applications and they can be readily optimized with a special TTL or ECL microprocessor design. Another advantage of a MSI TTL or ECL microprocessor is high speed. A standard MOS microprocessor may be too slow for the application. The fastest available MOS microprocessor can execute a single instruction in approximately 2 microseconds. The simpler and less sophisticated MOS microprocessors have instruction execution speeds in the 10 to 50 microsecond region. With a special TTL or ECL MSI microprocessor, execution speeds in the nanosecond region are easily obtained.

Figure 10-25
A microprocessor made with TTL MSI and SSI integrated circuits. This machine is more powerful than the typical LSI microprocessor but less powerful than a full minicomputer. (Photo courtesy Computer Automation Inc.).

In order to use a standard single chip microprocessor, some form of external memory must be used. The program to be executed by the microprocessor is generally stored in a ROM. Data is stored in RAM. Other external components needed to support a microprocessor are an external clock circuit, input-output registers, and peripheral devices.

All single chip microprocessors incorporate a data bus through which all external data transfers take place. This may be a 4 or 8-bit bi-directional bus over which all data transfers between the memory and input-output devices communicate with the CPU. A bus design of this type greatly minimizes the number of interconnections required to connect the microprocessor to the external devices. The limiting factor of such interconnections is the number of pins on the integrated circuit package. The bus organizations keeps the pin count to a minimum, but at the same time requires time sharing of the bus. Since all data transfers between the memory and CPU and between the CPU and the peripheral devices must use the same input-output lines, each operation must take place at a different time.

The input-output devices used with most microprocessors are quite different from those used with larger digital computers. Most larger computers are connected to input-output devices like CRT terminals, teletypewriters, paper tape readers and punches, card readers and line printers. On the other hand, microprocessors are interfaced to devices such as keyboards, 7-segment LED displays, thumbwheel switches, relays, analog to digital and digital to analog converters, temperature sensors, and other such components.

Applications of Microprocessors. Microprocessors are used primarily for dedicated functions. Rarely are microprocessors used to implement a general purpose digital computer. The program of a microprocessor is usually stored in the read only memory. This means that the program is fixed and dedicated to the specific application.

There are two broad general applications for modern LSI microprocessors. They can be used as replacements for minicomputers or as replacements for random hard-wired logic. The development of the minicomputer enabled many engineers to design digital computers into special control systems. The minicomputer was dedicated to the control application and its programmable flexibility offered many benefits. But its cost was very high. Some microprocessors have nearly as much computing power and capability as a minicomputer and can replace the minicomputer in many systems. A microprocessor has the advantage of smaller size, lower cost, and lower power consumption.

Another common use for the microprocessor is as an alternative to standard hard-wired digital logic circuits. Equipment customarily constructed with logic gates, flip-flops, counters, and other SSI and MSI circuits can often be implemented with a single microprocessor. All of the standard logic functions such as Boolean operations, counting and shifting can be readily carried out by the microprocessor through programming. The microprocessor will execute instructions and sort subroutines that perform the same logic functions.

Many benefits result from using the microprocessor in replacing hard-wired random logic systems. Some of these advantages are: (1) reduced development time and cost; (2) reduced manufacturing time and cost; (3) enhanced product capability; (4) improved reliability.

Development time and cost can be significantly reduced when a microprocessor is used. The design procedures used with standard logic circuits are completely eliminated. Much of the breadboarding, cut-and-try and prototype construction is completely eliminated. Design changes can be readily incorporated and new functions implemented by simply changing the program. With a microprocessor, the logic and control functions are implemented with programs. The program can be written and entered into memory and then tested. System changes are easy to make by simply rewriting the program. Unique functions can be readily added by increasing the size of the program. In many cases, the system can be made self-checking by programming special diagnostic routines. Development time is further reduced because usually a single integrated circuit microprocessor replaces many other integrated circuits. This reduces wiring and interconnections and simplifies printed circuit board layout. Often the printed circuit board will be significantly smaller with a microprocessor system. Power comsumption and cooling are also usually simplified. The benefit of reduced development time and cost, of course, is that the product can come to market or be applied sooner.

Manufacturing costs are also reduced as a result of replacing random logic with a microprocessor. Fewer integrated circuits and smaller printed circuit boards are required to construct the system. Therefore, less time and materials are required to assemble the equipment. The programmed nature of the microprocessor system also makes it easier to test and debug than an equivalent hard-wired system.

Enhanced product capability is another benefit of using the microprocessor to replace hard-wired logic.The power of a digital system implemented with a microprocessor is limited strictly by the imagination of the designer. Many unique features and capabilities can be incorporated into the design by simply adding to the program. The incremental cost for adding such features to a microprocessor system is small compared to that of a hard-wired logic system. The ROM used to store the program usually contains extra room for program additions. Therefore, it is very easy to add special features. Many of these special or unique features would be difficult to incorporate in a random hard-wired logic design because of the extra design time, the complexity, and the additional cost. When a microprocessor is used, no additional parts or significant amount of design time are required to add them. The more unique and special features that a product can incorporate the better it performs and the more competitive it will be in the marketplace.

Another benefit of using the microprocessor to implement digital systems is increased reliability. Whenever the number of integrated circuits and wiring interconnections are reduced in a system, reliability increases significantly. Most system failures result from the failure of an integrated circuit or from an interconnection. The number of integrated circuits and interconnections are reduced many orders of magnitude in going from a standard hard-wired logic system to a microprocessor system. Increased reliability means fewer failures and leads to a corresponding reduction in both warranty and service costs.

The benefits of using a microprocessor are so significant that they will soon replace most random hard-wired logic designs. But the biggest present disadvantage of using a microprocessor is the designer's lack of programming knowledge. Very little circuit or logic design is required to implement a system with a microprocessor. Instead, the primary skill required is digital computer programming. Most engineers and digital designers were not trained in this subject, and therefore, initial design attempts with microprocessors may be slow and frustrating. However, as microprocessors are more widely used and their benefits recognized, engineers and designers will learn programming and begin to implement their systems with these devices.

Where are Microprocessors Used? There are so many applications for microprocessors that it is difficult to classify and list them. However, to give you a glimpse at the many diverse uses for these devices, consider some of the applications where they are now being used or being considered.

1. Electronic Cash Registers
2. Electronic Scales
3. Electrical Appliance Controls
4. Automotive Controls
5. Traffic Signal Controllers
6. Machine Tool Controls
7. Programmable Calculators
8. Automatic Test Equipment
9. Data Communications Terminals
10. Process Controllers
11. Electronic Games
12. Data Collection

These are only a few of the many applications presently implemented with microprocessors. Just keep in mind that the microprocessor can be used in any other application where hard-wired standard logic systems are now used. In addition, microprocessors can also be used as the CPU in a small general purpose microcomputer or minicomputer.

A typical application for a microprocessor is illustrated in Figure 10-26. Here the microprocessor is used to implement an electronic scale for use in a grocery market. The item to be weighed is placed on the scale. A transducer and analog-to-digital converter convert the weight into a binary word that is read into the CPU under program control. A clerk enters the price per pound via the keyboard. This too is read into the CPU. Then the CPU computes the price by multiplying the weight by the

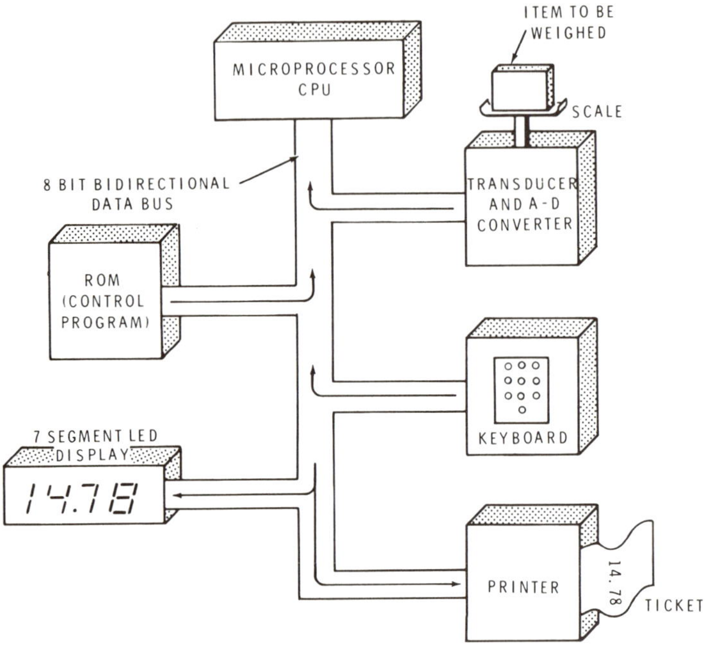

Figure 10-26
Typical application of a microprocessor in an electronic scale.

price per unit of weight. Then the total price is displayed on a 7-segment LED readout and printed on a ticket. All of this takes place under the control of the dedicated program stored in the ROM. Note the single 8-bit bi-directional data bus over which all data transfers take place.

Designing with Microprocessors. As indicated earlier, microprocessors can be used in two general ways. First, they can be used to replace minicomputers for dedicated control functions. Second, microprocessors can be used to replace standard hard-wired random logic systems. This section provides you with some guidelines to help you decide when and where a microprocessor should be used.

Microprocessors are generally much slower and less sophisticated than the typical minicomputer. But despite these limitations, microprocessors can often be used to replace minicomputers in some systems. The reason for this is that most minicomputers used in control systems are not used to their full capability. In a sense, they are a case of over-kill. Many control systems used the minicomputer simply because of the ease with which the control can be changed by modifying the program. The significantly higher cost has been traded off for the convenience of system modification. In these applications, the microprocessor can usually handle the control functions as well as the minicomputer. A careful study must be made in such designs to see when a microprocessor can replace a minicomputer. There are many trade-offs to consider (speed, cost, etc.). Keep in mind that microprocessor development is in its infancy. Many technological improvements will be made over the years causing the microprocessor to futher approach the capabilities of today's minicomputer.

The microprocessor is a design alternative which should be considered in the early design stages of any digital system. The benefits of a microprocessor over standard hard-wired designs is significant in the larger, more sophisticated digital systems. As a general guideline, a microprocessor can be used beneficially if it will replace from thirty to fifty standard MSI and SSI TTL integrated circuits. If a preliminary design indicates that this many TTL integrated circuits must be used a microprocessor should be considered. Unless the speed limitation of the microprocessor is a factor, all of the benefits mentioned earlier will result from the use of the microprocessor.

Another way to equate a microprocessor design with the more conventional hard-wired logic design, is to compare the number of gates in a hard-wired design with the number of bits of memory required by a microprocessor system. It has been determined that it takes approximately 8 to 16 bits of memory in a microprocessor system to replace a single gate. Since most read only memories used to store the program for a microprocessor can contain as many as 16,384 bits, such a memory can replace from 1000 to 2000 gates. Depending upon the number of gates per SSI or MSI package, this can represent a replacement of hundreds of integrated circuit packages. A 16,384 (16K) bit ROM in a single 40-pin IC package, for example, can replace one hundred to four hundred 14, 16, or 24-pin SSI and MSI packages. This is a significant saving.

At this point you may still have some doubts about the ability of a microprocessor to replace standard hard-wired logic functions. It may be difficult for you to imagine how a microprocessor can perform the functions you are so used to implementing with SSI and MSI packages. To ease your mind about this, let's consider all of the standard logic functions and illustrate how a microprocessor can perform them.

The microprocessor can readily perform all of the standard logical functions such as AND, OR, and Exclusive OR. It usually does this by executing the instructions designed for this purpose. Logical operations are generally performed on data stored in memory and in the accumulator register, with the result appearing in the accumulator. Suppose that you wanted to perform the NAND function on two 8-bit words. Using the instruction set in Table I, we could write the following program. Assume that the two words to be NANDed are stored in locations 8 and 9.

```
0    LDA  (8)
3    AND  (9)
6    CMP
7    HLT
```

The first instruction loads the first word into the accumulator. The second instruction performs the AND function with the word in the accumulator and the word in location 9. The result appears in the accumulator. Finally, this result is complemented to form the NAND function. This simple example illustrates the procedure you use to implement any Boolean function.

Arithmetic operations are also readily performed by a microprocessor. Special adders, subtractors and other arithmetic circuits are not required because all microprocessors can perform arithmetic operations through programming. Multiplication and division operations are carried out by subroutines. Even the higher math functions such as square root, trigonometric functions, and logarithms can be computed with subroutines. Many special algorithms have been developed for solving these higher mathematical functions with digital computers. To handle very large or very small numbers or to improve the accuracy of computation, multiple precision arithmetic subroutines are also available. Number size is limited by the number of bits in the basic computer data word. However, several computer words can be used to represent a quantity as large or as small as needed. Special programs can then be written to manipulate this data just as if it were represented by a single smaller word.

An example of a programmed arithmetic operation is shown in Figure 10-27. This flow chart illustrates the procedure for multiplying two positive numbers, A and B, by repeated addition. A is added B times to produce the product. The program to implement this algorithm is given below. The numbers to be multiplied are stored in locations 31 and 32. The product or answer is stored in location 33.

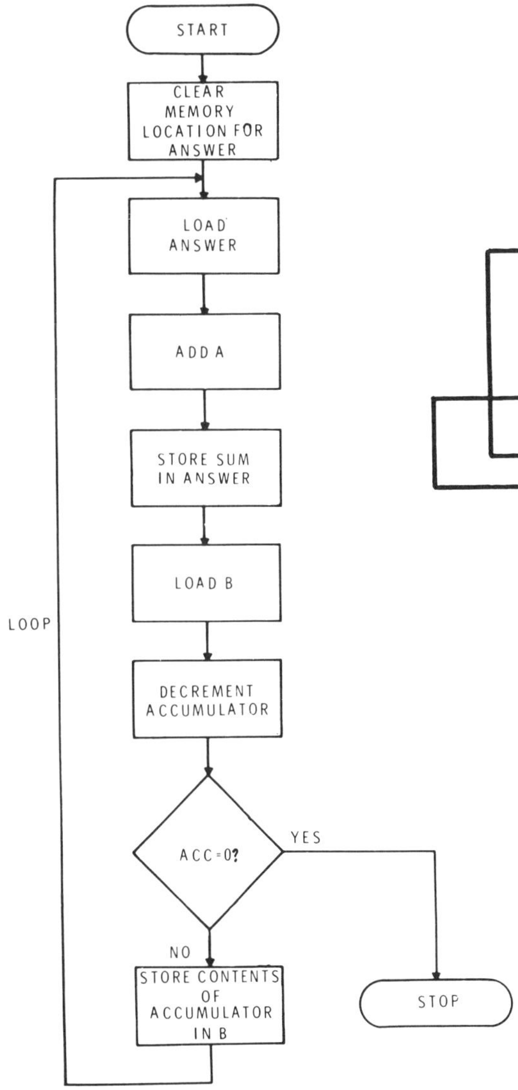


```
 0   CLA
 1   STA  (33)
 4   LDA  (33)
 7   ADD  (31)
10   STA  (33)
13   LDA  (32)
16   DCR
17   JMZ  (26)
20   STA  (32)
23   JMP  (4)
26   HLT
31   A
32   B
33   ANSWER
```

Figure 10-27
Flow chart of multiplication subroutine.

The first two instructions are used to clear the memory location where the ANSWER is to be stored. CLA resets the accumulator to zero and the STA instruction writes zeros in memory location 33. Next, the LDA instruction loads the contents of 33 (zero) into the accumulator. Then we add A to it with the ADD instruction. We then restore the partial product in location 33. We then load the content of location 32 (B) into the accumulator and subtract one from it with the decrement instruction DCR. We use a jump on zero (JMZ) instruction to see if the accumulator is zero. If it is not, we restore the accumulator content in location 32. The jump instruction creates a loop that returns us to the LDA (33) instruction. The entire sequence is then repeated. This continues until A has been added B times. Each time we add A to the answer, we subtract one from B. This permits us to keep track of how many times A has been added. When A has been added B times, the content of location 32 is again reduced by one producing a zero result. The JMZ instruction tests for zero. The correct product is contained in location 33 at this time. The JMZ causes the program to branch to location 26 where the HALT instruction is executed to stop the program.

Microprocessors can also be used to make decisions. For example, the microprocessor can compare two binary numbers and determine if they are equal or if one is greater than or less than another. This decision making function permits the microprocessor to evaluate information as it is developed and to modify its operation according to the values of the data.

The flow chart in Figure 10-28 illustrates one algorithm for comparing two binary numbers. Here, one number is subtracted from the other. A test for zero is then made. If the remainder is zero, of course, the numbers are equal. The program below implements this algorithm. The numbers to be compared are stored in locations 15 and 16.

```
     0    LDA  (15)
     3    SUB  (16)
  ┌─ 6    JMZ  (23)
  │  9    next instruction (A ≠ B)
  └►23    next instruction (A = B)
```

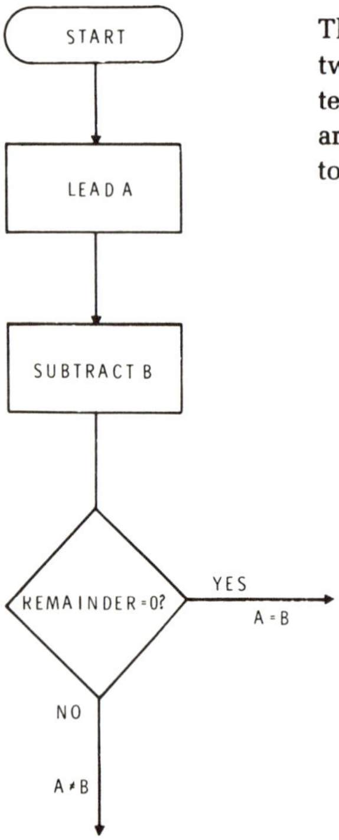

Figure 10-28
Flow chart of a procedure for comparing two numbers.

If the numbers are equal, the program branches to location 23. If the numbers are not equal, the program continues in its normal, sequential manner.

Another common logic function that is readily implemented with a microprocessor is counting. The microprocessor can count external events or a frequency standard. External events are counted by applying them to the interrupt line on the microprocessor. As each event occurs, an interrupt is generated with the microprocessor. This causes the microprocessor to jump to a subroutine that will increment the accumulator register or add one to some memory location. Up or down counters are readily implemented with the increment and decrement accumulator instructions. Decision-making techniques can be used to detect when a specific count is reached or to count quantities larger than the computer word size permits. For example, with an 8-bit data word in a microprocessor, the maximum count that the accumulator can handle is 1111 1111 or 255. To count to higher values, a program can be written to indicate each time the counter overflows.

The program below illustrates a method of detecting a count of 153. The flow chart in Figure 10-29 shows the approach.

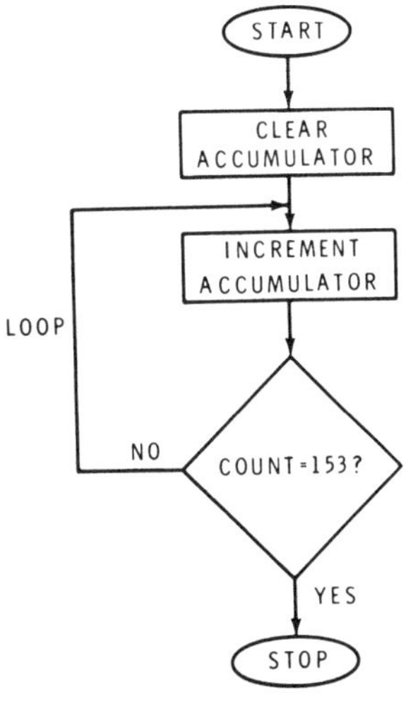

Figure 10-29
Flow chart showing a method of detecting a count of 153.

```
 0   CLA
 1   INC
 2   STA  (18)
 5   SUB  (19)
 8   JMZ  (21)
11   LDA  (18)
14   JMP  (1)
18   Count
19   153
21   LDA  (18)
24   HLT
```

The first instruction clears the accumulator. The accumulator is then incremented by the INC instruction, and the count is stored in location 18. The count is then compared by subtracting 153 and testing for zero. If a non-zero result occurs, the count is retrieved with the LDA instruction and the program loops back to the increment instruction. This loop continues until a count of 153 is reached. When the JMZ instruction detects the zero condition, the program branches to location 21 where the count is loaded and the program halts.

To count to numbers higher than 255, the program below can be used. See the flow chart in Figure 10-30 for an explanation of the procedure. This program counts in multiples of 256. Note that the program has two loops. The inner loop determines when a count of 256 occurs, while the outer loop determines the number of times that the inner loop occurs. The total count then is the product of the number of times the inner loop occurs and the count in location 37, in this case 5. The program halts on a count of 5 $\times$ 256 or 1280.

```
       0    CLA
       1    STA   (36)
       4    INC
       5    JMZ   (11)
       8    JMP   (4)
inner  11   LDA   (36)
loop   14   INC
       17   STA   (36)
       20   SUB   (37)
       23   JMZ   (30)
       26   CLA
       27   JMP   (4)
outer  30   HLT
loop   36   COUNT
       37   5
```

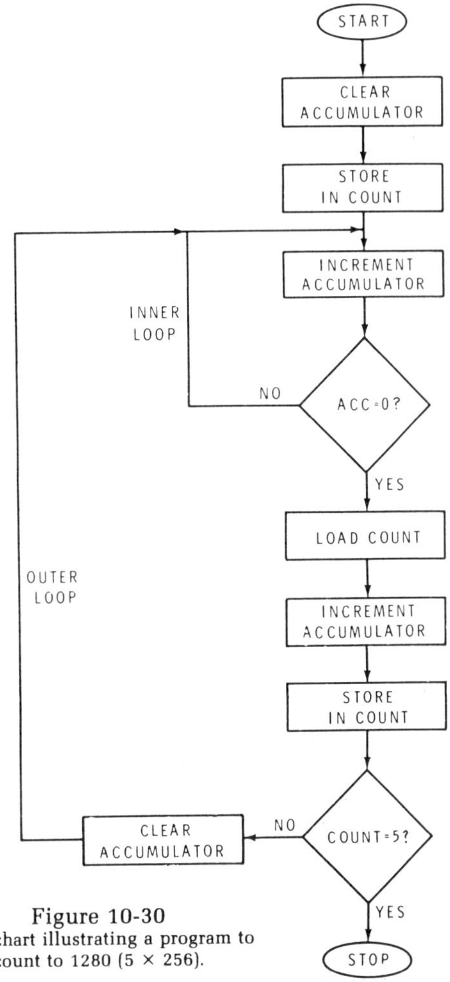

Figure 10-30
Flow chart illustrating a program to count to 1280 (5 × 256).

The first instruction clears the accumulator. The next instruction writes zero into memory location 36 which we call COUNT. The content of location 36 tells us how many times the inner loop is repeated. These first two instructions initialize the circuitry prior to starting the count. The program then begins the count by executing the increment instruction in location 4. The JMZ (11) then tests for zero. If the accumulator is not zero, the JMP instruction is executed creating a loop that returns the program to the INC instruction. The loop is repeated 255 times at which time the accumulator is 1111 1111. The INC instruction is executed a 256th time and the accumulator recycles to 0000 0000. Then the JMZ (11) instruction again tests for zero. This time the program branches to location 11 where the LDA (36) instruction is performed. This loads COUNT (which is initially zero). COUNT is then incremented to indicate that a count of 256 has occurred. COUNT is then restored by the STA (36) in location 17. Next, the program tests to see if COUNT is 5. It subtracts 5 from COUNT. If the remainder is not zero, the accumulator is cleared and the program loops back to the beginning where the inner loop is again repeated. When COUNT becomes 5, the inner loop has been repeated 5 times indicating a count of 1280. The program then branches via the JMZ (30) instruction to a HLT.

The microprocessor can also generate timed output pulses. It can do this in several ways. The simplest method is to store a series of binary numbers in sequential memory locations with the proper bit designations. These can then be read out of memory, one at a time, and sent to the output data bus. As the binary words change, the output bits change and generate any desired sequence of timing pulses. The rate of occurence of these pulses depends upon the speed of the microprocessor. Longer timed output pulses can also be generated by producing internal timing delays. This can be done by programming counting loops like the ones just illustrated. The time delays are a product of the instruction execution speed and the desired count. With a count of 153 and an instruction execution speed of 12.5 microseconds, the total delay would be $153 \times 12.5 = 1912.5$ microseconds. The binary words in memory could be outputted every 1.9125 milliseconds.

As you can see, microprocessors can perform all the same functions as hard-wired digital logic systems. The logic designer no longer spends his time in designing circuits or in minimizing Boolean expressions. Instead, he writes programs. For that reason, it is highly desirable for the digital designer to learn programming. As microprocessors become more widely used, the digital engineer will become more of a programmer than a circuit or logic designer.

Index

E

F

G

I

N

O

S

T

U

W

X